T0344678

Practical Creativity and Innovation
in Systems Engineering

WILEY SERIES IN SYSTEMS ENGINEERING AND MANAGEMENT

William Rouse, Series Editor
Andrew P. Sage, Founding Editor

A complete list of the titles in this series appears at the end of this volume.

Front Cover
Iconic of Japanese gardens, the Zen-Style Garden is designed to invite contemplation and seclusion. This image is part of a dry rock garden consisting of gravel and massive boulders placed by Hoichi Kurisu. Photo courtesy of Frederik Meijer Gardens & Sculpture Park, Grand Rapids, Michigan, USA.

Practical Creativity and Innovation in Systems Engineering

Avner Engel

Registered Office
John Wiley & Sons, Inc., 111 River Street, Hoboken, NJ 07030, USA

Editorial Office
111 River Street, Hoboken, NJ 07030, USA

For details of our global editorial offices, customer services, and more information about Wiley products visit us at www.wiley.com.

Wiley also publishes its books in a variety of electronic formats and by print-on-demand. Some content that appears in standard print versions of this book may not be available in other formats.

Library of Congress Cataloging-in-Publication Data

Names: Engel, Avner, author.
Title: Practical creativity and innovation in systems engineering / by Avner Engel.
Description: Hoboken, NJ : John Wiley & Sons, 2018. | Series: Wiley series in systems engineering and management |
Identifiers: LCCN 2018007226 (print) | LCCN 2018012978 (ebook) |
    ISBN 9781119383352 (pdf) | ISBN 9781119383383 (epub) | ISBN 9781119383239 (cloth)
Subjects: LCSH: Creative ability in technology. | Systems engineering–Technological innovations–Management.
Classification: LCC T49.5 (ebook) | LCC T49.5 .E525 2018 (print) | DDC 620/.0042–dc23
LC record available at https://lccn.loc.gov/2018007226

Cover design: Wiley
Cover image: © Frederik Meijer Gardens & Sculpture Park

Set in 10/12pt Warnock by SPi Global, Pondicherry, India

Printed in the United States of America

V10004335_090618

To Rachel

Contents

Preface

The aim of this book is to acquaint systems engineers with the practical art of creativity and innovation. The concept of creativity has evolved throughout history. The Greeks considered poetry the only legitimate creative practice. That is, poets, as opposed to artisans, merchants, and even nobility could create poetry freely with no restrictions or rules. Later, the Romans considered the visual arts as a creative practice, too. However, during the Middle Ages, creativity evolved to strictly mean God's creations. Therefore, the concept of creativity was no longer applicable to any human activity. Thereafter, during the Renaissance and beyond, creativity slowly progressed to imply freedom of expression in the arts. Only at the turn of the twentieth century did the concept of creativity began to be applied to science and engineering.

The basic premise of this book is that creative abilities of human beings are not fixed, inborn traits but, rather, are changing over their lifetime. For example, researchers show that children exhibit remarkable abilities to look at problems and come up with new, different, and creative solutions. However, as they grow to adulthood, these abilities diminish substantially. Fortunately, creative skills can also be learned. Many studies show that well-designed training programs enhance creativity across different domains and criteria. Hopefully, engineers adopting some of the creative methods discussed in this book will achieve improved creative skills as well.

Another premise of this book is that many creative engineers are stalled in their innovative efforts by organizations that claim to promote innovation but that, in fact, crush such efforts. Indeed, it is the author's impression (as well as other researchers) that, beyond boasting, the vast majority of companies and other organizations are creativity-averse. Naturally, creative engineers working for such organizations are frustrated and discouraged. Not less important are the accumulated losses for the organizations themselves as well as to society at large from neglecting many creative ideas without due consideration. The book attempts to explore this phenomenon and offer practical advice to organizations as well as to the multitudes of demoralized engineers. In particular, engineers are advised to expand their professional and intellectual horizons, seek to reduce risks inherent in their new ideas, and learn to obtain colleagues' support as well as deal with reactionary management. In short, adopt a more entrepreneurial attitude.

This book is organized in five parts: Part I: Introduction, contains material about the principles of the book and its content. Part II: Systems Engineering, describes basic systems engineering concepts and a partial and abbreviated summary of Standard 15288 systems' life cycle processes. In addition, this part includes a recommended set of creative methods for each life cycle process. Finally, this part provides some philosophical thoughts about engineering. Part III: Creative Methods, the heart of the book, provides an extensive repertoire of practical creative methods engineers may use. Part IV: Promoting Innovative Culture, deals with ways and means to enhance innovative culture within organizations. In addition, this part provides advice to creative engineers employed by non-creative organizations. Finally, Part V: Creative and Innovative Case Study, presents an exemplary creative and innovative research and implementation undertaking.

Fundamentally, this book is written with two categories of audience in mind. The first category is composed of practicing engineers in general and system engineers in particular as well as first- and second-line technical managers. These people may be employed by various development and manufacturing industries (e.g. aerospace, automobile, communication, healthcare equipment, etc.), by various civilian agencies (e.g. NASA, ESA, etc.) or with the military (e.g. Air Force, Navy, Army, etc.). The second category is composed of faculties and students within universities and colleges who are involved in Systems, Electrical, Aerospace, Mechanical, and Industrial Engineering. This book may be used as a supplemental graduate level textbook in creativity and innovation courses related to systems engineering. Selected portions of the book may be covered in one or two semesters.

Finally, readers should note that this book does not pursue new theories or theses with regards to creativity and innovation. To the contrary, the author seeks to acquaint systems engineers with well-established facets of creativity and innovation. In order to achieve this objective, the author drew upon his engineering experience, communicated with many people, and collected information from many sources, books, articles, internet blogs, and the like (giving credit where credit's due). Bibliographies at the end of each part of the book identify invaluable sources for deeper understanding of the various subject matters discussed in the book. The author gained much knowledge from these resources and is indebted to the individuals, researchers, and experts who created them.

Acknowledgments

Many people have generously contributed to the writing of this book. To all of them, I would like to express my sincere gratitude and appreciation.

In particular, I wish to thank Shalom Shachar, formerly from the Israel Aerospace Industries, and Professor Tyson Browning from the Texas Christian University, tireless colleagues and friends, much of whose scientific and engineering writings and words of wisdom are embedded in this book.

The AMISA project, funded by the European Commission, focused my attention onto the value of creativity and innovation within systems engineering. My appreciation goes to all the consortium members and in particular to Professor Yoram Reich of the Tel Aviv University for his steadfast support and advice and also to Michael Garber of Adi Mainly Software (AMS), who developed the DFA-Tool software package, which embodies the Architecture Option model.

Two people had direct impact on the manuscript of the book. Professor Shulamith Kreitler of the Tel Aviv University encouraged my book project and advised me on its structure. Professor Cecilia Haskins of the Norwegian University of Science and Technology volunteered to review the manuscript and contributed numerous and valuable suggestions to improve it. Also, I would like to thank my good friend, Menachem Cahani (Pampam) for contributing two caricatures to the book. I am indebted to them both. I would also like to express my deep appreciation to the dedicated and tireless Wiley editing team, especially to Victoria Bradshaw and Grace Paulin, as well as to Cheryl Ferguson for her diligent assistance with preparing the manuscript.

Several researchers empowered me to share their research with the readers of this book, and I am beholden to them all: Professor Sang Joon Lee of the Pohang University of Science and Technology, South Korea on biomimicry engineering, Professor Emeritus Ravi Jain of the University of the Pacific, California, on managing research, development, and innovation, Professor Christian Richter Østergaard of the Aalborg University, Denmark, on innovation and employee diversity, Inger Danilda of Quadruple Learning, Sweden, and Jennie Granat Thorslund of the Swedish Governmental Agency for

Innovation Systems (VINNOVA) on gendered innovation, and lastly, Professor T.K. DAS of the City University of New York on cognitive biases.

My deep appreciation also goes to the Standards Institution of Israel (SII), which permitted me, on behalf of the International Organization for Standardization (ISO), to reproduce a partial and abridged portion of the International Standard ISO/IEC/IEEE 15288. My thanks also go to the Royal Academy of Engineering, London, United Kingdom, for permission to reproduce intriguing portions of papers presented during seminars on the philosophy of engineering held at the academy in June 2010.

Most of all, my deepest thanks go to my wife, Rachel, and my sons, Ofer, Amir, Jonathan, and Michael, who encouraged my book efforts with advice, patience and love.

Avner Engel
Tel-Aviv, Israel

Part I

Introduction

"One must still have chaos in oneself to be able to give birth to a dancing star."

Friedrich Nietzsche (1844–1900)

1.1 Introduction to Part I

The aim of this book is to acquaint engineers in general and systems engineers in particular with the practical art of creativity and innovation. Systems engineers are people with a capacity to understand many engineering, scientific, and management disciplines. In addition, systems engineers tend to examine issues in a holistic way considering the total system life cycle. This capacity is obtained through formal education, as well as experience in leading multidisciplinary teams in creating, manufacturing, and maintaining complex systems within sustainable environments.[1]

The basic premise of this book is that creative abilities of human beings are not fixed, inborn traits but, rather, change over their lifetime. For example, in the late 1960s and early 1970s, George Land tested the level of creativity among children and adults.[2] The results, presented in Figure 1.1, are quite shocking. According to the study, 98% of five-year-old children could be categorized as geniuses in terms of their abilities to look at problems and come up with new, different, and creative solutions. This percentage drops to 2% within the average adults' population. Land and Jarman (1998) concluded from this longitudinal study that non-creative behavior is learned.

---

1 Adapted from: Urban Dictionary, http://www.urbandictionary.com/define.php?term=super-systems-engineer. Accessed: July, 2017.
2 See: TEDxTucson George Land, The failure of success, https://www.youtube.com/watch?v=ZfKMq-rYtnc. Accessed: July, 2017.

Practical Creativity and Innovation in Systems Engineering, First Edition. Avner Engel.
© 2018 John Wiley & Sons, Inc. Published 2018 by John Wiley & Sons, Inc.

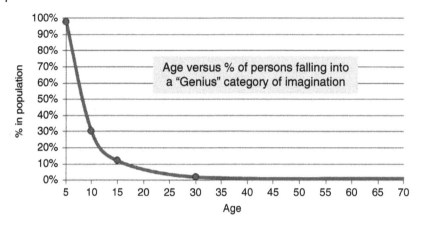

Figure 1.1 Age versus imagination

Fortunately, creativity skills can also be learned. For example, Scott et al. (2004) analyzed some 70 studies related to creativity training and concluded that well-designed training programs promoted distinct creativity performance gains across different domains and criteria. Hopefully, engineers adopting some of the creative methods discussed in this book will achieve improved creative skills as well.

Another premise of this book is that many creative engineers are stalled in their innovative efforts by organizations that claim to promote innovation but, in fact, consistently crush such efforts. Indeed, it is the author's impression (as well as other researchers[3]) that, beyond boasting, the vast majority of companies and other organizations are creativity-averse. Naturally, creative engineers working for such organizations are frustrated and discouraged. Not less important are the accumulated losses for the organizations themselves as well as to society at large from neglecting many creative ideas without due consideration. The book attempts to explore this phenomenon and offer practical advice to organizations as well as to the multitudes of demoralized engineers. In particular, engineers are advised to expand their professional and intellectual horizons, seek to reduce risks inherent in their new ideas, and learn to obtain colleagues' support as well as deal with reactionary management. In short, adopt a more entrepreneurial attitude.

Beyond this introductory chapter, Part I of this book provides some key points and a short outline related to the other four parts of the book, namely: (1) systems engineering, (2) creative methods, (3) promoting innovative

---

3 See for example T. Amabile, "How to Kill Creativity," https://hbr.org/1998/09/how-to-kill-creativity. Accessed: July, 2017.

culture, and (4) creative and innovative case study. In addition, Part I closes with a relevant bibliography. Figure 1.2 depicts the overall structure and contents of the entire book.

Part I: Introduction, true to its name, provides an introductory material to this book. Part II: Systems Engineering, describes basic systems engineering concepts as well as a partial and abbreviated depiction of Standard 15288 systems' life cycle processes. In addition, for each process, the book identifies a relevant small set of recommended creative methods. Finally, this part presents some intriguing philosophical insights about engineering. Part III: Creative Methods, provides an extensive repertoire of practical creative methods. Part IV: Promoting Innovative Culture, describes ways and means to enhance innovative culture within organizations. In addition, this part provides advice to creative engineers employed by non-creative organizations. Part V: Creative and Innovative Case Study, describes an exemplary creative and innovative case study. Lastly, the back matter of the book contains relevant appendices.

The book contains a massive number of visuals. This is because the author believes engineers (and probably other people) tend to focus on visuals as their

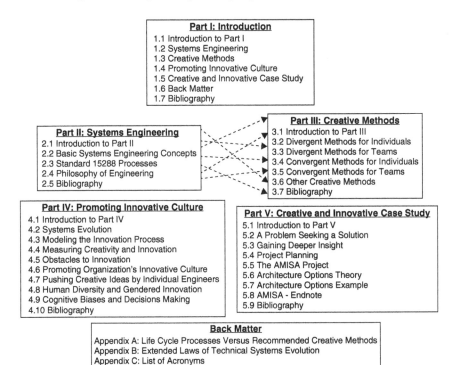

Part I: Introduction
1.1 Introduction to Part I
1.2 Systems Engineering
1.3 Creative Methods
1.4 Promoting Innovative Culture
1.5 Creative and Innovative Case Study
1.6 Back Matter
1.7 Bibliography

Part II: Systems Engineering
2.1 Introduction to Part II
2.2 Basic Systems Engineering Concepts
2.3 Standard 15288 Processes
2.4 Philosophy of Engineering
2.5 Bibliography

Part III: Creative Methods
3.1 Introduction to Part III
3.2 Divergent Methods for Individuals
3.3 Divergent Methods for Teams
3.4 Convergent Methods for Individuals
3.5 Convergent Methods for Teams
3.6 Other Creative Methods
3.7 Bibliography

Part IV: Promoting Innovative Culture
4.1 Introduction to Part IV
4.2 Systems Evolution
4.3 Modeling the Innovation Process
4.4 Measuring Creativity and Innovation
4.5 Obstacles to Innovation
4.6 Promoting Organization's Innovative Culture
4.7 Pushing Creative Ideas by Individual Engineers
4.8 Human Diversity and Gendered Innovation
4.9 Cognitive Biases and Decisions Making
4.10 Bibliography

Part V: Creative and Innovative Case Study
5.1 Introduction to Part V
5.2 A Problem Seeking a Solution
5.3 Gaining Deeper Insight
5.4 Project Planning
5.5 The AMISA Project
5.6 Architecture Options Theory
5.7 Architecture Options Example
5.8 AMISA - Endnote
5.9 Bibliography

Back Matter
Appendix A: Life Cycle Processes Versus Recommended Creative Methods
Appendix B: Extended Laws of Technical Systems Evolution
Appendix C: List of Acronyms
Appendix D: Permissions to Use Third-Party Copyright Material
Index

Figure 1.2 Book's overall structure

immediate and primary source of understanding. Many of these visuals require permission to use third-party copyright so, in order to reduce clutter and ease the reading process, these permissions are provided in Appendix D.

Finally, readers should note that this book does not pursue new theories or theses with regards to creativity and innovation. To the contrary, the author seeks to acquaint systems engineers with well-established facets of creativity and innovation. In order to achieve this objective, the author drew on his engineering experience, communicated with many people, and collected information from many sources, books, articles, internet blogs, and the like (giving credit where credit's due). Sections on further reading at the end of individual chapters, as well as the bibliographies at the end of each part of the book, identify invaluable sources for deeper understanding of the various subject matters discussed in this book. The author gained much knowledge from these resources and is indebted to the individuals, researchers, and experts who created them.

1.2 Systems Engineering

There are many books dedicated to the art of systems engineering, and it is not the purpose of this book to devote much space to this subject. Therefore, the intent of Part II is to construct scaffolding, bridging the gap between the domain of systems engineering and the domains of creativity and innovation. This is done by identifying some basic systems engineering concepts and then describing some 30 systems' life cycle processes in accordance with an abridged International Standard ISO/IEC/IEEE 15288. Each life cycle process is then associated with a specific and relevant set of recommended creative methods. Systems engineers can use these and other creative methods described in Part III to expand their creative skills and enhance their engineering output. Finally, this part provides some philosophical thoughts about engineering.

Chapter 2.2 describes basic systems engineering concepts. More specifically, it includes four basic concepts, namely: (1) organizations and projects concepts, (2) system concepts, (3) life cycle concepts, and (4) process concepts.

Chapter 2.3 describes systems life cycle processes harmonized with Standard 15288. The standard clusters these life cycle processes into four groups: (1) agreement process group, (2) organizational project-enabling process group, (3) technical management process group, and (4) technical process group.

Chapter 2.4 describes some key issues in philosophy of engineering. This includes: (1) engineering and truth, (2) The logic of engineering design, (3) the context and nature of engineering design, (4) roles and rules and the modeling of socio-technical systems, and (5) engineering as synthesis – doing right things and doing things right.

1.3   Creative Methods

Creativity may be defined as "The ability to transcend traditional ideas, rules, patterns, relationships, or the like, and to create meaningful new ideas, forms, interpretations, etc."[4] According to Teresa Amabile (1998), creativity is composed of three components: expertise, creative thinking, and motivation (Figure 1.3). Expertise consists of everything a person knows. Among others, this includes technical, procedural, and intellectual knowledge a person may possess. Creative thinking refers to ones' abilities to create meaningful new ideas and blend existing ideas together in new structures. Lastly, motivation determines what people will actually do. Extrinsic motivation comes from outside a person by way of offering person amenities like money, promotion, and the like. Intrinsic motivation, on the other hand, stems from a person's internal desire to pursue one's passion and interest.

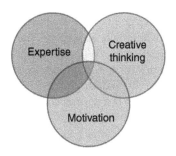

Figure 1.3 Three components of creativity

Fundamentally, creative methods may be partitioned along two axes: (1) divergent versus convergent creative methods and (2) creative methods primarily used by individuals versus teams. Along the first axis, divergent creative methods help in generating multiple creative solutions, whereas convergent creative methods help in trimming the number of creative solutions. Along the second axis, some creative methods are primarily appropriate for individuals, whereas other creative methods are primarily appropriate for teams.

Chapter 3.2 describes divergent methods for individuals, including: (1) lateral thinking, (2) resolving contradictions, (3) biomimicry engineering, and (4) visual creativity.

Chapter 3.3 describes divergent methods for teams, including: (1) classic brainstorming, (2) six thinking hats, (3) SWOT analysis, (4) SCAMPER analysis, and (5) focus groups.

Chapter 3.4 describes convergent methods for individuals, including: (1) PMI analysis, (2) morphological analysis, (3) decision tree analysis, (4) value analysis / value engineering, and (5) Pareto analysis.

Chapter 3.5 describes convergent methods for teams, including: (1) Delphi method, (2) SAST analysis, (3) cause-and-effect diagrams, (4) Kano model analysis, and (5) group decisions.

---

4 See Dictionary.com, http://www.dictionary.com/browse/creativity. Accessed: July 2017.

Chapter 3.6 describes other creative methods, including: (1) process map analysis, (2) nine screens analysis, (3) technology forecasting, (4) Design Structure Matrix analysis, (5) failure mode effect analysis, (6) anticipatory failure determination, and (7) conflict analysis and resolution.

1.4 Promoting Innovative Culture

Innovation may be defined as "the process of translating an idea or invention into a good or service that creates value or for which customers will pay."[5] Now readers can appreciate the fundamental difference between creativity and innovation. Whereas creativity refers to conceiving new and unique ideas, innovation implies introducing new systems, artifacts, processes, and the like into the market.

There is a natural conflict between the creative person, the dreamer, and the innovative person, the top-notch, down-to-earth leader and manager. This is why individuals rarely combine these two attributes within one person. Towering above them all is Moses (approx. 1390–1270 BC) the biblical prophet who presented the most fundamental and concise laws of ethics and worship and brought forth the creative idea of releasing the Israelites (and by extension, all mankind) from slavery in Egypt (Figure 1.4). Then, during 40 years of wandering in the desert, he forged a nation and a culture that propagated throughout the world to this day.

Innovation culture is the work environment that engineers and leaders cultivate within organizations in order to nurture individualistic thinking and give fair hearing to the implementation of new and often unorthodox ideas. Accordingly, this chapter examines the following issues: How do systems evolve? How should we model the innovation process? How is creativity and innovation measured? What are the obstacles to innovation? How can an organization promote its innovative culture? and, most importantly, how can an organization push creative ideas by individual engineers? Finally, this part of the book discusses human diversity and gendered innovation as well as cognitive biases and decisions making.

Chapter 4.2 describes systems evolution, including: (1) modeling systems evolution, S-curve, and (2) laws of systems evolution.

Chapter 4.3 describes modeling the innovation process, including: (1) classes and types of innovations, (2) technological innovation process, and (3) innovation funding.

---

5 See BusinessDictionary.com, http://www.businessdictionary.com/definition/innovation.html. Accessed: July 2017.

Figure 1.4 Moses, a creative and innovative giant

Chapter 4.4 describes measuring creativity and innovation, including: (1) defining innovation objectives, (2) measuring the innovation process, and (3) innovation capability maturity model.

Chapter 4.5 describes obstacles to innovation, including: (1) human habits factors, (2) cost factors, (3) institutional factors, (4) knowledge factors, (5) market factors, and (6) innovation obstacles and classes of innovations.

Chapter 4.6 describes promoting organization's innovative culture, including: (1) innovation and leadership, (2) innovation and organization, (3) innovation and people, (4) innovation and assets, (5) innovation and culture, (6) innovation and values, (7) innovation and processes, (8) innovation and tools, and (9) ascent to innovation – practical steps.

Chapter 4.7 describes pushing creative ideas by individual engineers, including: (1) large organizations seldom innovate, (2) characteristics of innovative engineers, and (3) innovation advice to creative engineers.

Chapter 4.8 describes human diversity and gendered innovation, including: (1) human diversity, (2) shift in gender paradigm, (3) gender disparity and innovation implications, (4) advancing gendered innovation, and (5) gendered innovation example.

Chapter 4.9 describes cognitive biases and decisions making, including: (1) cognitive biases, and (2) cognitive biases and strategic decisions.

1.5   Creative and Innovative Case Study

The purpose of Part V is to tell the story of an exemplary creative and innovative case study that started in 2003 and continued to 2016. The research question that emerged over time was: "How can adaptability[6] be designed into systems so that they will provide maximum lifetime value to stakeholders?"

The WOW factor was revealed in two papers (Engel and Browning, 2006, 2008) proposing to solve the problem using well-accepted economic theories: the financial option theory (FOT) and the transaction cost theory (TCT). Transforming each theory into the engineering domain and then blending them together was dubbed the architecture option (AO).

Transitioning to the innovation portion, a consortium (AMISA[7]) composed of two universities, four industries, and two small and medium enterprises (SMEs) were created. Then, a three-year funding request for a research project costing €4 million was submitted to the European Commission (EC), and after approval, the AMISA project started on April 2011, ending three years later.

End-project reports from the AMISA participants confirmed that the AO approach was indeed helpful, allowing participants to increase product adaptability, cost-efficiency, lifespan, and overall value. According to a post-project review by the EC, this research delivered a "step-change" in the performance of European industry, characterized by a higher reactivity to market needs and more economically compatible products and services. Bottom line, the AMISA partners seem to understand the importance of designing systems for future unforeseen upgrades, yet no partner truly incorporated this approach into their day-to-day systems' design operations.

Chapter 5.2 describes the problem at hand, including: (1) the problem and its inception and (2) initial funding effort.

---

6 According to the Merriam-Webster dictionary, to adapt means "to make fit, often by modification" (...from the outside). Adaptability is distinguished from "Flexibility," which is derived from the Latin word *flexus* and literally refers to what is capable of withstanding stress without injury and figuratively to what may naturally adjust itself as needed.

7 AMISA stands for: "Architecting Manufacturing Industries and Systems for Adaptability."

Chapter 5.3 describes how the people involved in this undertaking gained deeper insight. It includes: (1) the problem and the approach, (2) main ideas of the proposed work, (3) measurable project objectives, (4) basis for predicting the objectives, and (5) systems adaptability – state of the art.

Chapter 5.4 describes the project planning, including: (1) project planned activities, (2) detailed work package descriptions, (3) risks and contingency plans, (4) management structure and procedures, (5) project participants, and (6) resources needed.

Chapter 5.5 describes the AMISA project, including: (1) AMISA initiation, (2) identifying the DFA state of the art, (3) establishing requirements for AMISA, (4) implementing a software support tool, (5) developing six pilot projects, (6) generating deliverables, (7) planning exploitation beyond AMISA, (8) disseminating project results, (9) assessing the AMISA project, (10) consortium meetings, and (11) EC summary of the project.

Chapter 5.6 describes architecture options theory, including: (1) financial and engineering options, (2) transaction cost and interface cost, (3) architecture adaptability value, (4) Design Structure Matrix, and (5) dynamic system value modeling.

Chapter 5.7 describes an AO example, including: (1) general architecture option process, and (2) AO example – solid state power amplifier (SSPA).

Chapter 5.8 provides a summation of the AMISA project.

Note
Readers mostly interested in the creative aspects of this case study are advised to focus on chapters 5.2, 5.3, 5.6, and 5.7. Readers primarily interested in the innovative aspects of this case study are advised to consider chapters 5.4 and 5.5.

1.6 Back Matter

The back matter part of the book contains several relevant appendices and the index.

Appendix A: Depicts a table associating systems' life cycle processes with recommended creative methods.

Appendix B: Provides an extended set of technical systems evolution laws.

Appendix C: Provides a list of relevant acronyms.

Appendix D: Provides permissions to use third-party copyright material.

An index of important terms.

1.7 Bibliography

Amabile (1998). How to kill creativity. *Harvard Business Review* (September–October).

Cropley, D.H. (2015). *Creativity in Engineering: Novel Solutions to Complex Problems*. Academic Press.

Engel, A., and Browning T. (2006). Designing systems for adaptability by means of architecture options, INCOSE-2006, the 16th International Symposium, Florida, USA (July 09–13, 2006).

Engel, A., and Browning R.T. (2008). Designing systems for adaptability by means of architecture options. *Systems Engineering Journal* 11 (2): 125–146.

Kasser, J.E. (2015). *Holistic Thinking: Creating Innovative Solutions to Complex Problems*, 2nd ed. CreateSpace Independent Publishing Platform.

Land, G., and Jarman B. (1998). *Breakpoint and Beyond: Mastering the Future Today*. Leadership 2000 Inc.

Ruggiero, V.R. (2014). *The Art of Thinking: A Guide to Critical and Creative Thought*, 11th ed. Pearson.

Scott, G., Leritz, L.E., and Mumford M.D. (2004). The effectiveness of creativity training: a quantitative review. *Creativity Research Journal* 16 (4): 361–388.

Part II

Systems Engineering

"All you need in this life is ignorance and confidence; then success is sure."
Mark Twain (1835–1910)

2.1 Introduction to Part II

The assumption embodied in this book is that there are sufficient books and other means from which to learn about system engineering, i.e. the author expects readers to be reasonably familiar with this art. The key motivation for including Part II in this book is to launch systems engineers onto the focal point of this book, namely, Part III: Creative Methods and Part IV: Promoting Innovative Culture. This is done by characterizing some basic systems engineering concepts and then providing a condensed and abridged description of systems life cycle processes as defined in the International Standard ISO/IEC/IEEE 15288.[1] Each life cycle process is then associated with a specific set of recommended creative methods, exemplifying potential benefits that systems engineers may attain by using creative methods.

Among others, standard 15288 provides a total of 30 processes covering the life cycle of virtually any engineered system. These processes are applicable at the system level and express a coherent and cohesive set that satisfies a variety of needs. In addition, the standard provides conformance criteria that users can easily understand and apply. Finally, the standard supports tailoring by adding or subtracting processes or their constituents, making these processes widely applicable, yet adaptable to individual needs.

---

1 A partial and abridged portion of the International Standard ISO/IEC/IEEE 15288, Systems and software engineering System life cycle processes, was reproduced with permission from the Standards Institution of Israel (SII) on behalf of the International Organization for Standardization (ISO). Copyright remains with ISO.

Practical Creativity and Innovation in Systems Engineering, First Edition. Avner Engel.
© 2018 John Wiley & Sons, Inc. Published 2018 by John Wiley & Sons, Inc.

Engineering standards exhibit several advantages. First, they are developed by experts as well as practitioners. As such, they capture widespread communal engineering knowledge and experience. Second, engineering standards define common terminology, thus reducing confusion as well as communication problems. Lastly, engineering standards provide an effective tool guiding the various engineering processes in a methodic and organized manner.

Be that as it may, readers involved in developing, manufacturing, maintaining, or disposing engineered systems are urged to utilize the authentic standard 15288 and not substitute it with the following partial and abridged variation of the standard.

Some philosophical discussion about engineering and science in general concludes this part of the book.

Part II: Systems Engineering is composed of five chapters (Figure 2.1).

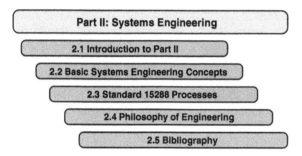

Figure 2.1 Structure and contents of Part II

Chapter 2.1 Introduction to Part II. This chapter describes the contents and structure of Part II.

Chapter 2.2 Basic Systems Engineering Concepts. This chapter describes the basic concepts of systems engineering, including (1) organizations and projects concepts, (2) system concepts, (3) life cycle concepts, and (4) process concepts.

Chapter 2.3 Standard 15288 Processes. This chapter summarizes portions of standard 15288 processes within the following four categories: (1) agreement processes, (2) organizational project-enabling processes, (3) technical management processes, and (4) technical processes.

Chapter 2.4 Philosophy of Engineering. This chapter provides illuminating philosophical thoughts about engineering, including: (1) engineering and truth, (2) the logic of engineering design, (3) the context and nature of engineering design, (4) roles and rules and the modeling of socio-technical systems, and (5) engineering as synthesis – doing right things and doing things right.

Chapter 2.5 Bibliography. This chapter provides bibliography related to Part II topics.

2.2 Basic Systems Engineering Concepts

2.2.1 Essence of Systems Engineering

According to the International Council on Systems Engineering (INCOSE), systems engineering[2] is "an interdisciplinary approach and means to enable the realization of successful systems. It focuses on defining customer needs and required functionality early in the development cycle, documenting requirements, and then proceeding with design synthesis and system validation while considering the complete problem."

INCOSE further upholds that: "Systems engineering integrates all the disciplines and specialty groups into a team effort forming a structured development process that proceeds from concept to production to operation. Systems engineering considers both the business and the technical needs of all customers with the goal of providing a quality product that meets the user needs."

Based on standard 15288, the author considers systems engineering as resting on four basic pillars: (1) organization and project concepts, (2) system concepts, (3) life cycle concepts, and (4) process concepts.

2.2.2 Organizations and Projects Concepts

System engineering is a human-intensive endeavor. Success or failure depends on individual scientists, engineers, managers and professional staff. Therefore, organizations and projects-related processes constitute a major part of standard 15288.

According to the BusinessDictionary (BD),[3] *organization* is: "A social unit of people that is structured and managed to meet a need or to pursue collective goals. All organizations have some type of management structure that determines relationships between the different activities and the members, and subdivides and assigns roles, responsibilities, and authority to carry out different tasks. Organizations are open systems. They affect and are affected by their environment." The reader should note that a part of an organization (e.g. a department within an organization) as well as a single person constitutes, by definition, an organization. The same source defines a project as: "A planned set of interrelated tasks to be executed over a fixed period and within certain cost and other limitations."

Again, the reader should note that organizations make agreements with other organizations in order to acquire / supply products or services (Figure 2.2). When an organization enters into such an agreement, it is sometimes called a

---

2 See: INCOSE, What Is Systems Engineering? http://www.incose.org/practice/whatissystemseng. aspx. Accessed: July 2017.
3 See: http://www.businessdictionary.com/definition/organization.html. Accessed: July 2017.

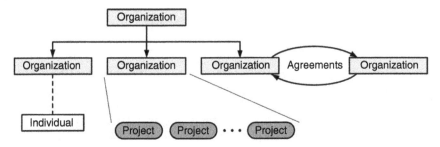

Figure 2.2 Organizations and projects concepts

party to the agreement. A party that is responsible for certain aspect of the project is usually referred to by the name of that responsibility. For example, the organization that supplies certain raw material or subsystems to the project is often called the *supplier*.

2.2.3 System Concepts

The term *system* in this book refers, in fact, to manmade systems or engineered system rather than to any system (e.g. the human body). INCOSE adopted Eberhardt Rechtin's definition of system, which states: "A system is a construct or collection of different elements that together produce results not obtainable by the elements alone. The elements, or parts, can include people, hardware, software, facilities, policies, and documents; that is, all things required to produce systems-level results. The results include system level qualities, properties, characteristics, functions, behavior, and performance. The value added by the system as a whole, beyond that contributed independently by the parts, is primarily created by the relationship among the parts; that is, how they are interconnected"[4] (Figure 2.3).

Engineered systems are characterized by the following attributes:

- Engineered systems exist within a given environment with which the system interacts in one way or another. The environment could be one or more defined entities like other systems, people, and the like.
- Engineered systems have boundaries that delineate and separate the system from its environment. A system engineer can define such boundary between a system and its environment at will.
- Engineered systems are composed of various entities, typically subsystems or parts that carry out processes and functions, and there are specific relations among these systems' entities.

---

4 See: Maier and Rechtin (2009).

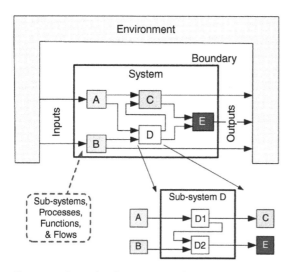

Figure 2.3 Example of an engineered system

- Entities within engineered systems have inputs and outputs supporting internal flow of materials, energy, and/or information among them. Similarly, virtually all engineered systems have external input and output flow, consisting of materials, energy, and/or information between the system and its environment.
- Any entity within an engineered system may be comprised of a hierarchically structured set of subordinate system entities.

Systems engineers distinguish between operational products and enabling products; both are constituents of engineered systems (Figure 2.4).

Operational products are the elements that, in total, perform the operational functionalities of the system. Enabling products do not contribute directly to the operational functioning of the system but provide essential services to the

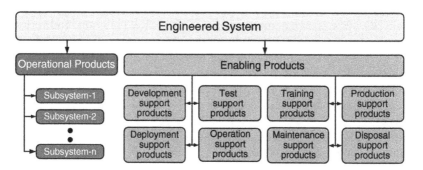

Figure 2.4 Engineered system: example of operational and enabling products

system at different stages of the life cycle, thus facilitating the progression of the system toward its designers' goals. Each enabling product plays its role in one or more stages of the system's life cycle. For example, Figure 2.5 depicts a wind tunnel testing at NASA's Langley Research Center, supporting preparations for deep space missions of the second-generation Space Launch System, or SLS. This activity, verifying broad system design concepts, may take place during the late system design stage.

Figure 2.5 Concept testing at NASA's wind tunnel facility (NASA photo)

2.2.4 Life Cycle Concepts

A system life cycle is a structured approach to create, maintain, and retire engineered systems. Many life cycle models have been defined based on analysis of typical tasks undertaken by engineers within different domains. The objective of this approach is to make those tasks amenable to traditional techniques of management planning and control. For example, different organizations such as the US Department of Defense (DoD), the National Aeronautics and Space Administration (NASA), and various industries (e.g. automobile, electronics, telecommunication, aerospace, etc.) define different system life cycle models.

Typically, a life cycle model is composed of stages. Each one represents a major period in the life of a system. More specifically, a life cycle stage defines its unique contribution to the cradle-to-grave progression of an engineered system. Standard 15288, by the way, does not advocate any particular life cycle model.

Figure 2.6 and Table 2.1 depict a generic and practical life cycle model, adopted for this book. This generic model incorporates the V-model proposed

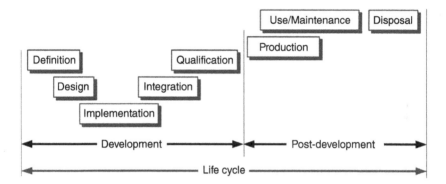

Figure 2.6 Generic system's life cycle model

Table 2.1 Generic system life cycle model

	Stage	Purpose
Development	Definition	Formulate the system operational concepts and develop the system requirements.
	Design	Create a technical concept and architecture for the system.
	Implementation	Build the elements (e.g. subsystems, components, etc.) of the system. Each element is built or purchased, then tested to ensure its stand-alone compliance with its allocated requirements.
	Integration	Connect the implemented elements into a complete system.
	Qualification	Perform formal and operational tests on the completed system to assure that the system performs its intended functionality and ensure the quality of the system as a whole.
Post development	Production	Produce the completed system in appropriate quantities.
	Use/ Maintenance	Operate the system in its intended environment in order to accomplish intended functionality, maintain the system, and correct any defects.
	Disposal	Properly dispose of the system and its elements upon completion of its life.

by Barry Boehm in the 1970s, describing the development portion of the systems' life cycle. The left-hand side of the V-model corresponds to satisfying stakeholders' requirements and the design of the desired system and its components. The right-hand side of the V-model consists of building the individual components, integrating them, and then verifying and validating the whole system. In addition, the generic model describes the post-development portion of the life cycle.

2.2.5 Process Concepts

Standard 15288 defines a framework of processes that reflects the life cycle evolution of engineered systems. More specifically, these processes represent a comprehensive superset providing: (1) organizational context to enable projects, (2) project management processes for any stage in a system's life cycle as well as (3) all of the technical processes for the entire life cycle of the system. Users (e.g. organizations, projects, etc.) can adopt and apply appropriate set of processes, defining individualized systems' life cycle practices appropriate to their purpose and doctrine. Each process within standard 15288 is described in terms of the following attributes.

Process. A process is a set of interrelated or interacting activities that transform inputs into outputs.

Title. A title summarizes the scope of the process as a whole.

Purpose. The purpose describes the high-level objective for performing a given process.

Inputs. Inputs consist of information, artifacts, or services that are used by processes in order to generate specified outcomes.

Activity. An activity encapsulates a set of cohesive process tasks.

Task. A task defines a set of requirements or recommendations or actions intended to contribute to the achievement of one or more outcomes of a process.

Outcome. An outcome is an observable result, emanating from a successful process performance.

Figure 2.7 shows a concept map depicting elements of a standard 15288 processes. The image should be "read" as follows: "Standard 15288 process has a Title, has a Purpose, Receives Inputs, is Composed of Activities, which are further composed of Tasks, producing Outcomes which fulfills Expected process outputs."

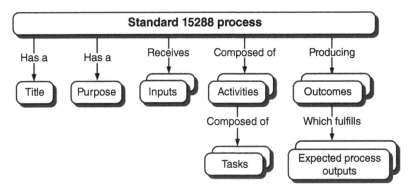

Figure 2.7 Elements of a process: presented by way of a concept map

2.2.6 Further Reading

- Blanchard and Blyler, 2016.
- Blanchard and Fabrycky, 2010.
- Boehm, 1979.
- de Weck et al., 2011.
- Holt et al., 2016.
- INCOSE, 2015.
- ISO/IEC/IEEE, 2015.

- Kossiakoff et al., 2011.
- Maier and Rechtin, 2009.
- Micouin, 2014.
- Morse and Babcock, 2013.
- NASA, 2007.
- Sage and Rouse, 2014.
- Wasson, 2015.

2.3 Standard 15288 Processes

The following summarizes portions of standard 15288[5] processes. The standard breaks systems' life cycle processes into four process groups: (1) agreement processes, (2) organizational project-enabling processes, (3) technical management processes, and (4) technical processes. Within this summary, each individual life cycle process is described in terms of: (1) the purpose of the given process and (2) the activities and tasks needed to achieve this purpose. In addition, for each process, the book identifies a relevant small set of recommended creative methods (Also see Appendix A: Life Cycle Processes versus Recommended Creative Methods).

---

5 As mentioned above, a partial and abridged version of the International Standard ISO/IEC/IEEE 15288, Systems and software engineering, System life cycle processes, was reproduced with permission from the Standards Institution of Israel (SII) on behalf of the International Organization for Standardization (ISO). Copyright remains with ISO.

The reader should note that most creative methods, described in Part III, could help systems engineers in performing the majority of these life cycle processes. Nevertheless, the book provides individual sets of recommended creative methods, which are considered more applicable for specific life cycle process.

2.3.1 Agreement Process Group

This section summarizes the first life cycle processes group of standard 15288, the *agreement processes*. This process group discusses how to establish agreements among internal or external organizations. It is composed of two parts as follows (Figure 2.8).

2.3.1 Agreement Processes

• Acquisition Process
• Supply Process

Figure 2.8 Individual processes within the agreement processes

2.3.1.1 Acquisition Process

Purpose. The purpose of this process is to obtain products or services in accordance with agreed contracts.

Activities and tasks. The acquirer shall implement the following activities and tasks:

I) Prepare for the acquisition by defining the acquisition parameters and then prepare an appropriate request for proposal (RFP).

II) Select a supplier for the required product or service.

III) Establish and maintain an acquisition agreement by developing, negotiating, and maintaining the agreement as necessary.

IV) Monitor the agreement by assessing its execution and resolving open issues in a timely manner.

V) Accept the said product or service by confirming that it complies with the agreement as well as providing payment or other agreed compensation.

Recommended creative methods

• 3.2.2 Resolving Contradictions	• 3.3.1 Classical Brainstorming
• 3.2.4.1 Concept Map	• 3.3.5 Focus Groups
• 3.2.4.2 Concept Fan	

2.3.1.2 Supply Process

Purpose. The purpose of this process is to supply products or services in accordance with agreed contracts.

Activities and tasks. The supplier shall implement the following activities and tasks:

I) Prepare for the supply by determining the existence and identity of relevant acquirers and define an appropriate supply strategy.

II) Respond to an RFP or a tender by determining its feasibility and the appropriate response to it. Then, prepare a response that satisfies its requirements.

III) Establish and maintain a supply agreement by developing, negotiating and maintaining the agreement as necessary.

IV) Execute and assess the agreement according to the established project plan.

V) Deliver and support the product or service by providing assistance to the acquirer in accordance with (IAW) the agreement. In addition, accept and acknowledge payment or other agreed compensation.

Recommended creative methods

• 3.2.2 Resolving Contradictions	• 3.3.1 Classical Brainstorming
• 3.2.4.1 Concept Map	• 3.3.5 Focus Groups
• 3.2.4.2 Concept Fan	

2.3.2 Organizational Project-Enabling Process Group

This section summarizes the second life cycle process group of standard 15288, the *organizational project-enabling processes*. This process group encompasses processes that help ensure the organization's capability to acquire and supply products or services through the initiation, support, and control of projects. It is composed of six parts as follows (Figure 2.9).

2.3.2 Organizational Project-Enabling Processes	
• Life Cycle Model Management Process	• Human Resource Management Process
• Infrastructure Management Process	• Quality Management Process
• Portfolio Management Process	• Knowledge Management Process

Figure 2.9 Individual processes within the organizational project-enabling processes

2.3.2.1 Life Cycle Model Management Process

Purpose. The purpose of this process is to define, maintain, and assure availability of policies, life cycle processes, life cycle models, and procedures for use by the organization.

Activities and tasks. The organization shall implement the following activities and tasks:

I) Establish a process that will include: (1) definition of policies and procedures for process management and deployment, (2) definition of processes that implement standard 15288 and also be consistent with the organization strategies, (3) definition of the stakeholders' roles, responsibilities and accountabilities, (4) definition of business criteria that will control projects' progression through their life cycle, and (5) creation of standard life cycle model for the organization.

II) Assess the process by monitoring its execution across the organization as well as conducting periodic reviews of the life cycle models used by the projects with the intent, among other things, of identifying improvement opportunities.

III) Improve the process by prioritizing and planning improvement opportunities as well as implementing them.

Recommended creative methods

• 3.2.4.3 Mind-Mapping	• 3.5.1 Delphi Method
• 3.3.2 Six Thinking Hats	• 3.6.1 Process Map Analysis
• 3.4.4 Value Analysis/Value Engineering	• 3.6.3 Technology Forecasting

2.3.2.2 Infrastructure Management Process

Purpose. The purpose of this process is to provide the infrastructure and services, including facilities, tools, communications and information technology and so forth, needed to support the organization throughout projects' life cycles.

Activities and tasks. The organization shall implement the following activities and tasks:

I) Establish the infrastructure by defining project's infrastructure requirements as well as identifying and providing needed infrastructure resources.

II) Maintain the infrastructure by evaluating the suitability of the existing infrastructure resources relative to projects' needs as well as identifying and providing improvements or changes to the infrastructure resources when projects' requirements change.

Recommended creative methods

• 3.2.4.3 Mind-Mapping	• 3.5.3 Cause-and-Effect Diagram
• 3.3.2 Six Thinking Hats	• 3.6.5 FMEA Analysis
• 3.3.5 Focus Groups	

2.3.2.3 Portfolio Management Process

Purpose. The purpose of this process is to initiate and sustain necessary, sufficient, and suitable projects in order to meet the strategic objectives of the organization.

Activities and tasks. The organization shall implement the following activities and tasks:

I) Define and authorize new business initiatives by: (1) identifying potential new or modified projects, (2) prioritizing and establishing new business opportunities, (3) defining projects' accountabilities, and authorities, (4) identifying the expected goals, objectives, and outcomes of each project, (5) identifying and allocating resources for the achievement of projects' objectives, (6) identifying any multiproject dependencies to be managed by each project, (7) specifying the project milestones and reporting requirements, and (8) authorizing each project to commence execution.

II) Evaluate the portfolio of projects by assessing each project in order to confirm ongoing viability. In addition, act to continue successful projects or redirect or terminate failed ones.

Recommended creative methods

• 3.2.1 Lateral Thinking	• 3.5.1 Delphi Method
• 3.3.1 Classical Brainstorming	• 3.5.3 Cause-and-Effect Diagram
• 3.3.2 Six Thinking Hats	• 3.6.3 Technology Forecasting
• 3.4.5 Pareto Analysis	• 3.6.7 Conflict Analysis and Resolution

2.3.2.4 Human Resource Management Process

Purpose. The purpose of this process is to provide the organization with necessary human resources and to maintain their competencies, consistent with business needs.

Activities and tasks. The organization shall implement the following activities and tasks:

I) Identify strategic skills based on current and expected projects as well as identifying and recording existing skilled personnel within the organization.

II) Develop skills by: (1) establishing skills development strategy, (2) obtaining or developing training and education, (3) providing planned skill development, and (4) maintaining records of skill development.

III) Acquiring and providing skills by: (1) obtaining qualified personnel, (2) maintaining a pool of skilled personnel, (3) making project assignments based on staff-development considerations, (4) motivating personnel through career development, and (5) controlling multi-project management interfaces to resolve personnel conflicts.

Recommended creative methods

• 3.2.4.3 Mind-Mapping	• 3.6.3 Technology Forecasting
• 3.4.3 Decision Tree Analysis	• 3.6.7 Conflict Analysis and Resolution
• 3.4.5 Pareto Analysis	

2.3.2.5 Quality Management Process

Purpose. The purpose of this process is to assure that products and services meet organizational and project quality objectives.

Activities and tasks. The organization shall implement the following activities and tasks:

I) Plan quality management (QM) by: (1) establishing QM policies, objectives and procedures, (2) defining responsibilities and authorities for implementation of QM, (3) defining quality evaluation criteria and methods, and (4) providing resources and information for QM.

II) Assess QM by: (1) gathering and analyzing quality data, (2) assessing customer satisfaction, and (3) conducting periodic reviews and monitoring project quality data.

III) Perform QM corrective and preventive action by planning preventive and corrective actions as well as monitoring these actions to completion and informing relevant stakeholders.

Recommended creative methods

• 3.2.3 Biomimicry Innovation	• 3.5.4 Kano Model Analysis
• 3.3.1 Classical Brainstorming	• 3.6.2 Nine-Screens Analysis
• 3.3.4 SCAMPER Analysis	

2.3.2.6 Knowledge Management Process

Purpose. The purpose of this process is to create the capability and assets that enable the organization to exploit opportunities by reapplying existing knowledge.

Activities and tasks. The organization shall implement the following activities and tasks:

I) Plan knowledge management by: (1) defining a knowledge management strategy, (2) identifying the knowledge, skills, and knowledge assets to be managed, and (3) identifying projects that can benefit from this knowledge.

II) Share knowledge and skills throughout the organization by: (1) establishing and maintaining a classification for capturing and sharing knowledge across the organization, (2) capturing or acquiring knowledge, and (3) sharing knowledge across the organization.

III) Manage knowledge assets by: (1) maintaining knowledge assets, (2) monitoring and recording the use of knowledge assets, and (3) reassessing periodically the market needs vis-a-vis knowledge assets.

Recommended creative methods

• 3.2.1 Lateral Thinking	• 3.3.5 Focus Groups
• 3.2.4.3 Mind-Mapping	• 3.4.4 Value Analysis/Value Engineering
• 3.3.3 SWOT Analysis	• 3.6.4 DSM Analysis

2.3.3 Technical Management Process Group

This section summarizes the third life cycle process group of standard 15288, the *technical management processes*. This process group is utilized to establish and evolve plans, to execute the plans, to assess actual achievement and progress against the plans and to control project's execution through to fulfillment. It is composed of eight parts, as follows (Figure 2.10).

2.3.3 Technical Management Processes

• Project Planning Process	• Configuration Management Process
• Project Assessment and Control Process	• Information Management Process
• Decision Management Process	• Measurement Process
• Risk Management Process	• Quality Assurance Process

Figure 2.10 Individual processes within the technical management processes

2.3.3.1 Project Planning Process

Purpose. The purpose of this process is to produce and coordinate effective and workable project plans.

Activities and tasks. The project shall implement the following activities and tasks:

I) Define the project by: (1) identifying the project objectives and constraints, (2) defining the project scope, (3) defining and maintaining appropriate life cycle model, (4) establishing appropriate work breakdown structure (WBS), and (5) defining and maintaining appropriate formal processes.

II) Plan the project and technical management by: (1) defining and maintaining a project schedule based on management and technical objectives and work estimates, (2) defining life cycle decision gates, delivery dates and dependencies on external inputs, (3) defining and planning appropriate budget, (4) defining roles, responsibilities, accountabilities, and authorities, (5) defining the infrastructure and services required, (6) planning the acquisition of materials, enabling products and services supplied from outside the project, and (7) generating and communicating a plan for project and technical management and execution.

III) Activate the project by obtaining internal authorization for the project as well as obtaining management commitments for the necessary resources to implement the project.

Recommended creative methods

• 3.2.4.1 Concept Map	• 3.5.2 SAST Analysis
• 3.4.2 Morphological Analysis	• 3.6.4 DSM Analysis
• 3.4.4 Value Analysis/Value Engineering	• 3.6.6 Anticipatory Failure Determination

2.3.3.2 Project Assessment and Control Process

Purpose. The purpose of this process is to assess the status of the project and help ensure that its performance meet plans, schedules, budgets, and technical objectives.

Activities and tasks. The project shall implement the following activities and tasks:

I) Plan for project assessment and control by defining the project assessment and control strategy.

II) Assess the project by: (1) assessing alignment between project objectives and plans versus project technical status, (2) assessing management and technical plans against project objectives, (3) assessing the adequacy of roles, responsibilities, accountabilities, and authorities as well as project resources, (4) assessing progress using measured achievement and milestone completion, (5) conducting required management and technical reviews, audits, and inspections, (6) measuring and analyzing critical processes as well as new technologies and making appropriate recommendations, (7) recording and disseminating project status, and (8) monitoring process execution within the project.

III) Control the project by: (1) initiating necessary actions needed to address identified issues, especially with regards to contractual changes to costs, time, or quality due to impact of an acquirer or supplier request, (2) authorizing the project to proceed toward the next milestone or event, if justified.

Recommended creative methods

• 3.2.2 Resolving Contradictions	• 3.5.5, 3.5.6 Group Decisions (theory
• 3.4.1 PMI Analysis	& practice)
• 3.5.2 SAST Analysis	• 3.6.7 Conflict Analysis and Resolution

2.3.3.3   Decision Management Process

Purpose. The purpose of this process is to provide a structured framework for objectively identifying, characterizing, and evaluating decision alternatives, and selecting the most beneficial course of action.

Activities and tasks. The project shall implement the following activities and tasks:

I) Prepare for decisions by defining a decision management strategy as well as identifying the circumstances whereby decisions-making processes should involve relevant stakeholders.

II) Analyze the decision information by: (1) selecting and declaring the decision management strategy for each decision, (2) determining a desired outcomes and measurable selection criteria, (3) identifying the alternative possible decision, and (4) evaluating each alternative against the selected criteria.

III) Make and manage decisions by: (1) determining preferred alternative for each decision, and (2) recording the selected decision, its rationale, and relevant assumptions, as well as tracking and evaluating past decisions.

Recommended creative methods

• 3.3.3 SWOT Analysis	• 3.5.4 Kano Model Analysis
• 3.3.4 SCAMPER Analysis	• 3.6.1 Process Map Analysis
• 3.4.2 Morphological Analysis	

2.3.3.4 Risk Management Process

Purpose. The purpose of this process is to continually identify, analyze, treat, and monitor all relevant risks.

Activities and tasks. The project shall implement the following activities and tasks:

I) Plan risk management by defining the risk management strategy as well as recording the context of the risk management process.

II) Manage the risk profile by: (1) defining and recording the risk thresholds and conditions under which risk level may be accepted, (2) establishing and maintaining a risk profile, and (3) periodically providing the applicable risk profile to relevant stakeholders for assessment.

III) Analyze risks by: (1) identifying risks within categories described in the risk management context, (2) estimating the likelihood of occurrence and consequences of each identified risk, (3) evaluating each risk against its risk thresholds, and (4) defining and recording recommended treatment strategies for each risk.

IV) Treat risks by implementing the most appropriate risk treatment alternatives. This shall be done after coordinating with relevant stakeholders and obtaining their concurrence.

V) Monitor risks by (1) continually monitoring all risks for changes and evaluating risks when their state has changed, (2) implementing and monitoring measures to evaluate the effectiveness of risk treatments, and (3) continually monitoring for the emergence of new risks throughout the life cycle.

Recommended creative methods

• 3.2.3 Biomimicry Innovation	• 3.5.4 Kano Model Analysis
• 3.4.3 Decision Tree Analysis	• 3.6.5 FMEA Analysis
• 3.5.3 Cause-and-Effect Diagram	• 3.6.6 Anticipatory Failure Determination

2.3.3.5 Configuration Management Process

Purpose. The purpose of this process is to manage and control system elements and ensure consistency between products and their associated configuration definition.

Activities and tasks. The project shall implement the following activities and tasks:

I) Plan configuration management (CM) by defining a CM strategy as well as defining the archive and retrieval approach for configuration items and other pertinent artifacts.

II) Perform configuration identification by: (1) identifying the system elements and information items that are configuration items, (2) identifying the hierarchy and structure of all configuration items, (3) establishing system, system element, and information item identifiers, (4) defining baselines through the life cycle, and (5) obtaining acquirers and suppliers agreement to this baseline.

III) Perform configuration change management by: (1) identifying and recording requests for change (RFC) and requests for variance (RFV), (2) coordinating, evaluating, and disposing such RFCs and RFVs, (3) submitting requests for review, and approval, and (4) tracking and managing approved changes to the baseline.

IV) Perform configuration status accounting by developing and maintaining the CM status information as well as capturing, storing, and reporting CM status.

V) Perform configuration evaluation by: (1) identifying the need for CM audits and schedule the events, (2) verifying whether the product configuration meets the configuration requirements, (3) monitoring the status of the approved configuration changes, (4) assessing whether the system meets baseline functional and performance capabilities as well as whether it conforms to the operational and configuration information items, and (5) recording the CM audit results and disposition action items.

VI) Perform release control by approving system releases and deliveries as well as tracking and managing system releases and deliveries.

Recommended creative methods

• 3.4.3 Decision Tree Analysis	• 3.6.4 DSM Analysis

2.3.3.6 Information Management Process

Purpose. The purpose of this process is to generate, obtain, confirm, transform, retain, retrieve, disseminate and dispose of information, to designated stakeholders. Information includes, among other, technical, project, organizational, agreement, and user information.

Activities and tasks. The project shall implement the following activities and tasks.

I) Prepare for information management by: (1) defining the strategy for information management, (2) defining the items of information that will be managed, (3) designating the authorities and responsibilities for information management, (4) defining the content, formats and structure of the information items, and (5) defining information maintenance actions.

II) Perform information management by: (1) obtaining, developing, or transforming the identified items of information, (2) maintaining the information items, (3) publishing, distributing, and providing access to information items to designated stakeholders, (4) archiving designated information, and (5) disposing unwanted or invalid information.

Recommended creative methods

• 3.2.3 Biomimicry Innovation	• 3.5.4 Kano Model Analysis
• 3.4.1 PMI Analysis	• 3.5.5, 3.5.6 Group Decisions
• 3.5.1 Delphi Method	(theory & practice)
	• 3.6.4 DSM Analysis

2.3.3.7 Measurement Process

Purpose. The purpose of this process is to collect, analyze, and report objective data and information to support effective management as well as providing quantitative estimates regarding the quality of the products, services, and processes.

Activities and tasks. The project shall implement the following activities and tasks:

I) Prepare for measurement by: (1) defining the measurement strategy, (2) describing the characteristics of the organization that are relevant for measurement, (3) identifying and prioritizing the information needs, (4) selecting and specifying measures that satisfy the information needs, (5) defining data collection, analysis, access, and reporting procedures, (6) defining criteria for evaluating the information items and the measurement process, and (7) identifying and planning for the necessary enabling systems or services to be used.

II) Perform measurement by: (1) integrating procedures for data generation, collection, analysis, and reporting into the relevant processes, (2) collecting, storing, and verifying data, (3) analyzing the data, and (4) recording results and inform relevant to stakeholders.

Recommended creative methods

• 3.3.2 Six Thinking Hats	• 3.6.5 FMEA Analysis
• 3.3.5 Focus Groups	• 3.6.6 Anticipatory Failure Determination
• 3.5.2 SAST Analysis	

2.3.3.8   Quality Assurance Process

Purpose. The purpose of this process is to help ensure the effective application of the organization's quality management (QM) process to the project. This will increase confidence that quality requirements will be fulfilled.

Activities and tasks. The project shall implement the following activities and tasks:

I) Prepare for quality assurance (QA) by defining a QA strategy and insuring the independence of the QA team from other life cycle processes.

II) Perform product or service evaluations by: (1) evaluating products and services for conformance with established criteria, contracts, standards, and regulations, and (2) performing verification and validation of the outputs of the life cycle processes to determine conformance to specified requirements.

III) Perform process evaluations by: (1) evaluating project life cycle processes for conformance, (2) evaluating tools and environments that support the conformance process, and (3) evaluating supplier processes for conformance to process requirements.

IV) Manage QA records and reports by: (1) creating records related to QA activities, (2) maintaining and distributing these records to relevant stakeholders, and (3) identifying incidents and problems associated with products, services, and process evaluations.

V) Treat incidents and problems by: (1) analyzing, classifying and resolving problems, (2) prioritizing the treatments of problems and tracking their resolution, (3) noting and analyzing trends in the occurrence of incidents and problems, and (4) informing relevant stakeholders regarding the status of incidents and problems.

Recommended creative methods

• 3.2.3 Biomimicry Innovation	• 3.5.4 Kano Model Analysis
• 3.3.1 Classical Brainstorming	• 3.5.5, 3.5.6 Group Decisions
• 3.3.2 Six Thinking Hats	(theory & practice)
	• 3.6.5 FMEA Analysis

2.3.4   Technical Process Group

This section summarizes the fourth life cycle process group of standard 15288, the *technical processes*. The technical processes are used to define the requirements for a system, to transform the requirements into an effective system, to support consistent reproduction of the system where necessary, to utilize the system in order to provide the required services, to sustain the provision of those services and finally, to dispose of the system when it retires from service. It is composed of 14 parts as follows (Figure 2.11).

<table>
<tr><td colspan="2" align="center">**2.3.4 Technical Processes**</td></tr>
<tr><td>

• Business or Mission Analysis Process
• Stakeholder Needs and Requirements Definition Process
• System Requirements Definition Process
• Architecture Definition Process
• Design Definition Process
• System Analysis Process
• Implementation Process
</td><td>

• Integration Process
• Verification Process
• Transition Process
• Validation Process
• Operation Process
• Maintenance Process
• Disposal Process
</td></tr>
</table>

Figure 2.11 Individual processes within the "technical processes"

2.3.4.1 Business or Mission Analysis Process

Purpose. The purpose of this process is to define the business or mission problem or opportunity, characterize the solution space, and determine potential solution classes that could address a problem or take advantage of an opportunity.

Activities and tasks. The project shall implement the following activities and tasks:

I) Prepare for business or mission analysis by: (1) reviewing identified problems and opportunities in the organization strategy, (2) defining the organization's business strategy, (3) identifying and planning the enabling systems or services needed in support of business or mission analysis, and (4) obtaining access to such enabling systems or services.

II) Define the problem or opportunity space by analyzing relevant problems and opportunities as well as defining the mission, business, or operational problem or opportunity.

III) Characterize the solution space by defining preliminary operational concepts and other concepts in life cycle stages, as well as identifying candidate alternative solution classes.

IV) Evaluate alternative solution classes by assessing each of them and selecting the preferred one.

V) Manage the business or mission analysis by maintaining traceability of business or mission analysis and providing key information items that have been selected.

Recommended creative methods

• 3.2.1 Lateral Thinking	• 3.4.5 Pareto Analysis
• 3.2.3 Biomimicry Innovation	• 3.5.4 Kano Model Analysis
• 3.2.4.2 Concept Fan	• 3.6.2 Nine-Screens Analysis
• 3.3.3 SWOT Analysis	• 3.6.7 Conflict Analysis and Resolution
• 3.3.4 SCAMPER Analysis	

2.3.4.2 Stakeholder Needs and Requirements Definition Process

Purpose. The purpose of this process is to define the stakeholder requirements for a system and its environment.

Activities and tasks. The project shall implement the following activities and tasks:

I) Prepare for stakeholder needs and requirements by: (1) identifying the stakeholders who have an interest in the system throughout its life cycle, (2) defining the stakeholder needs and requirements, (3) identifying and planning the enabling systems or services needed to support the stakeholders and (4) obtaining or acquiring access to these enabling systems or services.

II) Define stakeholder needs by: (1) defining the context of use within the system life cycle, (2) identifying the actual stakeholder needs and rational, and (3) prioritizing and down-selecting these needs.

III) Develop the operational concept and other life cycle concepts by: (1) defining a representative set of scenarios, related to capabilities corresponding with anticipated operational and other life cycle concepts, and (2) identifying the interactions between users and the system.

IV) Transform stakeholder needs into stakeholder requirements by: (1) identifying the constraints on a system solution and (2) defining stakeholder requirements, consistent with life cycle concepts, scenarios, interactions, constraints, and critical quality characteristics.

V) Analyze stakeholder requirements by: (1) analyzing the complete set of stakeholder requirements, (2) defining critical performance measures that enable the assessment of technical achievement, (3) feeding back the analyzed requirements to applicable stakeholders in order to validate that their needs and expectations have been adequately captured, and (4) resolving stakeholder requirements issues.

VI) Manage the stakeholder needs and requirements definition by: (1) obtaining explicit agreement on the stakeholder requirements, (2) maintaining traceability of stakeholder needs and requirements, and (3) providing key information items that have been selected for a baseline.

Recommended creative methods

• 3.2.1 Lateral Thinking	• 3.3.4 SCAMPER Analysis
• 3.2.3 Biomimicry Innovation	• 3.4.5 Pareto Analysis
• 3.2.4.2 Concept Fan	• 3.5.4 Kano Model Analysis
• 3.3.3 SWOT Analysis	• 3.6.2 Nine-Screens Analysis

2.3.4.3 System Requirements Definition Process

Purpose. The purpose of this process is to transform the stakeholder and user-oriented view of desired capabilities into a technical view of a solution that meets the operational needs of the users.

Activities and tasks. The project shall implement the following activities and tasks:

I) Prepare for system requirements definition by: (1) defining the functional boundary between the system and its environment as well as its behavior and properties, (2) defining the system requirements definition strategy, (3) identifying and planning for the necessary enabling systems or services needed to support system requirements definition, and (4) obtaining or acquiring access to these enabling systems or services.

II) Define system requirements by: (1) defining each function that the system is required to perform, (2) defining all implementation constraints, (3) identifying system requirements that relate to risks, criticality of the system, or critical quality characteristics, and (4) defining system requirements and their rationale.

III) Analyze system requirements by: (1) analyzing the complete set of system requirements, (2) defining critical performance measures that enable the assessment of the technical achievement, (3) feeding back the analyzed requirements to applicable stakeholders for review, and (4) resolving system requirements issues.

IV) Manage system requirements by: (1) obtaining explicit agreement on the system requirements, (2) maintaining traceability of the system requirements, and (3) providing key information items that have been selected for the baseline.

Recommended creative methods

• 3.2.1 Lateral Thinking	• 3.4.2 Morphological Analysis
• 3.2.2 Resolving Contradictions	• 3.4.3 Decision Tree Analysis
• 3.3.1 Classical Brainstorming	• 3.6.1 Process Map Analysis
• 3.3.2 Six Thinking Hats	• 3.6.6 Anticipatory Failure Determination

2.3.4.4 Architecture Definition Process

Purpose. The purpose of this process is to generate system architecture alternatives as well as to select one or more alternative that meet system requirements expressing it in a consistent framework.

Activities and tasks. The project shall implement the following activities and tasks:

I) Prepare for architecture definition by: (1) reviewing pertinent information and identifying key drivers of the architecture, (2) identifying stakeholder needs, (3) defining the architecture approach and strategy, (4) defining evaluation criteria based on stakeholders requirements, (5) identifying and planning for the necessary enabling systems or services needed to support the architecture definition process, and (6) obtaining or acquiring access to these enabling systems or services.

II) Develop architecture viewpoints by: (1) selecting, adapting, or developing viewpoints and models based on stakeholder needs, (2) establishing or identifying potential architecture framework to be used in developing models and views, (3) capturing rationale for selection of framework, viewpoints, and models, and (4) selecting or developing supporting modeling techniques and tools.

III) Develop models and views of candidate architectures by: (1) defining the system context and boundaries as well as interfaces and interactions with external entities, (2) identifying architectural and relationships between entities that address key stakeholder requirements, (3) allocating concepts, properties, characteristics, behaviors, functions, or constraints to these architectural entities, (4) selecting, adapting or developing models of the candidate architectures of the system, (5) composing views from these models in order to addresses stakeholder requirements, and (6) harmonizing the architecture models and views with each other.

IV) Relate the system architecture to its design by: (1) identifying system elements that relate to architectural entities and the nature of these relationships, (2) defining the interfaces and interactions between the system elements with external entities, (3) allocating requirements to architectural entities and system elements, (4) mapping system elements and architectural entities to the system design, and (5) defining principles for the system design and evolution.

V) Assess architecture candidates by: (1) assessing each candidate architecture against constraints and requirements, (2) assessing each candidate architecture against stakeholder requirements using evaluation criteria, (3) selecting the preferred architecture and capturing key decisions and rationale, and (4) establishing the architecture baseline for the system.

VI) Manage the selected architecture by: (1) formalizing the architecture approach as well as specifying the roles, responsibilities, accountabilities, and authorities with respect to (WRT) its design, quality, security, safety, etc., (2) obtaining explicit acceptance of the architecture from the system's stakeholders, (3) maintaining completeness of the architectural entities and their architectural characteristics, (4) organizing, assessing, and controlling the evolution of the architecture models and views, (5) maintaining the architecture definition and evaluation strategy, (6) maintaining traceability of the architecture, and (7) providing key information items that have been selected for the baseline.

Recommended creative methods

- 3.2.1 Lateral Thinking
- 3.2.2 Resolving Contradictions
- 3.2.3 Biomimicry Innovation
- 3.2.4.1 Concept Map
- 3.4.1 PMI Analysis
- 3.4.4 Value Analysis/Value Engineering
- 3.5.3 Cause-and-Effect Diagram
- 3.6.2 Nine-Screens Analysis
- 3.6.3 Technology Forecasting

2.3.4.5 Design Definition Process

Purpose. The purpose of this process is to provide sufficient detailed data and information about the system and its elements to enable its implementation consistent with the architectural models and views.

Activities and tasks. The project shall implement the following activities and tasks:

I) Prepare for the design definition by: (1) determining technologies required for each system element, (2) defining the principles for evolution of the design, (3) defining the design strategy, (4) identifying and planning for the necessary enabling systems or services needed to support the design process, and (5) obtaining or acquiring access to these enabling systems or services.

II) Establish the design characteristics and design enablers related to each system element by: (1) allocating system requirements to system elements, (2) transforming architectural characteristics into the design concepts, (3) defining the necessary design enablers, (4) examining design alternatives, (5) defining the interfaces among the system elements as well as the interfaces between the system and its environment, and (6) establishing the design artifacts.

III) Assess alternatives for obtaining system elements by: (1) identifying non-developmental items (NDI) candidates and (2) determining the preferred alternative among these NDI solutions.

IV) Manage the design by: (1) mapping the design characteristics up to the system elements, (2) capturing the design and its rationale, (3) maintaining traceability of the design, and (4) providing key information items that have been selected for the baseline.

Recommended creative methods

• 3.2.4.1 Concept Map	• 3.5.1 Delphi Method
• 3.3.1 Classical Brainstorming	• 3.5.3 Cause-and-Effect Diagram
• 3.4.3 Decision Tree Analysis	• 3.6.1 Process Map Analysis

2.3.4.6 System Analysis Process

Purpose. The purpose of this process is to provide a rigorous basis of data and information for thorough technical understanding of the system in order to aid the decision-making process throughout the life cycle.

Activities and tasks. The project shall implement the following activities and tasks:

I) Prepare for system analysis by: (1) identifying the problems or questions that requires system analysis, (2) identifying the relevant stakeholders, (3) defining the scope, objectives, and level of fidelity as well as the strategy of the system analysis, (4) identifying and planning for the necessary enabling systems or services needed to support system analysis, (5) obtaining or acquiring access to these enabling systems or services, and (6) collecting the data needed for the analysis.

II) Perform system analysis by: (1) identifying and validating relevant assumptions, (2) applying the selected analysis methods to perform the required system analysis, (3) reviewing the analysis results for quality and validity, (4) establishing conclusions and recommendations, and (5) recording the results.

III) Manage system analysis by maintaining traceability of the analysis results and providing key information items that have been selected for the baseline.

Recommended creative methods

• 3.2.4.2 Concept Fan	• 3.6.2 Nine-Screens Analysis
• 3.4.1 PMI Analysis	• 3.6.3 Technology Forecasting
• 3.4.5 Pareto Analysis	• 3.6.4 DSM Analysis
• 3.5.3 Cause-and-Effect Diagram	

2.3.4.7 Implementation Process

Purpose. The purpose of this process is to realize a specified system element.

Activities and tasks. The project shall implement the following activities and tasks:

I) Prepare for implementation by: (1) defining an implementation strategy, (2) identifying constraints to the implementation strategy and implementation technology, (3) identifying and planning for the necessary enabling systems or services needed to support the implementation, and (4) obtaining or acquiring access to these enabling systems or services.

II) Perform implementation by: (1) realizing or adapting system elements, according to the strategy, constraints and defined implementation procedures, (2) packaging and storing the system element, and (3) recording objective evidence that the system element meets system requirements.

III) Manage the results of the implementation by recording any anomalies encountered and maintaining traceability of the implemented system elements.

Recommended creative methods

• 3.2.1 Lateral Thinking	• 3.5.2 SAST Analysis
• 3.2.2 Resolving Contradictions	• 3.6.1 Process Map Analysis
• 3.2.4.2 Concept Fan	• 3.6.3 Technology Forecasting
• 3.3.1 Classical Brainstorming	• 3.6.6 Anticipatory Failure Determination
• 3.4.2 Morphological Analysis	

2.3.4.8 Integration Process

Purpose. The purpose of this process is to integrate a set of system elements into a realized system or service that satisfies system requirements, architecture, and design.

Activities and tasks. The project shall implement the following activities and tasks:

I) Prepare for integration by: (1) identifying a set of check points for the correct operation of the assembled interfaces as well as the selected system functions, (2) defining the integration strategy, (3) identifying and planning for the necessary enabling systems or services needed to support integration, (4) obtaining or acquiring access to these enabling systems or services as well as materials to be used, and (5) identifying any relevant integration constraints to be incorporated in relevant system's documentation.

II) Perform integration by successively integrating system elements until the complete system is incorporated. This will be done by: (1) obtaining implemented system elements in accordance with agreed schedules, (2) assembling the implemented system elements, and (3) checking the interfaces, selected functions as well as the required quality characteristics.

III) Manage results of the integration by: (1) recording the integration results and any anomalies encountered, (2) maintaining traceability of the integrated system elements, and (3) providing relevant information items that have been selected for baselines.

Recommended creative methods

• 3.4.3 Decision Tree Analysis	• 3.5.3 Cause-and-Effect Diagram
• 3.4.4 Value Analysis/Value Engineering	• 3.6.1 Process Map Analysis
• 3.4.5 Pareto Analysis	• 3.6.4 DSM Analysis

2.3.4.9 Verification Process

Purpose. The purpose of this process is to provide objective evidence that system elements as well as the complete system fulfills its specified requirements and characteristics.

Activities and tasks. The project shall implement the following activities and tasks:

I) Prepare for verification by: (1) identifying the verification scope and corresponding verification actions, (2) identifying the constraints that may limit the feasibility of verification actions, (3) selecting appropriate verification methods and associated criteria for every verification action, (4) defining the verification strategy, (5) identifying and planning for the necessary enabling systems or services needed to support the verification, and (6) obtaining or acquiring access to these enabling systems or services.

II) Perform verification by defining the verification procedures for each system elements or the complete system and then performing these procedures.

III) Manage the results of the verification by: (1) recording the verification results and any anomalies encountered, (2) recording operational incidents and problems and track their resolution, (3) obtaining stakeholder agreement regarding the status of the system, (4) maintaining traceability of the verified system elements, and (5) providing key information items that have been selected for the baseline.

Recommended creative methods

• 3.2.2 Resolving Contradictions	• 3.5.5, 3.5.6 Group Decisions (theory & practice)
• 3.4.3 Decision Tree Analysis	
• 3.4.4 Value Analysis/Value Engineering	• 3.6.1 Process Map Analysis
• 3.5.4 Kano Model Analysis	• 3.6.5 FMEA Analysis
	• 3.6.6 Anticipatory Failure Determination

2.3.4.10 Transition Process

Purpose. The purpose of this process is to establish a capability for a system to provide services specified by stakeholder requirements in its operational environment.

Activities and tasks. The project shall implement the following activities and tasks:

I) Prepare for the transition by: (1) defining a transition strategy, (2) identifying any facility or site changes needed, (3) identifying and arranging training of operators, users, and other stakeholders necessary to ensure system utilization and support, (4) identifying system constraints vis-a-vis the transition process to be incorporated in the relevant system's documents, (5) identifying and planning for the necessary enabling systems or services needed to support the transition, (6) obtaining or acquiring access to these enabling systems or services, and (7) identifying and arranging shipping and receiving of all the relevant systems' operational and support products.

II) Perform the transition by: (1) preparing the site of the operation in accordance with installation requirements, (2) delivering the system for installation at the required location on time, (3) installing the system in its operational location and connecting it to its environment, (4) confirming the proper installation of the system, (5) providing training of the operators, users, and other stakeholders necessary for proper system utilization and support, (6) performing activation and check-out of the system, (7) confirming that the installed system is capable of delivering its required functions, (8) confirming that the functions provided by the system are sustainable by the enabling systems, (9) reviewing the system for operational readiness, and (10) commissioning the system for operations.

III) Manage results of the transition by (1) recording transition results and any anomalies encountered, (2) recording operational incidents and problems and tracking their resolution, (3) maintaining traceability of the transitioned system elements, and (4) providing key information items that have been selected for the baseline.

Recommended creative methods

• 3.2.4.3 Mind-Mapping	• 3.5.1 Delphi Method
• 3.4.2 Morphological Analysis	• 3.6.3 Technology Forecasting
• 3.4.4 Value Analysis/Value Engineering	

2.3.4.11 Validation Process

Purpose. The purpose of this process is to provide objective evidence that the system, when in use, fulfills its business or mission objectives and stakeholder requirements, achieving its intended use in its intended operational environment.

Activities and tasks. The project shall implement the following activities and tasks:

   I) Prepare for validation by: (1) identifying the validation scope and corresponding validation actions, (2) identifying the constraints that potentially limit the feasibility of validation actions, (3) selecting appropriate validation methods or techniques and associated criteria for each validation action, (4) defining the validation strategy, (5) identifying and planning for the necessary enabling systems or services needed to support validation, and (6) obtaining or acquiring access to these enabling systems or services.

   II) Perform validation by (1) defining the validation procedures, each supporting one or a set of validation actions, (2) performing the validation procedures in the defined environment, and (3) reviewing validation results to confirm that all stakeholders' requirements are met.

   III) Manage results of the validation by: (1) recording validation results and any anomalies encountered, (2) recording operational incidents and problems and tracking their resolution, (3) obtaining stakeholder agreement that the system meets the stakeholder needs, (4) maintaining traceability of the validated system elements, and (5) providing key information items that have been selected for baselines.

Recommended creative methods

• 3.2.2 Resolving Contradictions	• 3.5.5, 3.5.6 Group Decisions
• 3.4.3 Decision Tree Analysis	(theory & practice)
• 3.4.4 Value Analysis/Value Engineering	• 3.6.1 Process Map Analysis
• 3.5.4 Kano Model Analysis	• 3.6.5 FMEA Analysis
	• 3.6.6 Anticipatory Failure Determination

2.3.4.12 Operation Process

Purpose. The purpose of this process is to utilize the system in order to deliver its services.

Activities and tasks. The project shall implement the following activities and tasks:

   I) Prepare for ongoing operation by: (1) defining an operation strategy, (2) identifying system constraints vis-à-vis ongoing operation to be incorporated in the system's documentation, (3) identifying and planning for the necessary enabling systems or services needed to support the operation, (4) obtaining or acquiring access to these enabling systems or services, (5) defining training and qualification requirements for personnel needed for operating the system, and (6) assigning trained, qualified personnel to be operators.

II) Perform ongoing operation by: (1) using the system in its intended operational environment, (2) applying materials and other resources, as required, to operate the system and sustain its services, (3) monitoring the system operation, (4) identifying and recording when system service performance is not within acceptable parameters, and (5) performing system contingency operations, if necessary.

III) Manage the results of the ongoing operation by: (1) recording results of the operation and any anomalies encountered, (2) recording operational incidents and problems and tracking their resolution, (3) maintaining traceability of the operations elements, and (4) providing key information items that have been selected for the baseline.

IV) Support the customer by: (1) providing assistance and consultation to the customers as requested, (2) recording and monitoring requests and subsequent actions for support, and (3) determining the degree to which the delivered system services satisfy the needs of the customers.

Recommended creative methods

• 3.3.1 Classical Brainstorming	• 3.4.4 Value Analysis/Value Engineering
• 3.3.2 Six Thinking Hats	• 3.6.6 Anticipatory Failure Determination
• 3.3.5 Focus Groups	• 3.6.7 Conflict Analysis and Resolution

2.3.4.13 Maintenance Process

Purpose. The purpose of this process is to sustain the capability of the system to provide its expected service.

Activities and tasks. The project shall implement the following activities and tasks:

I) Prepare for maintenance by: (1) defining a maintenance strategy, (2) identifying system constraints from the maintenance point of view in order to be incorporated in the system's documentation, (3) ensuring that the system maintenance is affordable, operable, supportable, and sustainable, and (4) identifying and planning for the necessary enabling systems or services needed to support maintenance.

II) Perform maintenance by: (1) reviewing incident and problem reports in order to identify future corrective, adaptive, perfective and preventive maintenance needs, (2) recording maintenance incidents and problems and tracking their resolution, (3) implementing the procedures for correction of random faults or scheduled replacement of system elements, (4) acting to restore the system back to its operational status upon encountering random faults that cause a system failure, (5) performing preventive maintenance by replacing or servicing system elements prior to failure, according to planned schedules and maintenance procedures, (6) performing failure identification actions when a noncompliance has occurred in the system, and (7) identifying when adaptive or perfective maintenance is required.

III) Perform logistics support by: (1) performing acquisition logistics and operational logistics, (2) implementing any packaging, handling, storage and transportation needed during the life cycle, (3) confirming that logistics actions satisfy the required replenishment levels so that stored system elements meet repair rates along planned schedules, and (4) confirming that logistics actions include supportability requirements that are planned, resourced, and implemented.

IV) Manage results of maintenance and logistics by: (1) recording maintenance and logistics results and any anomalies encountered, (2) recording operational incidents and problems and tracking their resolution, (3) identifying and recording trends of incidents, problems, and maintenance and logistics actions, (4) maintaining traceability of the maintenance elements, (5) providing key information items that have been selected for the baseline, and (6) monitoring customer satisfaction with system and maintenance support.

Recommended creative methods

• 3.4.3 Decision Tree Analysis	• 3.6.1 Process Map Analysis
• 3.4.5 Pareto Analysis	• 3.6.5 FMEA Analysis
• 3.5.1 Delphi Method	• 3.6.6 Anticipatory Failure Determination

2.3.4.14 Disposal Process

Purpose. The purpose of this process is to end the existence of a system or system element for a specified intended use, appropriately handling replaced or retired elements as well as to properly attend to identified critical disposal needs.

Activities and tasks. The project shall implement the following activities and tasks:

I) Prepare for system disposal by: (1) defining a disposal strategy for the system, as well as each system element and any resulting waste products, (2) identifying system constraints vis-à-vis the disposal process in the system's documentation, (3) identifying and planning for the necessary enabling systems or services needed to support disposal, (4) obtaining or acquiring access to these enabling systems or services to be used, (5) specifying containment facilities, storage locations, inspection criteria, and storage periods, if the system is to be stored, and

(6) defining preventive methods to preclude disposed elements and materials that should not reenter into the supply chain.

II) Perform system disposal by: (1) deactivating the system or system elements to prepare them for removal, (2) removing the system as well as system elements, or waste material from use or production for appropriate disposition and action, (3) withdrawing impacted operating staff from the system or system element and recording relevant operating knowledge, (4) disassembling the system or system elements into manageable parts to facilitate its removal for reuse, recycling, reconditioning, overhaul, archiving, or destruction, (5) handling system elements and their parts that are not intended for reuse in a manner that will assure they do not get back into the supply chain, and (6) conducting destruction of the system elements, as necessary, to reduce the amount of generated waste or to make the waste easier to handle.

III) Finalize the disposal process by: (1) confirming that no detrimental health, safety, security, and environmental factors exist following the disposal, (2) returning the environment to its original state or to a state that specified by a relevant agreement, and (3) archiving information gathered through the lifetime of the system to permit audits and reviews in the event of long-term hazards to health, safety, security, and the environment.

Recommended creative methods

• 3.2.1 Lateral Thinking	• 3.3.2 Six Thinking Hats
• 3.2.2 Resolving Contradictions	• 3.4.3 Decision Tree Analysis
• 3.2.3 Biomimicry Innovation	• 3.5.4 Kano Model Analysis
• 3.2.4.2 Concept Fan	• 3.6.3 Technology Forecasting

2.3.5 Further Reading

• INCOSE, 2015.	• ISO/IEC/IEEE 15288, 2015.

2.4 Philosophy of Engineering

Science and engineering are among the most impressive activities that we have engaged in as a species, so it is well worth a few philosophical thoughts trying to understand a little better how engineering activities actually work. Toward this end, the author of this book adapted a condensed and abridged version of five philosophical papers delivered during a series of seminars held at the Royal Academy of Engineering in London during 2010.[6]

2.4.1 Engineering and Truth

The purpose of this paper is to discuss where the difference between science and engineering actually makes a difference. That is, where the philosophical issues change if we focus on engineering rather than on science. Although there is no clear demarcation between science and engineering, there seem to be real and relevant differences between pure and applied work that are of philosophical importance. Here are three candidate differences between science and engineering: a difference in output, a difference in knowledge, and a difference in drivers.

The first contrast, as many philosophers of science would have it, is that the ultimate output in science is theory – a set of propositions, a set of equations, a set of assertions. Perhaps that is a defensible view of science, but it certainly does not seem to do justice to engineering. Of course, one can't do engineering without generating propositions, for example in specifying a design, but the ultimate output is an artifact, not a statement. It is something physical and manufactured. One certainly would expect that this contrast in ultimate output should make a difference to the form that a proper philosophical analysis should take.

The second contrast concerns knowledge. Epistemologists[7] distinguish between "knowing that" and "knowing how." Knowing that is propositional knowledge – it is knowing that something is the case. It is knowing that a statement is true or that a hypothesis is correct. By contrast, knowing how is ability or a skill, e.g. knowing how to build a bridge. The contrast between knowing that and knowing how suggests a contrast between science and engineering that parallels the contrast between theory and artifact. Philosophers investigating scientific knowledge have concentrated on knowledge that; but if we want to do justice to engineering, it would appear that philosophers need to put considerably more weight onto knowing how.

The third contrast has a different character. It is motivated by the thought that the drivers for problem choice in science and in engineering tend to be different in kind. In pure science, the driver is often internal to the scientific community: scientists often get to choose their own problems. In engineering,

---

6 See: Philosophy of Engineering, Volume 1 of the proceedings of a series of seminars held at The Royal Academy of Engineering, London, June 2010. The authors of the original papers are: P. Lipton, T. Hoare, J. Turnbull, M. Franssen and C. Elliott. Reproduced with permission of the Royal Academy of Engineering, London, UK.

7 Epistemology, from Greek, episteme, meaning "knowledge," and logos, meaning "logical discourse," is the branch of philosophy dealing with the theory of knowledge.

by contrast, it is more common for the driver to be external to the community of practitioners. Engineers often do not get to choose their own problems, but have them chosen instead by governments, by industries, or by other external sources. Here again, we have a difference since the way problems are selected may make a difference to the way they are addressed, such that a philosophical account that is more or less suitable to science does not, as it stands, do justice to the realities of engineering practice.

2.4.2 The Logic of Engineering Design

The first question in engineering is: what does the system do? In greater detail: What are its properties and behavior? How does it interact with its users and its external environment? Here, the engineer often gives an answer more technical in detail than the average user of the system would be interested in. The answer is likely to contain scientific technical terms, like ohms and farads, which the average user will not understand. One might think that the engineer who actually designs and implements a system should find this question trivially easy to answer. Surprisingly, in the case of complex systems like a computer program, this is not so. Even people who have just finished writing a program are often quite puzzled about what it actually does: if you ask an awkward question, they will have to conduct an experiment, actually running the program in order to find out what it does. In the ideal, this should not be necessary. A specification of the behavior of a program can in principle be written in advance, perhaps at the beginning of the project, and its accuracy should be maintained throughout the design, implementation, manufacturing, and usage.

The second question is one that surely interests all engineers – indeed, it probably was the initial motivation for their choice of engineering as the subject for their study. It is: how does the system work? How does the engine actually function, and how does it drive the wheels of a car? How does the airplane fly? The answer to this question is usually given by describing the structure of the system and its components. It includes a description of the ways in which the components are connected together, and the methods that they use to interact with each other. Again, we know that many good software engineers are seriously challenged to answer questions like this about their own programs. A few weeks after writing a program, they no longer know how it works. This causes problems when attempting to diagnose and repair errors that come to light later. It causes even more severe problems when the need arises to produce the next version of their program, when needed to make it do something a little different or a little better. The programmers then have to find out, again by experiment, how the program actually works.

The two questions "what?" and "how?" are equally relevant to the pursuit of all branches of natural science. For example, a classificatory biologist may enquire, what does a newt or an axolotl do? How does it relate to its environment? And the next question asks how the creature is constructed: what are its limbs and organs, and how do they interact? Sciences that concentrate just on these two questions are often characterized as being merely descriptive.

2.4.2.1 Basic Questions of Science

The more mature branches of science are certainly based on an extensive foundation of accurate description; but then they go on to address some rather deeper questions. The first of these is: Why does the system work? What are the basic scientific principles, the equations, and the laws of nature on which the working of the system actually depends? So, the aeronautical engineer studies aerodynamics, which makes explicit the laws that explain why an airplane flies. On the basis of the laws, it is possible to make predictions about how the airplane will respond to its controls; the modern engineer exploits such laws to optimize the quality of systems, and to reduce their cost.

The final and most distinctive feature of modern science is its pursuit of certainty of knowledge. The goal of the scientist is to assemble a massive body of convincing evidence that the answers to all the previous questions are in fact correct. The engineer similarly uses a wide range of testing before delivery of a system, to gain confidence that it will meet stakeholders' needs and not fail after delivery. Testing is also used to detect and remove any remaining deficiencies in the product that appear often after implementation. In architecture, this is known as *snagging*,[8] and in programming it is called *debugging*.

2.4.2.2 The Logic of Engineering Design

One may argue that the correctness of a process of rational engineering design follows the same rules of the propositional calculus as a mathematical proof. However, the individual lines in the proof are much larger than the normal statement of mathematical theorems; each of them is some engineering description of the system, either in part or as a whole. The proof itself is also much longer than most mathematical proofs: it consists of the entire collection of engineering documents recording the entire design process for the system. These engineering documents describe the system in different ways from different perspectives, for different purposes, and at different levels of detail and abstraction. The most abstract documents are the overall system specifications, answering the question, "What does it do?" in terms of the properties of the system that is of interest to its users.

---

8 Snagging is a slang expression widely used in the construction industry to define the process of inspection necessary to compile a list of minor defects or omissions in building works for the contractor to rectify.

2.4.3 The Context and Nature of Engineering Design

The design engineer plays a pivotal role in shaping society and its lifestyle and values – particularly in this modern, technology-driven age – and yet, since the age of Telford[9], Stephenson[10] and Brunel[11], engineers seem to have retreated and become largely anonymous background figures.

2.4.3.1 What Is Engineering?

One definition of engineering is: "Engineering is the knowledge required, and the process applied, to conceive, design, make, build, operate, sustain, recycle or retire, something of significant technical content for a specified purpose; – a concept, a model, a product, a device, a process, a system, a technology." This definition is intended to be an all-embracing and comprehensive definition. But it's somewhat legalistic approach seems misleading. It does describe what many engineers do, but it masks the really core and fundamental engineering activity that is design.

If we dissect the definition, we can say that "applying knowledge and skill" could refer to any profession. The activities covered by "make, build, operate, sustain" are essentially management activities requiring skilled, well-trained personnel to work according to the designer's recipes and instructions. However, the conception and design of systems are the fundamentals of engineering.

2.4.3.2 What Is "Design"?

According to the Webster's dictionary, design is: "The process of selecting the means and contriving the elements, steps and procedures for producing what will adequately satisfy some need." However, one can obtain a fascinating insight into what many people think that design is about by listening to the advocates of intelligent design in their arguments against evolution. This insight is separate from the conclusion that is drawn. It seems that the supporters of intelligent design see in the Darwinian concept of natural selection a kind of random, unstructured process in which chance plays a far greater role than would be logical to create the dynamic, intricate pieces of life that make up the universe. They fear that this robs human existence of meaning, and they argue that the sheer complexity of the natural world demonstrates, or even proves, the existence of a "designer." Purpose, in their minds, deserves particular emphasis and seems to be missing from natural selection. Anyway, it is clear that they see design as a very high-level activity that can bring order and meaning to life. Engineers should say Amen to that!

---

9 Thomas Telford (1757–1834), a noted Scottish civil engineer, architect and stonemason as well as a road, bridge and canal builder.
10 Robert Stephenson (1803–1859), a well-known English railway and civil engineer.
11 Isambard Kingdom Brunel (1806–1859), a famous English mechanical and civil engineer.

2.4.3.3 Engineering Design Process

The crucial nature of design and its phases should be emphasized. These involve intensive two-way dialogue with all sorts of clients.[12] A close fit has to be developed between the clients' business plan and risk model and the strategic elements of the proposed design. Both the designer and the clients must agree on the boundaries of the intended system and this must be done, not only in terms of the topography, but also in terms of time. "For what period of time is this system expected to operate?" is a key question. Everyone involved need to share a common risk assessment and management process that takes into account the inevitable uncertainties. This is particularly important in terms of the technologies to be employed as well as the areas where the design will require creativity and judgment. Finally, it must include a financial model that expresses and tackles the key economic uncertainties from both the clients and designer viewpoint.

In summary, this is the period when the design engineers ensure that there is a clear understanding of the purpose of the system to be delivered by the project and their ability to deliver it. Like most professionals, engineers' failure or success can be measured on a scale between 0 and 100%. But engineers are usually subject to a very transparent and public test. Engineering designers must produce systems and processes that work. Everyone can observe, and suffer the consequences if a bridge collapses, if an airplane does not fly, and automobiles do not start. The consequences of error are much more obvious than in the work of, say, a neurosurgeon, lawyer, or accountant.

2.4.3.4 Societal Risk

The design engineer has to take account not just of technology and economics. There are significant nonfinancial benefits and disbenefits to take into account. In addition, engineers need to accept that various stakeholders will often have a completely different agenda from one another. They may see the boundaries of the system quite differently. This was exhibited in a dramatic fashion when Shell planned the de-commissioning of Brent Spar. Greenpeace had a completely different agenda from that of Shell and succeeded in derailing the original Shell design.[13]

2.4.3.5 Aesthetics and Utility

When discussing "design" in the wider community, engineering design does not arise as a first thought or example. One could invite his friends to name a

---

12 In fact, all the stakeholders.
13 Brent Spar was a North Sea oil storage and tanker-loading buoy in the Brent oilfield, operated by Shell UK. By 1991, this facility was considered obsolete and Shell planned to dispose of it in the Atlantic waters approximately 250 km from the west coast of Scotland, at a depth of around 2.5 km. Shell abandoned its original plans in the face of stiff public and political opposition orchestrated, mostly by the Greenpeace organization (Wikipedia).

designer, and they, most probably, will reply with names like Antoni Gaudi (architect and interior designer, 1852–1926), Christian Dior (fashion designer, 1905–1957), Jonathan Ive (industrial designer, 1967–), and the like.

For example, a bank recently wrote and offered a cash card, which it said had been "designed" by Stella McCartney. The author of this paper was quite impressed to find that a fashion designer had mastered the technology of polymeric materials and their embossing and lamination, not to mention imprinting a magnetic strip, incorporating a chip and the necessary encryption technology. However, on further reading, it was discovered that what the bank meant by "design" was that she had provided a pretty picture to put on the front of the card. Nevertheless, engineers cannot ignore or dismiss this interpretation, because good aesthetics are valued and respected. On the other hand, "utility" is taken for granted, mainly because engineers do such a good job. Architects combine utility and aesthetics, but can engineers? and should they? Of course, they should – and often, they do.

2.4.3.6 Engineering and Aesthetics

Automobiles are highly engineered systems, but we well know that they are bought as much for their appearance and style as for their performance. The Concorde aircraft is a dramatic example of how, even when designing for performance, something aesthetically satisfying can result. One may be astonished to find, even today, how many people talk in sentimental, nostalgic terms about that beautiful, wonderful machine. It really was such a beautiful airplane, they say, and they appreciate that. Of course, they appreciated its performance also, but what they really emphasize is its looks.

Another point in case is the Viaduct de Millau bridge (Figure 2.12). One may ask his friends who actually designed it and they will say it was designed by Norman Foster (1935–). Norman Foster, of course, made a magnificent contribution, in terms of the outline and the shape of the bridge, but the man who actually ensured that that bridge stands up and can take the traffic and resist the elements is called Michel Virlogeux (1946–). Undoubtedly, no one in the public at large has ever heard his name.

Figure 2.12 The Creissels and Viaduct de Millau bridge in southern France

2.4.3.7   Purpose and Other Values

However, when reflecting on purpose, apart from utility and making it work, and apart from aesthetics, some other values begin to appear. There are ethics, for sure; a social focus, because people are concerned with health, education, and care for the elderly; environmental responsibility and sustainable development; and engagement with the developing world.

Fascinatingly, of course, these are issues that are rising up the engineering agenda more and more today. The list is not complete, but these are issues that are certainly receiving the attention of engineers worldwide and an increasing number of reports and studies are being generated to address these values. Most probably, this trend will help to increase the public awareness and appreciation of engineering, because it addresses areas that count out there in society. By addressing them, engineers are seen to be responding to society's wider agenda.

2.4.3.8   Engineering's Social Dimension

The fact of the matter is that engineering should not be technology driven. Technology is an enabler, and it is the key component in the toolkit, but it is not the reason why one is an engineer. The real driver is social, because engineers want to improve the quality of life out in the community. However, what distinguishes engineering from other equally socially driven professions is its range of activities, its ability to combine science with judgment and intuition, and all of this within a disciplined, technical framework. Engineers have the skill and ability to design, according to well-tested rules, complex systems that work. They can combine science-based technologies with social insight, to improve the quality of life.

Finally, engineering formation needs to recognize and give much more room for the social skills that a professional engineer needs. In fact, to be taught that engineering is 100% technology is a travesty of the truth. The engineer's mission must surely be to serve the community, and to do this, the engineer needs to have the skills of communication and debate to engage the community and to address and educate its needs and aspirations.

2.4.4   Roles and Rules and the Modeling of Socio-Technical Systems

The term *system* is in itself not very informative. It was introduced into engineering in the 1940s, leading to the rise of systems engineering in the 1950s and 1960s, but at the same time obtained a central place in biology and other sciences as well. The common understanding in the literature is that a system is a complex whole consisting of elements or components that are related to each other. This makes almost any technical artifact a system.

Since a model of a system should contain all elements that are relevant to the functioning of the system, a model of a socio-technical system must include the operators. In the literature, two approaches can be distinguished as to how this is to be done. The first of these is often called hard-systems thinking, the second soft-systems thinking. Hard-systems thinking was the predominant (in fact, the only) form of systems analysis until the early 1970s. In that period, the emphasis shifted from hard-systems engineering to soft-systems engineering.

Figure 2.13 shows a model of an aircraft system as part of a larger air transport system, which is again related to a national transport supersystem. The characterization of the various subsystems that make up the aircraft system does not indicate in any way whether or not some of the operating and controlling is done by people instead of machines. One component of the aircraft system model, the crew, which is an all-human subsystem, is presented as completely on a par with the other all-hardware subsystems. The air-traffic control system in the enveloping air transport system is operated by people – air-traffic controllers – supported by hardware systems, but it cannot be seen from the model that this is so, or that this is so for the air-traffic control system but not for, e.g., the ticketing system, which could be a completely automated subsystem.

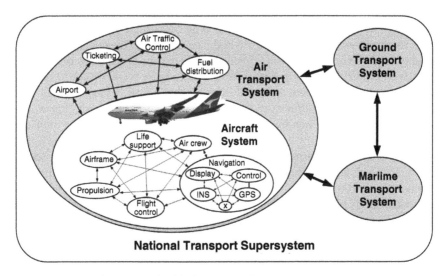

Figure 2.13 Aircraft system embedded in a national transport supersystem

Especially in the design of socio-technical systems, however, failing to take into account the fact that human beings are fundamentally different from machines may create serious difficulties. The point is not that human as biological organisms are fundamentally different from machines. That remains to be seen. The point is also not so much that people make mistakes, that is, that they choose the wrong action when a particular condition materializes, or that they fail to recognize a condition as one where they should take a particular action, and this is admitted by the operator.

Hardware malfunctions can also never be ruled out, due to our incomplete knowledge of nature. The point is that people can contest a judgment that actual circumstances are or were precisely equal to a specified condition in their list of instructions and can therefore contest whether they ought to choose or should have chosen a particular course of action. People do not coincide with the roles they fulfill. They are defined, rather, by their goals and desires, their beliefs and expectations, as individual persons. Their judgment will, therefore, involve a broader range of considerations than any list of instructions will contain. Finally, people are as individual persons, part of a social system. They have responsibilities, both in the roles they fulfill and as individuals, and they are held responsible for their deeds. This seriously affects which courses of actions they will choose. In conclusion:

- Socio-technical systems involve humans both in the role of operators and in the role of users. Operators are subsystems of the larger system in which they perform their operating work and are therefore included in the system. Users are not part of the system. They are free to use the system or, in the case of a socio-technical system, to participate in using it.
- A proper functioning of socio-technical systems requires the coordination of the actions of all people involved, both operators and users. This will usually be accomplished through rules, and the design of such rules is therefore an integral element of the task of designing a system.
- A human decision to follow a particular rule requires first of all a judgment that the situation is one where the rule applies. But even when an operator decides that a particular rule applies, he or she can also be expected to make a judgment whether or not it is in the person's interest to follow the rule.
- The history of technology consists to a large extent in attempts to remove the "friction" in the system that is caused by the freedom of operators, and many of these attempts have been successful. One approach to decrease such friction is to increase systems' automation and simply eliminate the operators as much as possible. Operators are everywhere and continuously being replaced by completely hardware systems. This option is, of course, no panacea: hardware systems can fail as well, even if differently. Additionally, there are institutional limits to this option, having to do with the distribution of responsibility, accountability, and liability.
- Finally, regardless of the extent of automation, the friction due to the interpretational and reflective freedom of the users of the system will remain. One can never automate the users of a system, because the system exists to serve their purposes; automated users have no purposes. Although a user cannot be considered part of a system, the person who constitutes the user is often present in the system in the role of operator, as is the case when an individual driver steers his or her car along the roads of a traffic network. What one can do is decouple as much as possible the user role and the

operator role. The increasing interest in the development of fully automated traffic, so that ultimately the user can sleep his way from A to B, is an example of this approach.

2.4.5 Engineering as Synthesis – Doing Right Things and Doing Things Right

2.4.5.1 Engineering as Synthesis

Fundamentally, it could be argued that engineering equals design. Everything else done under the label of engineering is either applied science and technology, or it is craft – making things. The element that makes engineering different from science and craft is design. That is not popular with university departments, which are often made up either of scientists or of people who are actually trying to pursue the craft of making things. Both groups deserve the greatest respect. Nevertheless, the thesis of this paper remains that engineering is design.

The notes in the flyer for this philosophy of engineering seminar say that "engineering is primarily a social rather than a technological discipline." This is not quite the case. Take an aeronautical example: if you are halfway across the Atlantic, do you want to know that the diameter of the bolt that holds the engine on was calculated and the materials chosen so that it is strong enough? Or that it is there because that is the correct social context for it? Engineering is about making things that work, and if they do not work – people die.

On a lighter note, here is a relevant quotation from one of the leading twentieth century philosophers of science, Douglas Adams (1952–2001). In a line from "The Restaurant at the End of the Universe," which is a part of the *Hitch-Hiker's Guide* series. It concerns a party of hairdressers and management consultants who are marooned on prehistoric earth, who have formed committees to invent things to make life better. They are having a review of their work: "What about this wheel thingy?" said the captain. "It sounds a terribly interesting project." "Ah," said the marketing girl, "we have a bit of difficulty there." "Difficulty?" exclaimed Ford, "what do you mean, difficulty? It is the single simplest machine in the entire universe." The marketing girl soured him with a look. "All right, Mr. Wise Guy," she said, "If you're so clever, you tell us what color it should be."

Douglas Adams is funny, but also, he makes many perceptive remarks. Getting so obsessed with the "what color should it be?" example, when you actually miss the point about whether it goes round and carries a load, seems to be a mistaken sense of priorities. Engineering, which is about making things that work, should never lose sight of the goal.

A popular engineering rule is that "form follows function." Alas, not always – a great example of where function followed form was the design of the Millennium Dome (Figure 2.14). The first decision was how big it would be in

Figure 2.14 The Millennium Dome, London, United Kingdom

square meters, followed by the choice of material to make the roof. Only then did somebody say, "That's pretty good – now what shall we do with it?" That is an archetypal example of letting the form dictate the function.

Everyone has their own definition of engineering and the author's is, "Changing the natural world to make it better meet the needs of mankind." Engineers are about reforming this place from what it was originally, so that it works better to meet the needs of at least a subgroup of mankind. If you want to be biblical, this is a very complex world to design and build in six days so engineers have to finish off what God left undone.

This brings us to the central message. Engineering, in practice, is of no use unless it is sensitive to what society wants and will use. If you are stuck on prehistoric earth, any sort of wheel is worth having. However, if you are trying to design a wheel for the next generation of expensive luxury car, it will not sell if it does not meet all the needs of prestigious cars customers! Engineering design has to be sensitive to the social context of what it designs and how it will be built.

2.4.5.2 The Engineering Design Process

It could be argued that design is the art of compromise. Since it was claimed that engineering equals design, people started to ask whether engineering actually means compromises. Apparently yes, because there is rarely a right answer, a right design, because there are so many stakeholders who have conflicting objectives – performance, delivery time, cost, risk, and many others. If the project becomes big enough to have a political dimension, you are talking about job security, national pride, and international relations. There are so many axes being ground in most engineering projects, and the engineer has to take them all into account.

One may describe the role of the engineering designer as finding the least bad compromise that all of the stakeholders can live with. Think of it as plotting their needs on a Venn diagram[14] and trying to find that little blob where they overlap, which everyone can live with. Of course, in almost all real engineering design challenges there is no overlapping blob. There is no common ground; the engineers have the diplomatic task of persuading someone to move his position (that is, redefine his needs) or the project is abandoned.

In engineering, there is the principle of iteration – that one puts up an idea and, if it does not fly or if the customer does not like it, one keeps tweaking it and working with all of the stakeholders until one comes up with something that can be done. If all else fails, one abandon it – and that is something which engineers are very bad at. Engineers persist, even when it does not make sense. The usual problem is that the customer wants a palace, until one tells him what it will cost, and then one starts again. It is especially true of institutional customers. Willy Messerschmitt (1898–1978) once said, "We can build any aircraft that the aviation ministry calls for, with any requirement satisfied. Of course, it will not fly." This is a global problem that always arises, with the customer requiring the impossible. The designer is left with trading off a whole load of benefits and constraints, to arrive at a compromise that everyone can work with – speed, reliability, maintainability, cost, timescale, mass, comfort, the list goes on.

Engineering design is a mass of disciplines, not all of which are purely technical. Most projects will involve a wide range of engineering disciplines – mechanics, electrics, electronics, computing, materials, etc. Then there is project management, including planning, construction, testing, operating, and disposing. To this we must add many subjective human issues, such as biomechanics, shape, color, and form.

Creativity is not a passive process and one does not just follow the rules. Analysis, going back to the bolt holding the engine must be based on hard calculation and not hand-waving. Judgment, one cannot look up answers for everything and, eventually, one must make a value judgment. Leadership is important; if the engineer is not going to lead the project, who is? It comes back to the marketing people in the prehistoric earth, and you would not want them leading it.

That sets the scene for the six principles, related to human as well as technical issues, engineers are trying to follow in their design efforts:

1) Debate, define, revise and pursue the purpose
2) Think holistic
3) Be creative
4) Follow a disciplined procedure
5) Take account of the people
6) Manage the project and the relationships

---

14 See: https://en.wikipedia.org/wiki/Venn_diagram. Accessed: February 2018.

2.4.5.3 Systems Engineers

People are often held to fall into one of two types, hedgehogs or foxes. Hedgehogs have one trick and they do it well – they have spikes. Foxes have many little tricks and they are cunning. The popular view is that engineers are hedgehogs and they are very good at something, while project managers are foxes, being quite good at a number of things.

Systems engineers, however, must be both; their resume is T-shaped: It has a lot of breadth and at least one deep piece: "There is at least one thing in this project for which I am the expert." If an engineer cannot say that, then apart from anything else, he will not have any credibility with the others. Those engineers have to be able to do one part of the project in detail and all of it in outline, which sets the agenda for their education. They have to know a lot of basic science and engineering – physics, chemistry, and mathematics, the science on which engineering is based. They also have to have an analytical spirit – one that tries to model problems, rather than just brainstorm them. They need an awareness of the many disciplines that contribute to the project. Finally, they need to be able to communicate with everybody – from the customer to the technician who assembles their design. The message we are trying to convey to engineering education is, please think about how you form engineers who fit that pattern. This is hard to do, and it is uncomfortable for traditional engineering thinking, but it is crucial if we are to be able to engineer complex systems that work.

That is where we come back to the title of this paper. It started by exploring and doing things right – one does not want the engine to be held on by goodwill of individuals – but, actually, doing the right things is the much wider context of engineering systems.

2.4.6 Further Reading

● Adams, 1995. ● RAENG, 2010.

2.5 Bibliography

Adams D. (1995). *The Hitchhiker's Guide to the Galaxy*. Del Rey: Mass Market Paperback.

Blanchard, B.S., and Blyler, J.E. (2016). *System Engineering Management*, 5th ed. Hoboken, NJ: John Wiley & Sons, Inc.

Blanchard, B.S., and Fabrycky, W.J. (2010). *Systems Engineering and Analysis*, 5th ed. Englewood Cliffs, NJ: Pearson.

Boehm, B.W. (1979). *Guidelines for Verifying and Validating Software Requirements and Design Specifications*. http://csse.usc.edu/TECHRPTS/1979/usccse79-501/usccse79-501.pdf.

de Weck, O.L., Roos, D., and Magee, C.L. (2011). *Engineering Systems: Meeting Human Needs in a Complex Technological World.* Cambridge, MA: MIT Press.

Holt, J., Perry, S., and Brownsword, M. (2016). *Foundations for Model-Based Systems Engineering: From Patterns to Models.* The Institution of Engineering and Technology.

Walden, D. D. (ed.) (2015). *INCOSE Systems Engineering Handbook: A Guide for System Life Cycle Processes and Activities,* 4th ed. Hoboken, NJ: John Wiley & Sons, Inc.

ISO/IEC/IEEE 15288:2015 (2015). *Systems and Software Engineering – System Life Cycle Processes,* International Organization for Standardization (ISO).

Kossiakoff, A., Sweet, W.N., Seymour, S.J. and Biemer, S.M. (2011). *Systems Engineering Principles and Practice,* 2nd ed. Hoboken, NJ: John Wiley & Sons, Inc.

Maier, M.W., and Rechtin, E. (2009). *The Art of Systems Architecting,* 3rd ed. Boca Raton, FL: CRC Press.

Micouin, P. (2014). *Model Based Systems Engineering: Fundamentals and Methods.* Hoboken, NJ: John Wiley-ISTE.

Morse, L.C., and Babcock, D. L. (2013). *Managing Engineering and Technology,* 6th ed. Englewood Cliffs, NJ: Pearson.

NASA (2007). *NASA Systems Engineering Handbook.* Washington, DC: National Aeronautics and Space Administration.

RAENG (2010). *Philosophy of Engineering, Volume 1* of the proceedings of a series of seminars held at The Royal Academy of Engineering, Royal Academy of Engineering (RAENG), London.

Sage, A.P., and Rouse, W.B. (2014). *Handbook of Systems Engineering and Management,* 2nd ed. Hoboken, NJ: John Wiley-Interscience.

Wasson, C.S. (2015). *System Engineering Analysis, Design, and Development: Concepts, Principles, and Practices.* Hoboken, NJ: John Wiley & Sons.

Part III

Creative Methods

"The true sign of intelligence is not knowledge but imagination."

Albert Einstein (1879–1955)

3.1 Introduction to Part III

Fundamentally, creative methods may be partitioned along two axes, (1) divergent versus convergent creative methods and (2) creative methods primarily appropriate to individuals versus creative methods primarily appropriate to teams. These general partitions are not absolute, but it is quite convenient to partition the creative methods into these four categories.

Along the first axis, divergent creative methods help in generating multiple creative solutions characterized by the following attributes: (1) fluency – the generation of many responses or ideas, (2) flexibility – changing the form or modifying information, or shifting perspectives, (3) originality – generating unusual or novel ideas, and (4) elaboration – augmenting ideas with details. Convergent creative methods differ qualitatively from divergent methods, in that they attempt to identify one or very few optimal solutions from a larger set of available ones for a given problem. Along the second axis, some creative methods are primarily appropriate for individuals, whereas some other creative methods are primarily appropriate for teams.

The following "Part III: Creative Methods," describes 21 creative methods organized in four chapters (Chapter 3.2 to Chapter 3.5) as depicted in Figure 3.1.

In addition, there are other creative methods that are not traditionally considered "Creative" but the author considers them as important tools in the creative engineer's arsenal. Seven such methods have been described in Chapter 3.6. Finally, Chapter 3.7 provides a detailed bibliography relevant to this part of the book (Figure 3.2).

As mentioned before, it is not the intent of this book to develop new creative methods. Similarly, it is not the intent to describe and teach each creative

Practical Creativity and Innovation in Systems Engineering, First Edition. Avner Engel.
© 2018 John Wiley & Sons, Inc. Published 2018 by John Wiley & Sons, Inc.

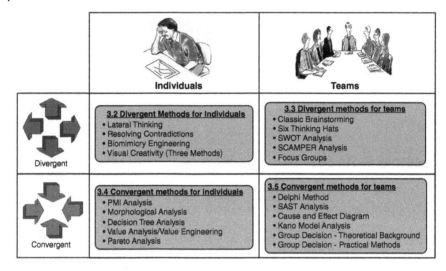

Figure 3.1 Structure of the first four chapters in Part III

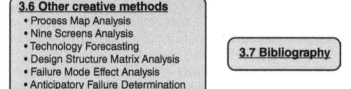

Figure 3.2 Structure of the last two chapters in Part III

method in exhaustive and minute details. The intent is mainly to acquaint systems engineers and engineers at large with relevant creative methods. Therefore, wherever possible, creative methods are described in a few pages covering four sections: (1) Theoretical Background, (2) Implementation Procedure, (3) Example, and (4) Further Reading. Some readers may find the level of details of a given creative method sufficient to understand and implement it. Others, interested in deeper understanding, may do so by following the links provided in the "Further Reading" section to the relevant bibliography, which is provided at the end of each part of the book.

Finally, any selection is arbitrary, and the author's is no exception. Creative methods were selected, first and foremost, on the basis of their applicability to engineers, particularly systems engineers. Another consideration was the feasibility of implementing each creative method by practicing engineers doing their business along the entire systems' life cycle. A final consideration relates to the maturity and robustness of each selected creative method, i.e. a method that has been thoroughly researched and implemented in many engineering and other projects.

3.2 Divergent Methods for Individuals

This chapter describes a collection of divergent creative methods for individuals as encapsulated in Figure 3.3.

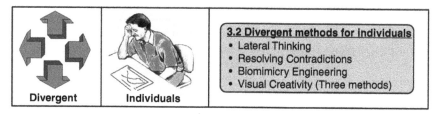

Figure 3.3 Divergent methods for individuals

3.2.1 Lateral Thinking

3.2.1.1 Theoretical Background

Education at the school systems, from the elementary to the university levels, instructs children and adults to think vertically. That is, to seek logical solution to problems on a step-by-step basis, identify a particular solution to a problem, and then explore the ramifications of that solution. A good example of vertical thinking is embodied in the way people play chess. At any time, one knows explicitly the positions of pieces involved in the game as well as the exact rules governing the game. On this basis one projects his or her future moves against any potential moves of the opponent. However, this is seldom the case in real life, and students are rarely taught other, overarching strategies to solve problems in this chaotic, unstructured world.

Lateral thinking (LT), the brainchild of Edward de Bono (1967), is a method for solving problems by way of an indirect and creative approach, using reasoning that is not immediately obvious. By and large, LT involves the generation of "horizontal" imagery, encouraging the generation of multiple solutions while ignoring, at least initially, their detailed implementation. LT exposes one's hidden assumptions and inner boundary conditions, bringing new insights into the problem-solution process. It penetrates through the confines of obsolete ideas, ushering creativity by generating new patterns. In short, LT is based on rearrangement of one's mental information. This, in turn, enables one to escape from the rigid brain-pattern established throughout one's life. The result of lateral thinking is an extensive set of alternative solutions, each of which can then be subjected to further vertical thinking.

Of the four thinking tool clusters defined by Edward de Bono (i.e. idea-generating tools, focus tools, harvest tools, and treatment tools), only the first one, idea-generating tool cluster, intended to break routine thinking patterns, is discussed in this book. Readers interested in the remaining thinking tools are encouraged to turn to the references at the end of this part of the book. We will use a brain-teasing example of how to cross a rickety bridge to examine the six LT strategies that follows.

3.2.1.2 Example: Four People on a Rickety Bridge

For ease of understanding, an example with a conventional solution is provided ahead of the lateral thinking method description.

Four people plan to cross a rickety bridge at night. They have one flashlight, and the bridge is too dangerous to cross without one. The bridge can support up to two persons at a time (Figure 3.4). The individuals crossing the bridge exhibit the following characteristics: person A requires 1 minute, person B requires 2 minutes, person C requires 7 minutes, and person D requires 10 minutes. What is the shortest time needed for all four persons to cross the bridge?

Figure 3.4 A rickety bridge example[1]

---

1 Photo: Drahtsteg pedestrian hanging bridge, in the Zillertal Alps, Austria.

The "correct," or conventional, solution is shown here. However, after reviewing the LT strategies, the reader may come up with some creative ideas, either individually or in a group.

Step 1. A and B cross the bridge ⇒ Total time so far: 2 minutes
Step 2. B returns with the flashlight ⇒ Total time so far: 4 minutes
Step 3. C and D cross the bridge ⇒ Total time so far: 14 minutes
Step 4. A returns with the flashlight ⇒ Total time so far: 15 minutes
Step 5. A and B cross the bridge ⇒ Total time so far: 17 minutes

3.2.1.3 Lateral Thinking Method

Lateral thinking may be implemented in various ways. This book suggests carrying out the following six LT strategies:

Strategy 1: Master the problem at hand. Strategy 4: Break the rules.
Strategy 2: Ask fundamental questions. Strategy 5: Broaden perspectives.
Strategy 3: Recognize and control Strategy 6: Use provocative ideas.
  assumptions.

We can use these strategies in order to think about possible solutions to the rickety bridge example.

Strategy 1: Mastering the Problem at Hand
The relevant maxim for this lateral thinking strategy is: "Look before you leap, think before you act." It advises that one should look deeply upon an issue at hand and think how it is to be done before beginning to act. It also hints of potential vulnerabilities for the ones who do not heed this advice. Mastering the issue entails (1) clarifying the problem and then (2) analyzing it. Clarifying the problem calls for identifying the known information about the problem. This should include but not be limited to the relevant facts, inferences, speculations, and opinions as well as identifying what is not known about the problem. Typically, analyzing the problem calls for answering relevant questions, such as:

- What is the problem?
- Why does the problem exist?
- Who is causing the problem?
- When did the problem first occur?
- What is the impact of the problem?

During this analysis, one can be fully aware of the problem and so avoid all kinds of pitfalls on the way to resolving it.

For the bridge example, one could analyze the problem and ask elemental questions like: "Why do these people need to cross the bridge? Could they take another route?"

Strategy 2: Asking Fundamental Questions

The next LT strategy is to ask a lot of nonthreatening, seemingly obvious questions that might bring about new insights into the true nature of the problem. The goal is to be able to challenge the basic perception of the problem and the issues around it. Hopefully, this will lead to fresh new ideas. Typical fundamental lateral thinking questions might be:

- Is it possible to restate the problem?
- Are we dealing with the real important problem?
- Is the problem caused by our operational procedures/strategy?

Asking fundamental questions provides mechanism to investigate the necessity, validity, and uniqueness of the problem at hand. As such, it can illuminate certain aspects of the problem that would otherwise be overlooked.

Regarding the rickety bridge example, one could ask seemingly unrelated questions like: Is the objective (i.e. minimize the overall bridge crossing time) truly important? Are there other, more important issues, to deal with (e.g. safety of the people involved or future travelers)?

Strategy 3: Recognizing and Controlling Assumptions

The origin of the word *assumption* is the Latin word *assumptionem*, meaning "a taking or receiving." An assumption may be defined as "a hypothesis that is taken for granted," Assumptions play important role in our lives and are created as a result of past experiences, societal culture and people and events that shape our lives. Furthermore, people are seldom aware of the effect of their inner assumptions on their actions. Assumptions often distort our reality by causing our brain to ignore information that does not correspond with what we have chosen to believe. Analyzing our assumptions identifies why we think and act in a certain way, while controlling assumptions enable us to set aside previous convictions and view problems in a more objective way.

Individuals and members of a group need to be cognizant of their own assumptions in order to be able to assess all the relevant issues pertinent to a given problem. In particular, psychologists observe common phenomenon called *the assumption of competence*. In essence, most people perceive themselves as more competent and infallible than they really are. In other word, our brain tends to assume we know more than we actually do. So what can be done about this?

- **Adopt some measure of doubt.** First, one must recognize that each and every person has ingrained assumptions about every situation. Healthy doubt allows one to consider he may be wrong and still feel comfortable with it. Doubt must be applied in a balanced way. Too much doubt leads us to inaction, while overconfidence can easily lead us astray.
- **Ask questions and listen.** Asking plenty of basic questions is probably the best mechanism to fill in the blanks and weed out fact from fiction. Listening

to a person talking, being truly present, is a necessary ingredient in the process of controlling one's assumptions. Also, true listening consists of accepting the integrity of the other person and waiting patiently, without interruption, for all of the evidence to be presented.

- **List the assumptions.** When confronted with a problem, it is often advisable to write out the assumptions inherent to the problem at hand, as well as the proposed solution. This process allows us to capture our understanding in a definitive way, thus gaining overt insight into the situation.

Regarding the rickety bridge example, one could ask a question about our assumptions: The stated problem was: "What is the shortest time needed for all four persons to cross the bridge?". We may assume the bridge must be crossed immediately but this is not necessarily the case. Can the bridge crossing operation be postponed to the daytime? In this case, two persons (A and B) will cross the bridge and then the other two persons (C and D) will cross the bridge, requiring a total time of only 12 minutes.

Strategy 4: Breaking the Rules

Some teachers love students who unquestioningly follow the rules and do not disturb the class. Many managers value employees who eagerly accept the dogma of the company, do not challenge authority or their assigned tasks and ask no questions. It is a human trait to resist change. We are often most comfortable in a steady and predictable environment. However, creativity and innovation are invariably related to changes in the status quo. If people never question the norm or challenge conventions, there would seldom be any advancement in this world.

According to Sternberg (1999), there are three major categories of creative contributors: ones that accept current paradigms, ones that reject current paradigms, and ones that attempts to integrate multiple current paradigms into a new one. Creative contributions that reject current paradigms are often based on breaking the rules and are the least likely to be accepted. Thus, these individuals may repeatedly be considered troublemakers. However, ideas created by them are often the most creative and innovative ones.

Within engineering, rules are put in place to establish standards, mostly aimed at ensuring the achievement of stated goals. So, based on our training, experience, and natural psychological tendencies, following rules is something most of us engineers are quite comfortable with. In summary, breaking rules can be tricky due to the risk involved, but the flip side of this outlook is that often people break "rules" that, in fact, are simply a set of conventions. Here are some suggestions for an enlightened rule-breaking process:

- **Rules identification.** A good starting point in problem solving is to list as many relevant rules as possible. The list should distinguish among laws of the land, explicit rules and standards, unwritten rules, implicit rules as well as day-to-day common practices.

- **Concept viability evaluation.** Deep evaluation must be undertaken regarding the rules to be broken and the proposed creative ideas. One should study carefully the nature of the said rules, their origin, raison d'etre and the level of resistance this approach may induce. Similarly, the pros and cons regarding the proposed creative ideas should be carefully analyzed vis-à-vis all stakeholders (colleagues, managers, customers, society at large, etc.).
- **Risks and consequences evaluation.** An engineer must be thoroughly familiar with the industry's laws and regulations before he proceeds. Being careless when it comes to industry's requirements can bear a risky outcome, sometimes with devastating effects on the public and the company as well as on the person's career. Therefore, one should carefully weigh the potential downside of breaking established rules and consider whether one's is willing to take responsibility for whatever may happen.
- **Ethics and value evaluation.** An engineer contemplating breaking established rules is advised to proceed only if he feels, true to himself, no ethical or moral objection. In other words, he should put firstly his obligations to colleagues, managers, clients, his company as well as his profession and others.

Regarding the rickety bridge example, one could break a rule and ask: "Is the stated bridge load limit (i.e. two persons at a time) well-founded? Maybe three persons could cross the bridge safely at a time?"

Strategy 5: Broadening Perspectives
Perspective, in the context of this book is "the faculty of seeing all the relevant data in a meaningful relationship," In other words, perspective is the ability to step back from an existing position and gain a broader understanding of the situation. This process of "reframing" the situation expands one's ability to generate new creative ideas. What can be done about this?

- **The obvious.** Broadening one's perspectives starts with the obvious: Read extensively, ask questions and listen, spend time with new people, and generally, open up to the world (e.g. surfing the internet, watching TV, visiting museums, theaters and concert halls, taking trips etc.). Finally, stay humble so all this deluge can actually penetrate one's mind and heart.
- **Practicing different perspectives.** Practice how other stakeholders regard a given situation (e.g. individual colleagues and managers, customers, suppliers, etc.).

Regarding the rickety bridge example, one could consider the perspective of other travelers trying to cross the bridge. They may come at night, not knowing the poor state of the bridge; they may not know the maximum allowable bridge load, etc. It may be more important to fix the bridge or install a clear warning sign rather than the mission presented in the original problem.

Strategy 6: Utilizing Provocative Ideas

In everyday language, provocation is an action or statement that is intended to make someone angry or distraught. However, in the context of this book, provocative ideas, radical and unrealistic as they may be, intend to forcibly cause one's mind to move out of its comfort zone, allowing one to consider potentially surprising solutions to a problem at hand.

Provocative ideas are intentionally quite absurd and often seem nonsensical. Such ideas are often different from ideas raised during brainstorming, where participants offer solutions without active judgment but those that are feasible within the problem domain. There are several provocation creative techniques like wishful thinking, exaggeration, reality distortion, and the like.

Provocative ideas could, in principle, solve the relevant problem but, by and large, are not realistic. Therefore, the reciprocal step needed is to undertake a mental transformation to convert the provocative ideas into sensible solutions. There are several ways to do this – for instance, performing transformation via principle. Here, one extracts key principles from the provocative ideas that are critical in solving the problem. Then one may devise a realistic solution based on those principles.

In summary, implementing creative solution based on provocative ideas entails the following:

- Create one or more provocative ideas that principally solve the problem at hand.
- Identify the principles that underlie the provocation.
- Move to a realistic solution based on the identified principles.

Another essential component of this technique is to seek more than a single provocative idea and underlying principle. The best strategy is to keep looking for reasonable number of sets containing provocative ideas as well as their reciprocal principles. It is reasonable to expect that several such sets will enable one to propose a better solution.

Regarding the rickety bridge example, one could propose a provocative idea involving building a large catapult that will hurl all four persons onto the other side of the bridge in one fell swoop. The principle that underlies this idea is the possibility of all four persons crossing the bridge simultaneously at great speed. One solution utilizing this principle is to fix or enlarge the bridge so it will be safe for four persons to cross it together at full running speed or, even, using a fast vehicle.

3.2.1.4 Further Reading

- De Bono, 2015.
- Richardson, 2016.
- Sloane, 2006.
- Sternberg, 1998.
- Sternberg, 1999.

3.2.2   Resolving Contradictions

3.2.2.1   Theoretical Background

TRIZ is the Russian acronym for "theory of inventive problem solving." It was developed by Genrikh Altshuller and his colleagues during the second half of the twentieth century in the former Soviet Union.[2] TRIZ is not based on intuitive creativity of individuals or groups but, rather, on meticulous study of some 3 million registered patents in order to identify specific patterns that predict breakthrough solutions to problems and then codifying it within TRIZ. More specifically, TRIZ proposes the following three hypotheses:

1) Problems and solutions are repeated across industries and sciences.
2) Patterns of technical evolution are repeated across industries and sciences.
3) Creative innovations may use scientific effects outside the field where they were developed.

These three hypotheses lead to a generalized problem-solution procedure depicted graphically in Figure 3.5. If a barrier blocks one's ability to solve a problem through a traditional approach, then one should: (1) analyze the problem and reformulate it into a general TRIZ problem, (2) use one or more of the TRIZ strategies and tools in order to produce a general TRIZ solution, and then (3) develop the specific solution by way of an analogy.

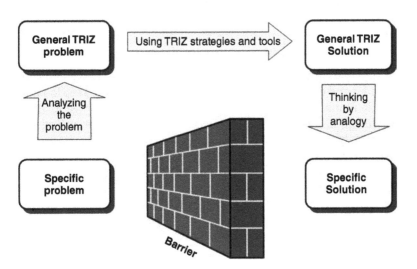

Figure 3.5 TRIZ generalized problem-solution procedure

---

2 See also: Brands R. (April 6, 2016). *An Introduction to TRIZ Theory for the New Product Development Process*. http://www.robertsrulesofinnovation.com/2016/04/new-product-development-process-triz-theory/. Accessed: November 2017.

TRIZ is a rich methodology consisting of a variety of tools and techniques that open new vistas on creative technical thinking. However, introducing newcomers to TRIZ is outside the scope of this book. Therefore, only selected parts of TRIZ are discussed in various chapters of this book.

Resolving contradictions is one of the most useful tools of TRIZ because it substantially broadens the solution space. It enables one to resolve contradictions, thereby improving one aspect of a system without compromising other aspects of the system. At the heart of most contradictions are two or more conflicting system requirements that one is expected to fulfill. For example, designing an aircraft that can (1) carry 500 passengers, (2) for a distance of 8,000 km, (3) within 5 hours flight duration, (4) using up to 1 liter of fuel per 100 km per passenger. Obviously, these are contradicting requirements, which are difficult or, perhaps, impossible to meet. In engineering, the common approach for dealing with such contradictions is to identify an optimal compromise amongst the conflicting requirements (Figure 3.6).

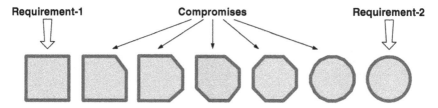

Figure 3.6 Two conflicting requirements and compromises to be selected

This is a frequent strategy, widely used throughout the industry. The problem with this strategy is twofold: First, selecting a compromise never fulfills the original requirements. Second, one always runs the risk of selecting an incorrect "optimal" compromise. In such cases, systems may fail in their ultimate testing grounds, the marketplace. Unfortunately, actual consequences of a risky compromise may only be discovered long after products have been designed, manufactured, and distributed with limited or costly corrective options. TRIZ offers numerous techniques to resolve existing contradiction with limited or no compromises. Therefore, the noncompromising solution space, available to problem solvers, is large and rewarding.

Some TRIZ experts employ a *contradiction matrix*, a table identifying optimal methods for resolving contradictions. However, this book will describe a set of *separation principles* that the author considers more effective and promising in resolving contradictions problem. Historically, TRIZ scholars defined three categories of separation principles: (1) separation in space, (2) separation in time, and (3) separation between parts and the whole. However, contemporary TRIZ researchers proposed several additional and unique separation principles.

3.2.2.2 Separation Principles and Examples

1) <u>Separation in time</u>

Under separation in time, one can resolve contradictions by identifying a time frame in which one function is performed and another time frame in which the other function is performed. This resolves the contradictions so long as the separations in time do not overlap. In some cases, separation of contradiction can be achieved under specific conditions without changing the system at hand. In other cases, one may be required to add new functions or elements to the system that may increase the complexity or cost of the said system, product, or service.

Example: Water faucet
The contradicting requirements for a water faucet are: (1) the water must be BLOCKED and (2) the water must FLOW. These contradicting requirements can be separated in time: (1) the water should be BLOCKED during the time the faucet is shut and (2) the water should FLOW during the time the faucet is open (Figure 3.7).

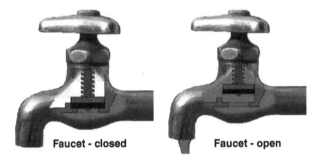

Faucet - closed   **Faucet - open**

Figure 3.7 Example: Water faucets

2) <u>Separation in space</u>

Separation in space can resolve contradictions in which, at the same time, two or more contradictory systems requirements must be fulfilled. That is, if an object is required to have property "A" and not have property "A," then one can separate the object into two objects, each with its own properties. Conversely, the contradiction may be resolved if one part of the object has property "A" and the other part does not have it.

Example: Safety matches
Different designs of matches were invented and patented during the early to mid-nineteenth century. Early designs were cumbersome and, most importantly, were susceptible to ignite spontaneously. The problem stemmed from two contradicting requirements: (1) a match SHOULD ignite and (2) a match SHOULD NOT ignite. The problem was solved by separating the reactive

chemicals between the match head and the striking surface located on the outside surface of the matchbox (Figure 3.8).

Figure 3.8 Example: Safety matchbox by Weltholzer

3) Separation between parts and whole

Resolving contradictions by way of separation between the parts and the whole may be employed when contradictory requirements state that a system exhibits specific properties and, at the same time and space, one or more of its parts exhibits opposing properties. One way to deal with this contradiction is to arrange them so that the parts can interact at their distinct and respective scales. That is, one property may be expressed at a larger, macro scale and the other property is expressed at a smaller, micro scale. As long as these two or more properties do not interfere with one another, the contradiction is resolved.

Example: Motorcycle chain
The contradicting requirements for a mechanical force transmission system between a motorcycle engine sprocket and a rear-wheel sprocket are: (1) the transmission system must be RIGID in order to transmit force and match properly with the two sprockets and (2) the transmission system must be FLEXIBLE in order to wrap around the two rotating sprockets (Figure 3.9).

Figure 3.9 Example: Motorcycle chain

The property expressed at the micro level (the chain links) is rigid when made to interface with the sprockets' teeth as well as each other by way of hinging pins. However, the overall system behavior expressed at the macro level (i.e. the motorcycle chain) is flexible.

4) Gradual separation

Resolving contradictions by way of gradual separation may be employed when contradictory requirements apply to a system in which properties shift gradually into other properties. In other words, there is no sharp region in which the system properties transform abruptly. Instead, the initial set of properties tends to change gradually until the full transformation to the new properties is exhibited.

Example: A screw

The contradicting requirements for a screw are: (1) a screw must be SHARP in order to drive it rapidly into a wooden slab and (2) a screw must be BLUNT in order to support the required mechanical load. A screw is designed in a conical shape in order to fulfill the above two contradicting requirements (Figure 3.10).

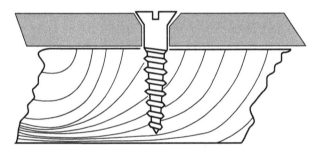

Figure 3.10 Example: Screw

5) Separation by directions

Resolving contradictions through separation by directions may be employed when contradictory requirements apply, within the same space and time, to different internal axes of a system, for which different properties are required. This is the case where a system exhibits one property in one direction, and the other property in another direction. Conversely, the system may be built or modified to do so. This phenomenon may be exploited to one's advantage.

Example: A rope

A rope is composed of a group of natural or synthetic fibers that are twisted together into a larger and stronger cable. The contradicting requirements for a rope are: (1) a rope must be STIFF in the direction of tension and (2) a rope must be FLEXIBLE when it is bent in any other direction. Due to its construction, a rope has a tensile strength but is flexible and cannot provide sideways strength. As such it fulfills the above two contradicting requirements (Figure 3.11).

Figure 3.11 Example: Rope

6) Separation by perspectives

Resolving contradictions through separation by perspectives may be employed when an object's emerging properties are dependent on perception, within the same space and time.

More specifically, the system under consideration does not change its properties but rather the exposed system's properties may be useful from one perspective and useless or even harmful from another perspective.

Example: A TV screen

The contradicting requirements for a Television screen are: (1) Due to its construction, a television screen exhibits a MEANINGLESS image when viewed from a short perspective and (2) a television screen must exhibit a MEANINGFUL image when viewed from a normal viewing perspective (Figure 3.12).

Figure 3.12 A TV screen

7) Separation by frame of reference

Resolving contradictions through separation by frame of reference may be employed when a system's properties depend on the specific frame of reference employed, within the same space and time. Therefore, resolving contradicting requirements for a system could be achieved by either changing the system at hand or changing the frame of reference used. Often, changing the system is expensive and time consuming, so changing the frame of reference could be an attractive resolution of the contradiction.

Example: Phone for seniors

Small (100–200 gram) hand-held mobile phones were introduced in Japan in the late 1990s. They had very limited memory and small screens and could support cellular (GSM) phone calls and FM radio reception. However, they had a large keypad and, today, are inexpensive to manufacture. Such devices have virtually no value in today's smart-phone mass market, so manufacturers of such phones must meet the following contradicting requirements: (1) sell a WORTHLESS phone in a market where (2) this phone is VALUABLE. For example, a firm produces a "Phone for seniors" and advertises such cell phones for under $30 (Figure 3.13).

Figure 3.13 Example: Phone for senior citizens

Under a conventional frame of reference (today's smart-phone mass market), this device has virtually no buyers; however, under a different frame of reference (market for older persons as well as people in developing countries such as parts of Asia, South America and Africa, etc.), similar device is quite valuable.

8) Separation by response of fields

In TRIZ, the concept of "Field" covers a verity of phenomena like: electromagnetic field (radio waves, microwaves, infrared light, visible light, ultraviolet rays, X-rays, gamma rays, etc.), electrostatic field, magnetic field, force (mechanical, gravity, centrifugal, inertial, friction, adhesion, coriolis, nuclear, etc.) and many more. Resolving contradictions by way of separation by response of fields may be employed when two or more fields respond differently to the properties of a system at hand, within the same space and time.

Example: UV-protected sunglasses

Extended exposure to the sun's ultraviolet (UV) rays has been linked to eyes damage (i.e. cataracts, macular degeneration, pingueculae, pterygia, and photokeratitis) that can lead to partial or full vision loss. Therefore, wearing UV-protected sunglasses will protect one's eyes from this harmful solar radiation. The contradicting requirements for such UV-protected sunglasses are: (1) the visible sunlight must PASS through the lens of the sunglasses and (2) the UV sunlight must be BLOCKED by the lens of the sunglasses. These contradicting requirements are resolved by the lens of the glasses.

The principle of separation by response of fields is applied as the sunlight spectrum is separated into different frequency regions. More specifically, a laminated glass totally blocks UV radiation, while the vast majority of the visible light passes through the lens (Figure 3.14).

Figure 3.14 Example: UV-protected sunglasses

9) <u>Separation between substance and field</u>

A substance is anything that has mass, occupies space, and has a specific composition and specific properties. Resolving contradictions by way of separation between substance and field may be employed when the substance from which a system is composed has one property and a field affecting the system, have conflicting properties, within the same space and time.

Example: A transformer

A transformer is a device that increases or decreases electrical potential (voltage) between its input and output by way of variable electromagnetic induction. More specifically, an alternating current in the transformer's primary coil creates a varying magnetic flux inside the transformer core, which, in turn, induces alternating potential at the secondary coil. The contradicting requirements for a transformer are: (1) the magnetic field created by the primary coil must FLUCTUATE in order to create variable magnetic flux within the core of the transformer and (2) the primary and secondary coils as well as the transformer core must be physically STATIONARY.

These contradicting requirements are resolved by the way the transformer is constructed and the effect of alternating current on the primary and secondary coils (i.e. time-varying magnetic field) (Figure 3.15).

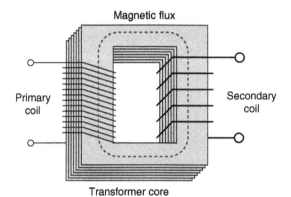

Figure 3.15 Example: A transformer

3.2.2.3 Further Reading

- Altshuller, 1996.
- Fey and Rivin, 2005.
- Salamatov, 1999.
- San, 2014.
- Savransky, 2000.

3.2.3 Biomimicry Engineering

3.2.3.1 Theoretical Background

The term *biomimicry* comes from the ancient Greek words *bios*, meaning "life" and *mimesis*, meaning "to imitate." Therefore biomimicry means "to imitate life." This term was made popular by Janine Benyus in her book *Biomimicry: Innovations Inspired by Nature*. The basic premise of this approach is that nature can be viewed as a role model and a teacher to mankind. Nature creates conditions conducive to maintaining living organisms, i.e. energy comes only from the sun and materials are created at ambient temperatures using local materials with no toxic pollution and minimal waste. This is particularly notable when one examines how the human race depletes Earth's natural resources, pollutes all corners of the globe, causes habitat loss as well as endangers, and eradicates many species at an alarming rate.

Nature has already solved many of the technological and sustainability problems that humans face, and we can learn from it. So the idea of biomimicry engineering is to design and build systems by emulating what nature has been doing for billions of years in a sustainable manner. This entails: (1) mimicking physical forms or designs of natural systems, (2) mimicking processes that take place in natural systems, and (3) mimicking ecosystems, i.e. integrating systems within their environment.

3.2.3.2 Implementation Procedure

When designers seek engineering solutions inspired by biological systems, they often collaborate with biologists who could possibly identify organisms that have solved similar problems. This process, called Biomimicry Design Spiral (BDS), was developed by Carl Hastrich in early 2000. BDS is a step-by-step process, often done iteratively in a nonlinear and dynamic manner such that output from a later phase frequently influences previous phases. This phenomenon necessitates iterative feedback and refinement loops. Biomimicry researchers have developed a procedure where engineers and biologists work together to identify and emulate natural solutions to engineering problems. This procedure uses the following steps (Figure 3.16).

Step 1: Identify the engineering challenge. In this step, one identifies the specific problem or human need as well as the required functionalities the system must fulfill.

Step 2: Interpret the design in biological terms. In this step, one explores the mechanism whereby natural organisms and biological systems are able to solve the stated problem.

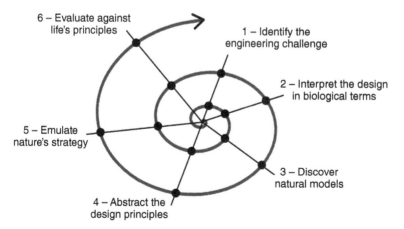

Figure 3.16 Biomimicry implementation procedure[3]

---

3 Adapted from Carl Hastrich (2005) via The Biomimicry Institute.

Step 3: Discover natural models. In this step, one identifies nature's models and strategies that meet the stated functionalities. In particular, one considers organisms whose survival depends on the stated function.

Step 4: Abstract the design principles. In this step, one identifies the engineering design principles that achieve the above functionalities.

Step 5: Emulate nature's strategy. In this step, one develops ideas and solutions based on the identified models and then convert them into the appropriate engineering domain. This process takes into consideration: (1) the structure or shape of the biological system, (2) the principle, process, strategy or mechanism of the biological system, and (3) the interactions between different organisms while meeting the stated functions.

Step 6: Evaluate against life's principles. In this step, one evaluates the design solution against the original problem or human need. In addition, one checks the design solution in the context of the following life's principles:[4] (1) ensure efficient use of resources, (2) use life-friendly chemistry (3) integrate development and growth, (4) be locally attuned and responsive, (5) adapt to changing conditions, and (6) evolve to survive. Lastly, one should strategize whether, and if yes, how he/she wants to use the next lap around the biomimicry design spiral.

3.2.3.3 Example: Sea Water Desalination

1) Example background

Large numbers of conventional seawater distillation and reverse osmosis systems have been built on industrial scales in order to overcome the scarcity of fresh water in many parts of the world. However, conventional methods suffer from severe limitations. In particular, high energy consumption, limited system durability, frequent needs to treat membrane fouling, and finally, the significant overall operating costs. The biomimetic example described here is based on a study conducted by a team of scientists from the Republic of Korea (Kim et al., 2016). The study provides insights into the mechanism underlying water filtration through halophyte roots. This method can also be used for the development of a bio-inspired desalination method.[5]

2) Example analysis

The study investigated the biophysical characteristics of seawater filtration in the roots of the mangrove *Rhizophora stylosa* (RS) plant. The idea was to explore a novel seawater desalination method utilizing RS plant hydrodynamic capabilities. Mangroves grow in saline water, and the plant is able to extract near-fresh water through filtration. The root possesses hierarchical,

---

4 Adapted from the Biomimicry Group, 2011.
5 The ideas discussed in the study constitute a basis for a patent application submitted by the scientists involved.

triple-layered pore structures that traps Na+ ions in the outermost layer. In addition, the second layer is composed of microporous structures that also facilitate Na+ ion filtration.

Figure 3.17 depicts a schematic of this water filtration process in mangrove roots: (a) The root is immersed in salted (NaCl) seawater solution and the outermost layer of the mangrove root is composed of three layers; (b) water passes through the outermost layer when a negative suction pressure is applied across this layer. The Na+ ions attach themselves to the outermost layer acting as a selective barrier to ionic diffusion, repelling the Cl– ions from entering the mangrove roots.

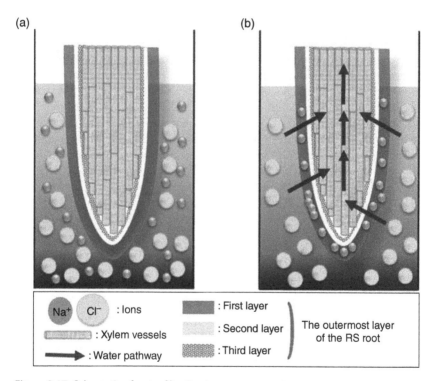

Figure 3.17 Schematic of water filtration in mangrove roots

3.2.3.4 Further Reading

- Bar-Cohen, 2005.
- Benyus, 2002.
- Kim et al., 2016.
- Lakhtakia and Martín-Palma, 2013.
- Passino, 2004.

3.2.4 Visual Creativity (Three Methods)

Simply stated, *creativity* is the ability to think of new ideas. Visual thinking is the practice of using pictures to solve problems, think through issues, and communicate clearly. More specifically, visual thinking is instrumental in: (1) seeing patterns and connections that are noticeable only in a graphical form, (2) analyzing complex problems, viewing their components and discerning their underlying effects, (3) spurring new possibilities and ideas, and (4) effectively communicating ideas with other people and building consensus. This section on visual creativity describes the following creative methodologies: (1) concept map, (2) concept fan, and (3) mind-mapping.

3.2.4.1 Concept Map

1) <u>Theoretical background</u>

Concept maps[6] were developed by Joseph D. Novak in the early 1970s. Concept maps are graphical means for organizing and representing information. They include nodes describing concepts, usually enclosed in circles or squares, as well as connecting directional edges indicating relationships between concepts. These edges are realized by way of connecting arrows and phrases specifying the nature of the relationship between concepts. Concept maps constitute an effective tool for capturing current knowledge as well as creating and integrating new knowledge. This new knowledge may include original insights, distinct concepts, and altogether unusual linkage to existing knowledge. Finally, concept maps are also an excellent means of communication between stakeholders, and thus they often employ a flexible vocabulary of concepts and linking words.

2) <u>Implementation approach</u>

Concept maps have specific characteristics:

Focus question. A good starting point in creating a concept map is to define a focus question. The intent here is to clearly specify the context of the concept map and what problem should it help to resolve. By and large, clear focus question can bring about a superior concept map. For example, the focus question for the example depicted in Figure 3.18 is: "What is the intended audience of this book? What is its philosophy and what is its content?"

Proposition format. Concept maps should express relationships between sets of concepts. Each relationship is depicted by way of linking phrases forming identifiable propositions. This means that every two concepts together with their linking phrases should form a short sentence. Thus, a concept map consists of a collection of graphical entities representing a set of propositions related to the focus question at hand. For example, in Figure 3.18 the

---

6 See also: A.J. Cañas and J.D. Novak, *What is a Concept Map?* Last updated September 2009: http://www.the-aps.org/APS-Storage/APS-Education/Pedagogy-Resources/Concept-Map.pdf. Accessed May 2017.

relationship between the concept "This book" and the concept "Practicing SEs & Engineering students" is defined through the linking words "Intended audience is," forming the proposition "This book's intended audience is practicing systems engineers and engineering students."

Hierarchical structure. By and large, concept maps are hierarchical structures beginning with a main concept (single root) that then branches out to show how each concept is linked to other concepts. So, usually the most general concept is drawn at the upper portion of the hierarchical graph and the more specific concepts are arranged below. As a result, concept maps are usually read from the top, progressing down toward the bottom. Therefore, within the framework of the concept map depicted in Figure 3.18, the concept "this book" is more general than the concepts "systems engineering," "creativity," and "innovation," Nevertheless, concept maps can have multiple initiation roots and, on occasions, may be organized along a circular structure (e.g. a concept map depicting butterflies' life cycle: egg, larva, pupa, and adult).

Cross-links. Cross-links are additional relationships between concepts in different segments of the concept map. They are often created after the first iteration of building a concept map has been completed and are emanated from creative leaps in which one may discover new propositions, not originally thought of. Bottom line, cross-links help us visualize how a concept in one segment is related to another concept in another segment of the concept map. For example, in Figure 3.18, the concept "this book" was originally linked to "Part III: Creative Methods." However, on further examination, a new link from the concept "System engineers could improve their skills by utilizing C&I knowledge" was added to "Part III: Creative Methods" through the proposition "by practicing."

Concept maps serve several purposes and provide distinctive benefits for creative engineers. They help engineers brainstorm an issue as well as discover concepts and propositions that connect them. In addition, this method encourages users to discover new relationships among concepts not originally thought of. Finally, concept maps provide an effective means for engineers to communicate ideas, thoughts, and information.

3) Example: This book

Figure 3.18 shows an example of a concept map depicting the intended audience, philosophy, and content of this book.

3.2.4.2 Concept Fan

1) Theoretical Background

The concept fan technique was proposed by Edward de Bono in his book *Serious Creativity* (1993). This method may be used when all obvious solutions to a problem fail to solve it. In such cases, the concept fan is a useful approach for widening the search for solutions as it broadens one's perspective by way of "taking a step back" in order to get a fresh view of the problem.

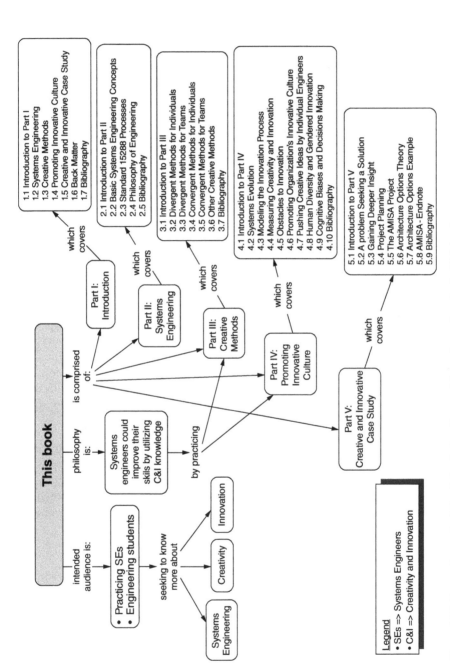

Figure 3.18 Concept map depicting intended audience, philosophy, and content of this book

2) Implementation Procedure

A concept fan is created by carrying out the following steps:

Step 1: Define the problem. Define the problem at hand within a circle (Figure 3.19).

Step 2: Identify the solutions. Next, draw one or more lines radiating outward in which one or more potential solutions should be identified (Figure 3.20).

Step 3: Step back. If none of the potential solutions are practical, desirable, or seem to solve the problem, then one should redefine the problem more broadly. This is done by drawing an arrow to the left of the original problem definition and then generating a broader problem definition in a circle attached to the arrow (Figure 3.21).

Step 4: Broaden the solution space. Once a broader view of the problem has been identified, one can create solutions to the new, broader problem. This, again, is done by drawing one or more lines radiating outward on which one or more partial solutions (in circles) and potential solutions should be identified (Figure 3.22).

Step 5: Repeat the process. Steps 3 and 4 may be repeated until one obtains a useful solution (Figure 3.23).

3) Example: Global warming

Global warming is a problem of too much heat-trapping carbon dioxide (CO_2), methane and nitrous oxide in the atmosphere. Most scientists believe this is a man-made phenomenon. For example, recent analyses suggests that New York City and other parts of the US Northeast are likely to be affected by regional rising sea levels, changes in ocean currents, increases in coastal flooding, storm surges, erosion, property damage, and loss of wetlands. More specifically, with regards to New York City, scientists conservatively estimate that if we do nothing to reduce carbon emissions, global sea level could rise as much as 2.3 feet (70 cm) by the end of this century.

Undoubtedly, global warming is one of the toughest problems facing the human race in the twenty-first century. The concept fan example presented in

Figure 3.19 Global warming: Step I

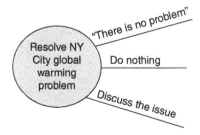

Figure 3.20 Global warming: Step II

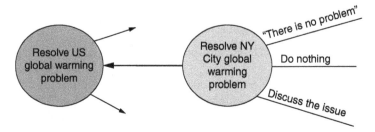

Figure 3.21 Global warming example: Step III

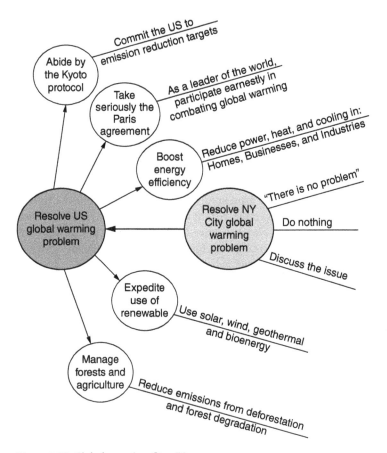

Figure 3.22 Global warming: Step IV

Figure 3.19 to Figure 3.23 addresses this problem and near-term actions needed to combat this problem.[7]

---

7 See more on global warming: Union of Concerned Scientists, http://www.climatehotmap.org/global-warming-locations/. Accessed: July 2017.

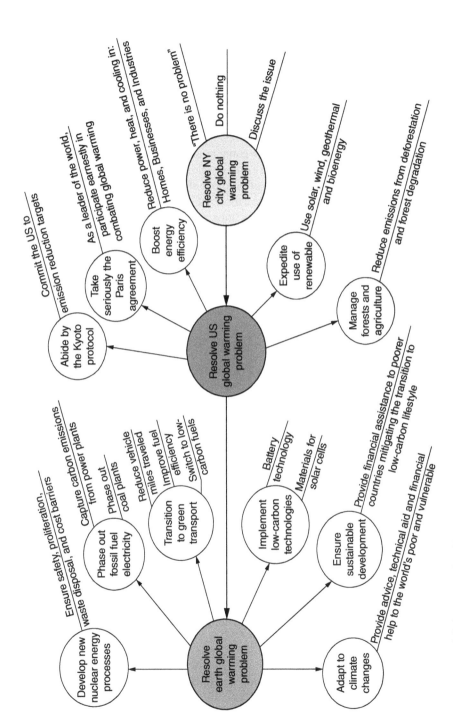

Figure 3.23 Global warming: Step V

3.2.4.3 Mind-Mapping

1) Theoretical Background

A mind map is a diagram used to visually organize information in a hierarchical and spatial structure. A key concept or problem is drawn in the center and a collection of relevant topics is arranged in a radial structure about this concept or problem. Mind maps were proposed by Tony Buzan during a 1974 BBC TV series he hosted, called *Use Your Head,* and a companion book bearing the same name.

Mind maps may be used to explore and develop ideas for resolving a specific concept or problem. A mind map is often more effective than linear note taking for several reasons. First, it is a graphical tool that may incorporate text, numbers, images, and colors. Second, it provides links and natural associations between individuals as well as groups of elements. Third, they may be created at different levels of granularity, providing optimal amount of information for a given situation. Fourth, it facilitates easy means of communication regarding new ideas and thought processes. Finally, mind maps present a very intuitive way of organizing one's thoughts, since mind maps mimic the way humans' brains function.

2) Implementation Procedure

Developing mind map includes the following steps.

Step 1: Define the main concept or problem. Identify the main concept or problem one would like to explore using a short phrase or sentence. One can also add a drawing or picture to represent the concept or the problem at hand.

Step 2: Define the primary branches. Create as many primary branches needed which directly relate to the main concept or problem, placing them in a radial hierarchy structure. Choosing the "right" primary branches is important, as this will facilitate the thinking at lower levels of the hierarchy. If these primary branches are too general or too specific, they may constrain ones' thinking and creativity.

Step 3: Define the sub-branches. Create as many sub-branches as needed. Sub-branches stem from the primary branches in order to further expand a given concept. This process may continue until one reaches the desired level of granularity.

At its most basic form, mind-mapping presents information in a graphical and hierarchical manner and could be drawn in any tree-shaped format. It is a beneficial learning and communication tool that helps users broadening their vision and think creatively.

3) Example: How to buy a used car?

Figure 3.24 is a typical mind-mapping example depicting an analysis of the question, "How to buy a used car? This example was generated using the free version of the XMIND tool (http://www.xmind.net/).

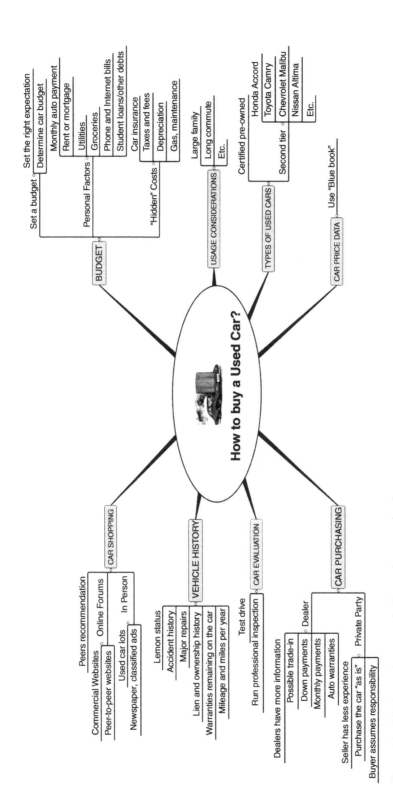

Figure 3.24 Mind-mapping example: How to buy a used car?

3.2.4.4 Further Reading

- Buzan, 2006.
- de Bono, 1993.
- Karl et al., 2009.
- Novak, 2009.
- Novak and Musonda, 1991.

3.3 Divergent Methods for Teams

This chapter describes a collection of divergent creative methods for teams as encapsulated in Figure 3.25.

3.3.1 Classic Brainstorming

Divergent   **Teams**

3.3 Divergent methods for teams
- Classic Brainstorming
- Six Thinking Hats
- SWOT Analysis
- SCAMPER Analysis
- Focus Groups

Figure 3.25 Divergent methods for teams

3.3.1.1 Theoretical Background

The idea of brainstorming, probably the best-known creative tool, was proposed by Alex Osborn in his 1953 book *Applied Imagination*. Brainstorming is a method for generating creative ideas, widely used in engineering circles and elsewhere. The method, intuitive and easy to implement, exploits the ability of the mind to solve problems by way of free association within a team setting. Akin to lateral thinking, the brainstorming process encourages people to contribute, among other things, strange and half-baked ideas, so this may trigger a cascade of fresh ideas by the participants. Therefore, the cardinal rule of brainstorming is to avoid criticizing or rewarding ideas that may stunt the idea-generation process and limit creativity. The objectives of brainstorming are twofold: (1) get as many ideas or solutions to a problem as possible and then (2) analyze and then combine, eliminate, and refine the results in order to distill the most suitable idea or solution to a problem.

The advantage of brainstorming is that it is usually conducted in a relaxed and relatively informal manner. Each team member may bring his/her diverse experience into the discussion, increasing the richness and diversity of emerging ideas. Also, engineers are more likely to accept ideas and solutions to problems if (1) they have been instrumental in shaping them and (2) these ideas have been discussed and arrived at in a quasi-democratic manner.

Brainstorming, however, is not a panacea; it has its own limitations. Sometimes the meeting may be hijacked by an overpowering individual who may reject other peoples' suggestions while being overprotective regarding his own ideas. Sometimes people are timid and may only present safe ideas. If too many people play it safe, say because of political reasons, the team may be bogged down, succumbing to a groupthink syndrome. Also, at times, the problem at hand is too intractable and more amenable to be solved by one or more specific individuals working in isolation.

3.3.1.2   Implementation Procedure

Typical brainstorming may follow these steps (Figure 3.26):

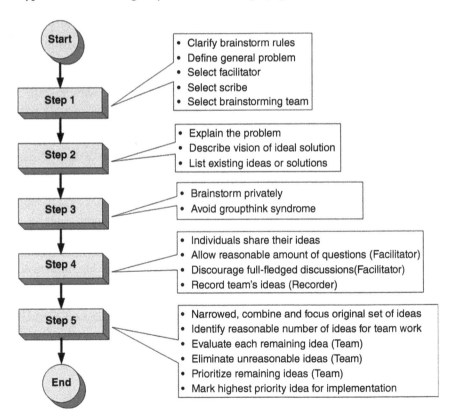

Figure 3.26 Brainstorming procedure

Step 1: Initially, define the general problem and designate the facilitator. The facilitator is responsible for selecting a scribe and brainstorming participants as well as leading the brainstorm process and enforcing the rules. The scribe's job is to record the ideas generated during the brainstorming process.

Step 2: Initiate a brainstorming session. Usually, such sessions will be most productive if they are preceded by a preliminary discussion in which the facilitator explains the rules governing the brainstorming processes. In addition, the team may share their understanding of the problem, its root causes, the barriers to realize change, the specifics of the present situation, and a vision of the ideal solution. Once the problem statement or issue is clearly defined, brainstorming usually starts as an inventory or listing of old, familiar ideas. Brainstorming often works best when the team starts by adapting, splitting, or combining old solutions creatively into new ones.

Step 3: Allow the team some interval of time in order to brainstorm privately. That is, write their ideas regarding the problem on a piece of paper. This is an effective way to captures one's own ideas. This technique is also helpful in avoiding a groupthink syndrome whereby the entire group goes off in one direction without exploring the full range of possibilities.

Step 4: Ask each member of the team to share his or her ideas with the other members of the team. As mentioned, the facilitator ensures that no criticism or cynical comments are expressed. However, a reasonable amount of questioning for better understanding of the ideas should be allowed. At the same time, the facilitator should discourage full-fledged discussion of these ideas. Usually one person (the recorder) notes the team's ideas on the board or on a laptop connected to a projector.

Step 5: Narrow down the set of ideas generated by the team by focusing and combining any redundant ideas. This activity could also generate new ideas on which the team may work. This could be achieved by means of team discussion as to the practicality and desirability of each idea. Some ideas will be considered outright unacceptable by the entire team and so be eliminated. The remaining ideas should be prioritized. One effective approach to prioritizing is based on a scheme whereby each member of the team rates each idea. A few ideas with the highest combined score will be discussed, further leading to a final decision on the optimal solution.

3.3.1.3 Example: Computer Memory Data Recovery

1) Example background

During the early 1970s, the International Business Machine (IBM) Corporation developed its first mainframe computer using all integrated-circuit (IC) memory (i.e. IBM-370/3147). A major challenge facing the engineering team was to maintain system integrity in case of main power loss (i.e. capture all perishable data and resume proper operations when electrical power is restored).

2) <u>Example analysis</u>

Several classical brainstorm meeting were held at IBM, and several ideas were raised and eventually, rejected due to technical or economic reasons. Eventually, one of the engineers exclaimed: "Eureka, I found it!" The principle of his solution is depicted in Figure 3.27. The three-phase alternating current is converted into direct current and, in the event of a power loss; a voltage drop detector identifies the event and activates recovery software that dumps the memory onto a (nonvolatile) magnetic disk. The magnetic disk continues to rotate for a while and a capacitor ensures the availability of electricity for the needed, less than a second, process time. Once power becomes available, the software uploads the entire memory and resumes operations automatically. IBM obtained a patent on this invention and implemented this scheme on all of its future mainframe computers.

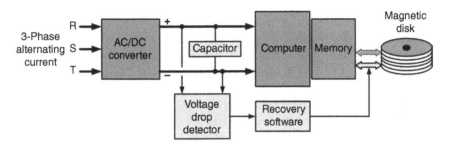

Figure 3.27 Computer memory recovery under power loss condition

3.3.1.4 Further Reading

- Gallagher, 2008.   • Osborn, 1979.
- Janis, 1972.

3.3.2 Six Thinking Hats

3.3.2.1 Theoretical Background

The Six Thinking Hats (STH) concept was developed by Edward de Bono in 1985. STH is mostly used to conduct effective workshops and group meetings. STH is based on the idea that human beings unconsciously construct their thinking along several and distinct dynamic modes of operation. Being aware of one's mode of operation at a given moment is quite helpful to an individual.

Furthermore, apprising a group regarding one's mode of operation yields substantial improvement in intergroup communication.

de Bono identified six distinct modes of operations and assigned a metaphoric colored hat to each one. Putting on a colored hat, either literally or symbolically, transmits a message to the entire group saying: "I transition into a specific mode of operation." An added bonus to this method is that it legitimizes modes of operations that otherwise might not be proper in certain setting. For example, in technical meetings, engineers speak freely about facts and figures but often are reluctant to express their feelings. It is simply not quite politically correct. However, under STH, one may openly say: "Wearing my red hat (emotion hat), I feel that this approach makes me mad."

3.3.2.2 Implementation Approach

Different organizations implement the Six Thinking Hats along different procedures but, in general, this technique could be quite instrumental in conducting efficient meetings. Wearing any specific colored hat permits one to discuss a problem, an idea, or a solution to a problem from one of six different perspectives.

1) White hat: Information seeking mode

In this thinking mode, one seeks to obtain factual, objective information related to the issue at hand. Typical white hat questions will be: "What do we know?" "What do we need to find out?" "How will we obtain the needed information?" "Who will be responsible for obtaining this data?" The leader or facilitator of the meeting should encourage the group to put on white hats (literally or symbolically) in the early phase of a meeting so the group could conduct a rational discussion, distinguishing facts from assumptions, speculations, and wishful thinking.

2) Green hat: Creativity mode

In this thinking mode one focuses on creativity, potential possibilities, alternatives, and new ideas. Typical green hat contributions will be out-of-the-box, provocative concepts, new perceptions, and solutions to the problem at hand.

3) Yellow hat: Logical-positive mode

In this thinking mode, one seeks to establish harmony and optimism within the group. Typical yellow hat remarks will illustrate logically the value, usefulness and benefits of a proposed idea or a solution to a problem. Yellow hat thinking compliments the black hat thinking, stressing the positive aspects of ideas and proposed solutions to problems.

4) Black hat: Logical-negative

In this thinking mode, one assumes the role of the devil's advocate in the group, identifying reasons to be cautious and conservative. Typical black hat remarks will explain logically why an idea or a proposed solution to a problem may not work. Black hat role is to warn the group regarding potential problems, difficulties, weaknesses, and hidden risks lurking in a given approach. Black hat thinking compliments the yellow hat thinking, stressing the downside of ideas and proposed solutions to problems.

5) Red hat: Emotional mode

In this thinking mode, one is encouraged to express emotions without having to justify it in a logical manner. Typical red hat remarks may be just hunches, instinctive gut reactions, or intuition, and may express feeling like hope, fear, love, hate, and so forth. Often, engineers find it difficult to express emotions in a productive way, although emotions play important role in the dynamics of meetings. Thus, a leader or facilitator of technical meetings should encourage members of the group to put on their red hats every now and then.

6) Blue hat: Manage and control mode

In this thinking mode, one focuses on planning, managing, and controlling the thinking process of the group. A typical blue hat contribution is to look at the big picture. That is to provide a control mechanism for the group in order to ensure that the meeting is well executed and the participants know and stick to its agenda and goals.

3.3.2.3 Example: Typical Six Thinking Hats Remarks

1) **White hat: Information seeking mode**
   - "What are the facts?"
   - "How many persons are involved?"
2) **Green hat: Creativity mode**
   - "We can fix the bridge instead of building a new one."
   - "Can we power the rocket using other propellant?"
3) **Yellow hat: Logical-positive mode**
   - "This team can solve the problem at hand."
   - "A solution based on electrochemical cell is feasible and cost effective."
4) **Black hat: Logical-negative**
   - "The proposed solution will not be acceptable to management."
   - "I am worried about the environmental impact of this solution."

5) **Red hat: Emotional mode**
   - "I like most the third approach."
   - "I am frustrated because the group repeatedly ignores my suggestions."
6) **Blue hat: Manage and control mode**
   - "We must cover three topics today."
   - "One more round of comments, please. Lunch is served in 30 minutes."

3.3.2.4 Further Reading

- de Bono, 1999.
- de Bono, 2017.

3.3.3 SWOT Analysis

3.3.3.1 Theoretical Background

The origins of SWOT (strengths, weaknesses, opportunities, and threats) analysis are a bit obscure, but its creation is often attributed to Albert Humphrey in the late 1960s. SWOT analysis is a qualitative as well as quantitative strategic method for assessing the value of a specified objective, system, project, or business venture. Qualitative SWOT is usually expressed graphically in a 2 × 2 matrix. The y-axis represents the internal and external factors and the x-axis expresses the helpful and harmful elements affecting the likelihood of attaining a desired objective. The internal factors include the strengths and weaknesses of the organization and the external factors include the opportunities and threats caused by the environment outside the organization (e.g. changes in competitive position as well as variation in the economy, technology, legislation, etc.). The four cells in the qualitative SWOT matrix represent the following perspectives (Figure 3.28).

	Helpful in achieving objective	Harmful in achieving objective
Internal factors (organization)	**Strength**	**Weakness**
External factors (Environment)	**Opportunities**	**Threats**

Figure 3.28 SWOT 2 × 2 matrix

Strengths. Characteristics of the planned objective that, within the organization, give it an advantage over other objectives.

Weaknesses. Characteristics of the planned objective that, within the organization, constitute a disadvantage relative to other objectives.

Opportunities. Elements outside the organization that the planned objective could exploit to its advantage.

Threats. Elements outside the organization that could hinder the success of the planned objective.

Traditional SWOT analysis is based on a qualitative approach, however; the literature proposes several quantitative SWOT methods that include various computation strategies augmenting the qualitative SWOT analysis. For example, one may create a weighted score matrix (WSM) for each of the four SWOT perspectives. Each WSM could define the following weighing scheme for each component within each of the four perspectives.

Weight. Indicates the relative importance of a given factor. One way to assign weights to individual factors is to use the range from 0.00 to 1.00, where zero indicates insignificant factor and one indicates very important factor. For simplicity, the total weight per single WSM should be 1.00.

Rating. Captures the likelihood of attaining a desired objective. For example, one can use rating on the scale from 1 to 3 (i.e. Minor = 1, Intermediate = 2, and Major = 3).

Weighted score. Computed by multiplying each factor's weight by its rating.

Total weighted score. Computed by summing up the weighted scores for each factor within a given perspective.

Finally, for each pairs of WSM, one can compute:

Internal factors ratio. Computed by dividing the strengths total weighted score by the weaknesses total weighted score.

External factors ratio. Computed by dividing the opportunities total weighted score by the threats total weighted score.

Higher ratios, for either internal or external circumstances, indicate greater likelihood to attain the desired objective. Quantitative SWOT analysis is important because it can indicate whether a group of people considers an objective to be feasible. If the objective is not feasible, then one may attempt to improve the ratio of the organization's strength to weakness. That is, enhance the strength of the organization or reduce the weakness of the organization or both. In the absence of these remedies, one should change the nature of the objective or give up the quest for attaining the said objective.

3.3.3.2 Implementation Procedure

The following is a procedure to conduct a qualitative and quantitative SWOT analysis (Figure 3.29).

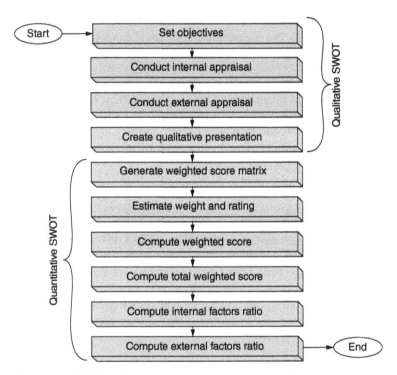

Figure 3.29 SWOT qualitative and quantitative implementation procedure

Step 1: Set objective. Formulate a specific objective to be accomplished by the organization.

Step 2: Conduct internal appraisal. Conduct an internal appraisal of the organization. This should include the identification of all the factors within the organization that contribute to the strength and weakness of the organization vis-à-vis the stated objective.

Step 3: Conduct external appraisal. Conduct an external appraisal of elements that may affect the organization. This should include the identification of all the factors outside the organization that contribute to the opportunities and threats to the organization vis-à-vis the stated objective.

Step 4: Create qualitative presentation. Place the information in a 2 × 2 SWOT matrix format in a concise form.

Expanding the procedure to deal with quantitative SWOT analysis is carried out by implementing the following steps:

Step 5: Generate weighted score matrix. Create a WSM for each of the four SWOT perspectives.
Step 6: Estimate weight and rating. For each factor within each WSM, estimate the values of the weight and rating.
Step 7: Compute weighted score. For each factor within each WSM, compute the weighted score by multiplying each factor's weight by its rating.
Step 8: Compute total weighted score. For each WSM, compute the total weighted score by summing up the weighted scores.
Step 9: Compute internal factors ratio. For each pair of WSMs, compute the internal factors ratio (IFR) by dividing the strengths total weighted score by the weaknesses total weighted score.
Step 10: Compute external factors ratio. For each pair of WSMs, compute the external factors ratio (EFR) by dividing the opportunities total weighted score by the threats total weighted score.

3.3.3.3 Example: Applying for a Research Funding

1) Example background

The objective of this example is to successfully apply for European Commission (EC) research funding involving a three-year development of a generic method and tools aimed at reducing the manufacturing lead-time of customized products by 50%. The objective was conceived by members of the engineering faculty of a large, well-established, university, having conducted two large, EC-funded, research projects in the past few years.

2) Example analysis

SWOT qualitative matrix. An example of SWOT qualitative matrix is presented in Table 3.1
SWOT quantitative matrices. The four SWOTs weighted score matrices are presented in Table 3.2 through Table 3.5.
Internal and external factors ratios. The maximal total weighted score is 3.00 and the minimal total weighted score is 1.00. Therefore, the maximal factor ratio is 3/1 = 3.00 and the minimal factor ratio is 1/3 = 0.33, so the breakeven factor ratio is 1.50. The internal and external factors ratios for this example are computed and shown below:

$$\text{Internal factor ratio} \left(\text{IFR}\right) = \frac{2.40}{1.95} = 1.23$$

$$\text{External factor ratio} \left(\text{EFR}\right) = \frac{2.60}{2.90} = 0.90$$

Table 3.1 SWOT qualitative matrix example

	Helpful in achieving objectives	Harmful in achieving objectives
Internal factors	**Strengths** • Familiarity with the research topic • Superb group of international partners • Availability of academic and industrial facilities • Availability of researchers and students • Experience in managing research projects	**Weaknesses** • Limited collaboration within an international project • Limited control within a consortium setup • Incompatible partners' internal goals • Inexperience in managing intellectual property rights • High level of administrative bureaucracy • Limited administrative support
External factors	**Opportunities** • EC funding is financially important to partners • Research will reduce manufacturing lead-time • Research will broaden university knowhow • University researchers could publish results	**Threats** • EC extreme competition limits likelihood of winning project • Potential financial issues may affect small partners • Inflation could reduce project's net funding • EC's funding policy changes may endanger project • Foreign currency usage hamper funding

Table 3.2 SWOT weighted score matrix: Strengths

Strengths factor	Weight	Rating	Weighted score
Familiarity with the research topic	0.30	3	0.90
Superb group of international partners	0.20	2	0.40
Availability of academic and industrial facilities	0.20	2	0.40
Availability of researchers and students	0.10	1	0.10
Experience in managing research projects	0.20	3	0.60
Total	1.00		2.40

Table 3.3 SWOT weighted score matrix: Weaknesses

Weaknesses factor	Weight	Rating	Weighted score
Limited collaboration in an international project	0.30	2	0.60
Limited control within a consortium setup	0.30	2	0.60
Incompatible partners' internal goals	0.20	3	0.60
Inexperience in managing intellectual property rights	0.10	1	0.10
High level of administrative bureaucracy	0.05	1	0.05
Limited administrative support	0.05	1	0.05
Total	1.00		1.95

Table 3.4 SWOT weighted score matrix: Opportunities

Opportunities factor	Weight	Rating	Weighted score
EC funding is financially important to partners	0.30	3	0.90
Research will reduce manufacturing lead-time	0.40	3	1.20
Research will broaden university knowhow	0.20	2	0.40
University researchers could publish	0.10	1	0.10
Total	1.00		2.60

Table 3.5 SWOT weighted score matrix: Threats

Threats factor	Weight	Rating	Weighted score
EC extreme competition limits likelihood of winning project	0.70	3	2.10
Potential financial issues may affect small partners	0.10	3	0.30
Inflation could reduce the project net funding	0.05	2	0.10
EC's funding policy changes may endanger project	0.10	3	0.30
Foreign currency usage hamper funding	0.05	2	0.10
Total	1.00		2.90

One can conclude that, from the organization's internal perspective and more so, from external perspectives, the likelihood of attaining the desired objective is quite low.

3.3.3.4 Further Reading

- Bensoussan and Fleisher, 2015.
- Fine, 2009.

3.3.4 SCAMPER Analysis

3.3.4.1 Theoretical Background

SCAMPER is a creative technique to stimulate thinking and helping generate new ideas. The original idea was proposed by Alex Osborne in 1953, identifying nine ways of manipulating a subject. Then in 1971, Bob Elerle rearranged these principles into the SCAMPER acronym, standing for the following seven active verbs: Substitute, Combine, Adapt, Modify, Put, Eliminate, and Reverse. SCAMPER analysis is achieved by asking questions about relevant topics, using each of the seven prompts above. Asking SCAMPER-directed questions helps one to come up with creative ideas, as follows:

Substitute. Asking questions related to substitutions of parts within an existing problem, system, or process in order to resolve a problem. One may ask: "Instead of using _____, can we use _____?"

Combine. Asking questions related to combining two or more elements within an existing problem, system, or process in order to solve a problem. One may ask: Can we bring together _____ and to achieve _____?"

Adapt or Adjust. Asking questions related to adapting or adjusting elements within an existing problem, system, or process in order to resolve a problem. One may ask: "Can we adapt or adjust _____ in a way _____ to achieve _____?"

Modify or Magnify. Asking questions related to modifying or distorting elements within an existing problem, system, or process in order to resolve a problem. One may ask: "Can we modify (distort) _____ in a way _____ to achieve _____?"

Put to other uses. Asking questions related to using differently elements within an existing problem, system, or process in order to resolve a problem. One may ask: "Can we reuse _____ in this way _____ by doing _____?"

Eliminate. Asking questions related to eliminating components from an existing problem, system or process in order to resolve a problem. One may ask: "Can we eliminate _____ by doing _____?"

Reverse or Rearrange. Asking questions related to reversing or rearranging elements from an existing problem, product, or system in order to resolve a problem. One may ask: "Can we rearrange _____ like this _____ so that _____?"

3.3.4.2 Implementation Procedure
The SCAMPER analysis may be implemented using the following procedure:

Step 1: Identify the problem. Identify the problem or product or process at hand in a group session.

Step 2: Generate SCAMPER questionnaire. Generate a SCAMPER questionnaire (Table 3.6). Preferably, each member of the group files the questionnaire and then the group integrates the data and creates a filled, group-wide SCAMPER questionnaire.

Table 3.6 SCAMPER questionnaire

Mnemonics	Questions
Substitute	
Combine	
Adapt or Adjust	
Modify or Magnify	
Put to another use	
Eliminate	
Reverse	

Step 3: Generate creative answers. Perform a brainstorm session and attempt to generate creative answers based on the SCAMPER group-wide questionnaire.

3.3.4.3 Example: Software Quality Delivered on Time

1) Example background

A real-time software department in a large industrial firm consistently creates software products containing unacceptable quantity of faults. In addition, the software is delivered, too often, behind schedule. The result of this situation is that software-intensive products miss their delivery target date and, in addition, the reputation of the firm is jeopardized. Management convenes a SCAMPER analysis meeting to deal with this problem.

2) Example analysis

SCAMPER questionnaire. Table 3.7 provides an example of a filled SCAMPER questionnaire related to the above scenario.
SCAMPER resolution. The following decisions were made attempting to ensure that the software products will contain an acceptable quantity of faults and be available on time:

Table 3.7 Example of SCAMPER questions

Mnemonic	Questions
Substitute	• Can we replace the software department head with a more capable manager? • Can we replace some of the software engineers? • Can we change some of the software department operational procedures?
Combine	• Can we transfer some of the software development load to another software group, combining the load across different company divisions?
Adapt or Adjust	• Can we make adjustment in the infrastructure surrounding the software department (e.g. interactions with other departments, management, internal guideline and procedures, physical structures, etc.)?
Modify or Magnify	• Can we increase the size of the software department? • Can we increase the size of the quality assurance department? • Can we improve the software department stuff moral, perhaps by monetary rewards for submission of high-quality software products on time?
Put to other uses	• Can we retrain software engineers in the department to develop software in accordance with more advanced software standards?
Eliminate	• Can we identify and let go the software engineers who consistently contribute the most to faults and delays in delivery of software?
Reverse or Rearrange	• Can we rearrange the current interfaces and procedures between the software departments and other related departments? • Can we undertake a major personnel rearrangement within the software department?

1) Retrain software engineers in the department to develop software in accordance with more advanced software standards.
2) Identify and let go software engineers who consistently contribute the most to software faults and delays in delivery of software.
3) Transfer some of the software development load to another software group, combining the load across different company divisions.
4) Make adjustment in the infrastructure surrounding the software department, in particular, modify the internal guideline for software development, and improve the computing infrastructure within the department.

3.3.4.4 Further Reading

• Brostow, 2015.   • Michalko, 2006.
• Eberle, 2008.

3.3.5 Focus Groups

3.3.5.1 Theoretical Background

Focus group is an idea-generation technique where stakeholders group, guided by a trained leader, discuss their point of view on a designated topic or problem. The concept of focus groups was introduced by the sociologist Robert Merton, after World War II in his book titled *The Focused Interview* published in 1956. The purpose of focus groups is either to (1) stimulate new ideas or (2) collect qualitative data about feelings and inclinations of participants regarding a specific issue at hand. Such attitudes tend to develop by interactions with other people in a positive, sharing group atmosphere. By and large, focus groups should be contemplated when one wants to explore an issue that can't easily be answered in a written document or survey.

The process starts with selecting a moderator to conduct the focus group as well as focus group participants. A moderator is expected to have the mental skills and group control abilities to handle such undertaking. Participants should be recruited on the basis of their familiarity with the subject matter as well as their similar demographics and psychographics characteristics. They should be sufficiently mature to be able to listen to others, freely express themselves, and learn from one another.

Within the sphere of this book, the focus group process is a qualitative engineering group study. It addresses questions that require depth of understanding that, by and large, cannot be resolved through quantitative methods. Overall, a focus group is a flexible process, providing a speedy result with high face validity at relatively low cost while the information obtained is expressed in participants' own words. In addition, focus groups will often generate creative and unexpected ideas.

On the other hand, assembling an effective focus group and leading a smooth process can be a challenge. Also, the skill of the facilitator is crucial to the dynamics of a focus group, and ultimately to the utility of the obtained information. In addition, analysis of the obtained data is not easy, requiring experienced analysts. Other drawbacks of focus groups are that the use of such data in order to generate meaningful conclusions must be done diligently in order to avoid outpacing the limits of the focus group itself. Other criticism leveled against the focus group method is that the process lacks anonymity, which means that participants cannot maintain their confidentiality. As a result, participant may adopt a groupthink attitude as the group aims to please the powers that be rather than offer their own opinions.

Finally, the results of the focus group should be communicated to management (or clients) by way of a formal report, detailing the process and the results obtained.

3.3.5.2 Implementation Procedure

Focus group process may be implemented using the following steps:

Step 1: Undertake activities prior to group meeting. Identify the subject matter that needs to be discussed and locate suitable moderator/leader to conduct the focus group. Then invite appropriate persons who will act as participants in the process. In parallel, prepare one or more questions to be discussed during the focus group session(s).

Step 2: Undertake activities during group meeting. Start the focus group by reviewing the group's purpose and goals. Then explain how the meeting should proceed and how participants are expected to contribute. Then proceed with one question at a time, making sure that each participant provides his/her opinion, and then anyone in the group has a chance to relate to these views. Also make sure that these conversations are recorded for later analysis. Finally, inform the group of any next steps that are scheduled as well as their responsibilities vis-à-vis the focus group.

Step 3: Undertake activities after group meeting. Examine the transcript or written summary of the focus meeting, looking for patterns, new ideas, or additional issues that should be brought before the group for further discussion. Then generate a concluding report and present it to the focus meeting for final approval. Thereafter, deliver the report to management (or customer) for concrete action.

3.3.5.3 Example: Sales Department Focus Group

1) Example background

The sales department in a large marketing concern is consistently failing to predict future sales targets. In addition, it establishes ineffective advertising strategy, proposes incorrect pricing policies relative to the competitors, and frequently receives antagonistic criticism from customers.

2) Example analysis

The vice president of marketing in this concern was selected to lead a focus group to investigate these problems. He chose an eight-member team to conduct the actual focus group process. This group included two representatives from marketing, one representative from finance, one representative from sales, one representative from operations, and one representative from the technology arm of the company. In addition, two outside consultants, representing customers' outlook, were added to the focus group.

Initially, members of the focus group interviewed each member of the sales department personnel and their associates in other bodies within and outside the company. Then, after several group deliberations, it was agreed that the problems originally identified were indeed present. In addition, it was agreed that the morale and interpersonal relationships within the sales department disrupted daily operations.

The focus group suggested overhauling and rejuvenating the entire sales department. The head of the sales department as well as several members of the department were transferred to other positions. The remaining members of the department undertook an extensive professional retraining and participated in several team cohesion exercises. As a result, the original problems greatly diminished or disappeared altogether.

3.3.5.4   Further Reading

- Merton, 1990.

3.4   Convergent Methods for Individuals

This chapter describes a collection of convergent creative methods for individuals as encapsulated in Figure 3.30.

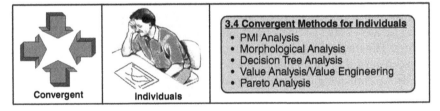

Figure 3.30 Convergent methods for individuals

3.4.1   PMI Analysis

3.4.1.1   Theoretical Background

One common problem with traditional decision-making processes is that people focus their attention on how to vindicate their original opinions. One approach to alleviate this tendency is to use the Plus-Minus-Interesting (PMI) analysis technique. PMI is a decision-making instrument proposed by Edward de Bono in the early 1990s. Relative to the traditional two-column pros and cons approach, under PMI one adds another column, "Interesting," which provides space for relevant information that does not fit easily into the other two columns. Essentially, PMI is a mechanism for making decisions and generating ideas about specific topics by considering as many aspects of a particular issue as possible with minimal preconceived notions or prejudices.

Under PMI, one identifies first the decision to be made. Next, under the label "Plus," one lists all of the possible positive consequences of making the decision, including anticipated benefits, outcomes, or results. Next, under the label "Minus," one lists all of the possible negative consequences of making the decision, including anticipated problems, harmful outcomes, or results. Then, under the label "Interesting," one lists relevant topics or interesting points as well as matters of curiosity or uncertainty that are not easily placed under either the plus or the minus columns. These interesting topics could lead one to examine other ideas that may be derived from the original plus and minus columns. At the same time, one could consider how to increase the impact of items in the plus column and vise-versa regarding the minus column.

So far, we have discussed handling a "go/no-go" type of decisions (e.g. "Should we move our production line to Michigan or stay put?"). However, sometimes decisions involve a selection from among several alternatives (e.g. "Should we move our production line to Michigan or to California or stay put?"). Under these circumstances, one could expand the traditional PMI table by dividing it into several sections, each one dealing with a different "Alternative" (creating an APMI table). Then, for each section one could develop a unique set of Plus-Minus-Interesting arguments.

Another expansion of the traditional PMI analysis is achieved by assigning a score (positive or negative) to each argument. One way to create these scores is to use a Likert scale, a widely used measure to quantify responses in various surveys. The method is especially pertinent within a group setting where team voting ensures that each individual voice will affect the final score values. One may use an extended Likert scale (Figure 3.31) in order to assign positive values to items in the Plus column and negative values to items in the Minus column. Likewise one could assign either positive or negative values (or none) to items in the Interesting column. Finally, one should add the scores of all the columns. A positive overall score implies that a positive decision should be taken while a negative one hints that a negative decision should be taken.

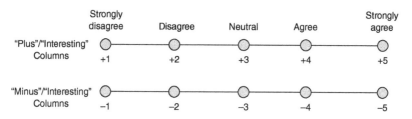

Figure 3.31 Extended Likert scale

3.4.1.2 Implementation Procedure
A procedure to implement a traditional PMI analysis includes the following steps (Figure 3.32).

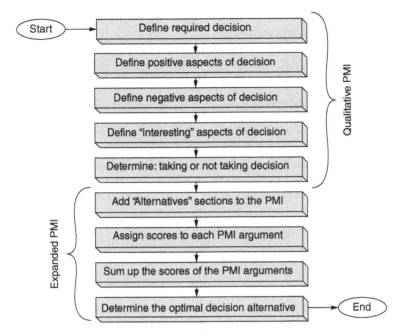

Figure 3.32 PMI implementation procedure

Step 1: Identify topic. Identify the topic or the issue that requires a decision.
Step 2: Identify positive aspects. Identify all of the positive aspects related to the issue at hand.
Step 3: Identify negative aspects. Identify all of the negative aspects related to the issue at hand.
Step 4: Identify interesting aspects. Identify all of the interesting points as well as matters of curiosity or uncertainty related to the issue at hand.
Step 5: Make a decision. Based on the previous steps, determine whether or not to take the decision.

Expanding the procedure to implement a quantitative and enlarged Alternatives-Plus-Minus-Interesting (APMI) analysis includes the following steps:

Step 6: Add alternative section. Expand the traditional PMI table by adding one or more "Alternatives" sections and then, for each of these decision alternatives, develop a unique set of Plus-Minus-Interesting arguments.
Step 7: Assign scores. For each decision alternative, assign a score (positive or negative) to each Plus-Minus-Interesting argument within the columns.

Step 8: Sum up scores. For each decision alternative, sum up the scores of the Plus-Minus-Interesting arguments.

Step 9: Make a decision. Select the decision alternative associated with the highest overall score.

3.4.1.3 Example: Unmanned Vehicle

1) <u>Example background</u>

A well-established corporation has been involved in the development and manufacturing of unmanned air vehicles (UAV) for many years. Management would like to expand this business area by developing and manufacturing either an unmanned ground vehicle (Figure 3.33) or an autonomous underwater vehicle (Figure 3.34).

Figure 3.33 Unmanned ground vehicle (UGV)

Figure 3.34 Autonomous underwater vehicle (AUV)

2) Example analysis

Table 3.8 and Table 3.9 depict APMI tables for this example.

As can be seen in the two APMI tables, the overall score for expanding the UGV business is (+3) whereas the overall score for expanding the AUV business is (−8). Clearly, expanding the business of unmanned ground vehicles (UGVs) is significantly more promising.

Table 3.8 APMI for expanding the UGV business

Plus	Minus	Interesting
The company has some UGV experience (+4).	Navigational accuracy requirements present substantial problems (−5).	The company could evaluate very precise navigation based on combining several independent global positioning systems (i.e. GPS, GLONASS, Galileo, BeiDou) (+5).
The commercial prospects for a UGV system are relatively good (+3).	UGV may lose communication under certain terrain conditions (−4).	
Physical conditions for conducting field tests are very good (+5).	Many operational issues must be resolved (−4).	The company could evaluate technical and commercial prospects for developing inexpensive UGV kits to be installed within commercial vehicles at nominal cost (+3).
	International competition in the UGV market is significant (−4).	
Overall score = (+3)		

Table 3.9 APMI for expanding the AUV business

Plus	Minus	Interesting
AUV is a potential new and promising business area for the company (+4).	The company has no AUV experience (−4).	The commercial prospects for an AUV system require further investigation (−2).
Physical conditions for conducting field tests are fairly good (+2).	AUV may operate independently. This will hamper its real-time control and prospects of recovery (−4).	
	Generally, AUV requires umbilical cable to maintain control, real-time communication and recovery operations (−4).	
Overall score = (−8)		

3.4.1.4 Further Reading

- Proctor, 2013.

3.4.2 Morphological Analysis

3.4.2.1 Theoretical Background

Morphological analysis was developed by Fritz Zwicky in the 1960s.[8] The method may be used in order to synthesize alternative overall system solutions in a multiattribute, often nonquantifiable, problem space. The analysis involves the creation of a two-dimensional matrix in which relevant attributes are listed across the matrix's top row and relevant parameters are listed beneath each attribute. For example, the analysis may involve placing systems functions across a matrix's top row and placing underneath each of these systems functions a set of design solutions (Figure 3.35). Next, an optimal and desirable system solution may be identified.

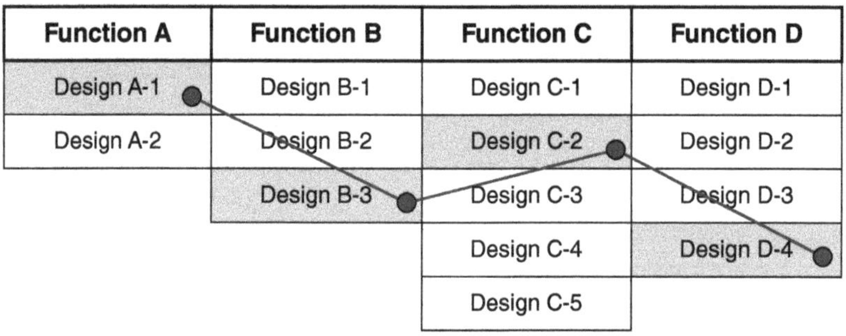

Function A	Function B	Function C	Function D
Design A-1	Design B-1	Design C-1	Design D-1
Design A-2	Design B-2	Design C-2	Design D-2
	Design B-3	Design C-3	Design D-3
		Design C-4	Design D-4
		Design C-5	

Figure 3.35 Typical morphological chart

As can be seen, there are, large numbers of theoretical systems' permutations (i.e. in this example there are a total of $2 \times 3 \times 5 \times 4 = 120$ systems solutions). But, of course, many systems solutions are impractical. Nevertheless, morphological analysis is a creative technique geared not only to solve problems but also to find innovative ideas for systems and services.

3.4.2.2 Implementation Procedure

The following steps should be used in order to create a morphological analysis:

Step 1: Identify the system. Identify the system or process to be analyzed.
Step 2: Define attributes. Define all the key attributes that affect the system under consideration.

---

8 Morphological analysis is an elaboration of the attribute listing technique pioneered by Robert Platt Crawford in the early 1930s.

Step 3: Create matrix. Create a two-dimensional matrix and place the attributes at the top row of the matrix.

Step 4: Define parameters. Place a set of relevant parameters beneath each attribute within the matrix.

Step 5: Determine possible solutions. Evaluate all the practical solutions that are part of the morphological space.

Step 6: Finalize solution. Select an optimal and desirable solution.

3.4.2.3 Example: T-38 Talon aircraft Avionics Upgrade

1) <u>Example background</u>

The T-38 Talon aircraft (originally, Northrop F-5) is part of a supersonic light fighter family, initially designed in the late 1950s by Northrop Corporation. Being a relatively smaller and simpler aircraft, the T-38 Talon cost less to procure and operate, making it a popular export aircraft to many countries all over the world (Figure 3.36).

Figure 3.36 The T-38 Talon aircraft

The defense department of one country decided to upgrade the avionics of its aged T-38 Talon fleet. This example describes a morphological analysis used to select an optimal avionics suite configuration for these aircrafts.

2) <u>Example analysis</u>

Figure 3.37 and Figure 3.38 depict the result of the T-38 Talon aircraft's avionics upgrade morphological analysis.

The chosen avionics upgrade suite is composed of (1) two displays, (2) head-up display, (3) no helmet display, (4) air data computer, (5) embedded GPS-INS, (6) radar system with movable antenna, (7) Flight planning capability

Display units	Head-up Display (HUD)	Helmet Display	Air Data System (ADS)	Navigation	Radar system
One Display	Not installed	**Not installed**	Altimeter + Speedmeter	Attitude, Heading, Roll, System (AHRS) + GPS	Low performance Radar
Two Display	**Install HUD**	Electro-optics-based helmet	Analog Air Data System (AADS)	Inertial Navigation System (INS)	**Moveable Antenna**
Glass Cockpit		Electromagnetic-based helmet	**Air Data Computer (ADC)**	**Embedded GPS-INS (EGI)**	Fixed (Phase Array) Antenna

Figure 3.37 Morphological analysis: Part I

Flight planning	Radio communication	Identification Friend or Foe (IFF)	Electronic Warfare (EW)	Forward Looking Infrared (FLIR)
GPS based system	Voice	Not installed	Not installed	**Not installed**
GPS + Mission Computer System (MCS)	**Voice + Data**	**Integrated existing IFF**	Passive - Threat detection system	Install FLIR
	Encrypted Voice + Data	Provide new IFF system	**Active - Threat Countermeasures technology**	

Figure 3.38 Morphological analysis: Part II

based on GPS and Mission Computer System, (8) radio communication supporting nonencrypted voice and data, (9) integrated existing identification friend or foe system, (10) electronic warfare system based on active threat countermeasure technology and (11) no forward-looking infrared system. This specific avionics upgrade configuration emanates from a total of $3 \times 2 \times 3 \times 3 \times 3 \times 3 \times 2 \times 3 \times 3 \times 3 \times 2 = 52{,}488$ theoretical avionics suits combinations.

3.4.2.4 Further Reading

- Eisner, 2005.
- Zwicky, 1969
- Parsch, 2016.

3.4.3 Decision Tree Analysis

3.4.3.1 Theoretical Background

Decision tree analysis is a support tool that uses a tree-like branched graph to implement possible outcome of either decisions or statistical probabilities.[9] Decision trees help people visualize and understand potential option and

---

9 See: John F. Magee, "Decision Trees for Decision Making, *Harvard Business Review* (July 1964), https://hbr.org/1964/07/decision-trees-for-decision-making. Accessed: May 2017.

weigh each course of action. More specifically, decision trees simplify the settlement of complex problems by slicing them into a series of independent decision points. In addition, the structure of decision trees allows users to observe the relationship between different events or decisions.

Decision trees are composed of three types of nodes connected by a dispersed set of edges or branches. The tree nodes may include (1) decision nodes where one must make a decision among several outcomes (commonly represented by squares), (2) chance nodes where the outcome depends on statistical probability (represented by circles), and (3) leaf nodes at the very end of a tree branch, representing certain action to be performed.

In addition, a decision tree can be linearized into a set of classification rules, whereby each path from the tree root to a given leaf node, represents a single rule. Typically, each rule will have the form: "If {Condition-1 AND Condition-2 AND… Condition-n} THEN execute the action defined in the relevant leaf node."

In summary, decision trees are simple to understand and interpret and have visual value even with limited hard data. They can provide important insights and help determine worst, best, and expected values for different scenarios. In addition, they provide the means to analyze quantitatively possible consequences of decisions.

3.4.3.2 Implementation Procedure

Decision tree analysis may be implemented using the following steps:

Step 1: Draw the decision tree. Normally, one starts creating a decision tree by identifying possible expected results to the left side of a page or a display screen, representing the root of the tree. From the root, one draws separate lines to the right representing different decisions to be made. At the end of each line, one may place either a decision node (square) or chance node (circle). If the result is a decision node, then the same procedure should be repeated. If the result is a chance node, then lines representing possible outcomes and their probabilities (totaling 1.00) are to be drawn. This process should be repeated as necessary.

Step 2: Calculate the values of chance nodes. In this step, one usually assigns a monetary value to each possible tree outcome. Then work backward and calculate the outcome value of all connected chance nodes (Figure 3.39).

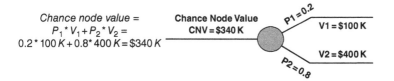

Figure 3.39 Example: Chance node value

Step 3: Calculate the value of decision nodes. In this step, one usually derives the value of a decision node from the value of the corresponding upstream chance node or decision node along the decision line. This is done by subtracting the relevant decision costs from the upstream chance node value (Figure 3.40).

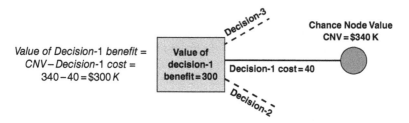

Value of Decision-1 benefit =
CNV − Decision-1 cost =
340 − 40 = $300 K

Figure 3.40 Example: Value of decision benefit

Step 4: Optimize the decision strategy. By applying this technique for all possible outcomes, one can optimize the decision strategy.

3.4.3.3 Example: Building a Car Assembly Plant

1) Example background

An automobile firm plans to build a new car assembly plant. Management must decide whether to build a relative large plant or a small plant that may or may not be expanded in the future. For simplicity, the example will assume a time horizon of 10 years with a possible plant expansion after 4 years. Additional simplifying assumptions are: (1) zero interest rate, (2) zero inflation rate, and (3) static yearly product's demand.

In summary, management must now make one and potentially two decisions: (1) between a large and a small plant, and if the company chooses to build a small plant and then finds the demand sufficiently high, management must decide in the future; (2) to expand or not expand its plant in 4 years' time. Figure 3.41 depicts the decision tree for the planned car assembly plant first phase.

2) Example analysis

The strategy for solving this problem is by way of the rollback method – that is, by resolving decision 2 first and simplifying the decision tree chart. Table 3.10 depicts this analysis. Accordingly, the total expected value of the "Expansion" alternative is $2,240K, whereas the total expected value of the "No-expansion" alternative is $600K. Obviously, the expansion alternative is preferable.

Once decision 2 is made, the decision tree chart can be simplified (Figure 3.42).

Next, Table 3.11 depicts the decision 1 analysis. Accordingly, the large plant expected value is $800K, whereas the small plant expected value is $1,212K. Obviously, the small plant option is preferable.

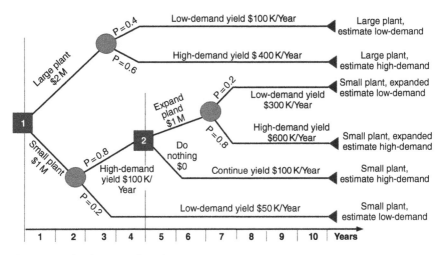

Figure 3.41 Decision tree: first phase

Table 3.10 Decision 2 analysis

Choice	Chance event	Probability	Yearly yield	Years	Expected value
Expansion	Six years low demand	0.2	$300,000	6	$360,000
	Six years high demand	0.8	$600,000	6	$2,880,000
	Plant expanding cost				−$1,000,000
	Total				**$2,240,000**

No expansion	Six years continuous demand	1	$100,000	6	$600,000
	Plant expanding cost				$0
	Total				**$600,000**

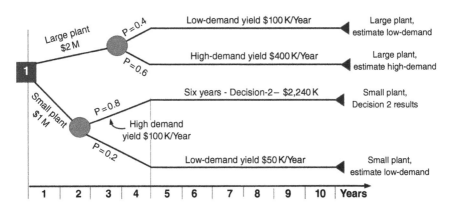

Figure 3.42 Decision tree: second phase

Table 3.11 Decision 1 analysis

Choice	Chance event	Probability	Yearly yield	Years	Expected value
Large plant	Ten years low demand	0.4	$100,000	10	$400,000
	Ten years high demand	0.6	$400,000	10	$2,400,000
	Large plant cost				−$2,000,000
	Total				**$800,000**

Choice	Chance event	Probability	Yearly yield	Years	Expected value
Small plant	Six years decision-2	0.8	$2,240,000		$1,792,000
	Four year high demand	0.8	$100,000	4	$320,000
	Ten years low demand	0.2	$50,000	10	$100,000
	Small plant cost				−$1,000,000
	Total				**$1,212,000**

In summary, according to the previous analysis, the first decision is to build a small plant and the second decision (given that the selected parameters correspond to future reality) is to expand the plant.

3.4.3.4 Further Reading

- Magee, 1964.
- Skinner, 2009.
- Quinlan, 1987.

3.4.4 Value Analysis/Value Engineering

3.4.4.1 Theoretical Background
Value is difficult to measure because it is used by different people in a variety of ways. Be it as it may, the most objective measure is perhaps, in monetary units.[10] The concept of *value analysis (VA)* was conceived by Lawrence Miles in the mid-1940s. In general, the intent of VA is to measure the value of an item by dividing its worth by its cost. The equivalent term in engineering is *value engineering (VE)*. The specific intent of engineers is to increase a system's value by improving the relationship between its worth to its stakeholders (i.e. functionality, performance and quality) to its life cycle's cost. Along the same line, the difference between worth and cost is known as a value gap. The larger the value gap, the more potential there is for value improvement. So, in summary, value engineering may be defined as "an organized study of a system's worth in order to satisfy users' needs and wants with a quality product at the lowest life cycle cost":

---

10 "There are seven classes of value that are recognized: Economic, Moral, Aesthetic, Social, Political Religious and Judicial" (Rumane, 2010).

$$\text{System value} = \frac{\text{System worth}}{\text{System cost}} \approx \frac{\text{Function} + \text{Performance} + \text{Quality}}{\text{Life cycle cost}}$$

One should be aware of the four ways a system's value may be increased: (1) increase system's worth, (2) decrease system's cost, (3) increase system's cost and increase even more system's worth, and (4) decrease system's worth and decrease even more system's cost.

Integrating value engineering-quality function deployment (VE-QFD) within a single matrix provides a powerful method for analyzing the value of a system. Under this scheme, one builds a matrix containing the functions of a system along with its subsystems and the cost and worth of all relevant combinations. Each function should be described in two words: verb and noun, placing the basic functions first followed by the supporting functions. Next, the subsystems supporting the various functions are listed and then cost and worth of each subsystem against each function is identified (Table 3.12).

Table 3.12 Integrated VE-QFD matrix

	Subsystems							
	Subsystem-1		Subsystem-2		....		Subsystem-n	
System functions	Cost	Worth	Cost	Worth	Cost	Worth	Cost	Worth
Function 1								
Function 2								
....								
Function *n*								
Total								
Total cost [$]								
Total worth [$]								
System value								

The entire value engineering process should commence as early as possible within a system's life cycle in order to maximize the cost saving potential (Figure 3.43).

3.4.4.2 Implementation Procedure
The following procedure may implement the Value analysis/Value engineering process.

Step 1: Identify the system and environment. Identify the system to be analyzed, its boundary, and its environment.

Cost

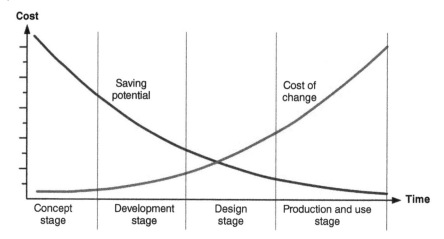

Figure 3.43 Value engineering saving potential versus cost of change

Step 2: Identify the subsystems. Identify the subsystems and list them across the header of the integrated VE-QFD matrix.

Step 3: Identify the functions of the system. List, down the leftmost column of the integrated VE-QFD matrix, the functions that constitute the main purpose of the system as well as its supporting or secondary functions.

Step 4: Assign costs to subsystems and functions. Determine the cost of each subsystem (including development manufacturing, distribution, maintenance, etc.) and assign them to the various functions performed by the given subsystem. Accordingly, update the integrated VE-QFD matrix.

Step 5: Assign worth to subsystems and functions. By way of simplification, one may assume that worth of stable system is equal to its price. Only when the system undergoes changes or modification (e.g. within a value engineering process) may its price and its worth diverge. Accordingly, update the integrated VE-QFD matrix.

Step 6: Compute system value. Sum up the cost and worth of each subsystem and then the total cost and worth of the entire system. Thereafter, compute the total system value and update the integrated VE-QFD matrix.

Step 7. Update cost and worth of subsystems and functions. Analyze each subsystem and identify any changes that can improve the worth versus cost ratio. Based on this analysis, insert updated cost and worth values into the integrated VE-QFD matrix.

Step 8: Compute upgraded system value. Sum up the cost and worth of each subsystem and then the total cost and worth of the entire system. Thereafter compute the total upgraded system value. Accordingly, update the integrated VE-QFD matrix.

3.4.4.3 Example: Hairdryer Design

1) Example background

A handheld hairdryer is an electromechanical device designed to blow hot air over damp hair in order to dry it (Figure 3.44). The system is composed of the following subsystems: (1) cord, (2) On/Off switch, (3) fan speed switch, (4) motor, (5) fan, (6) heat switch, (7) heating element, and (8) casing. The environment is composed of (1) electricity and (2) air.

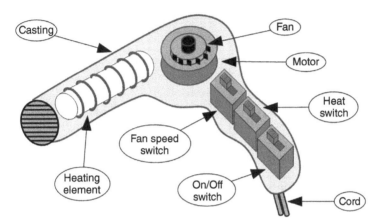

Figure 3.44 Handheld hairdryer and its components

This system is analyzed with the aim of improving its value relevant to the current design. The integrated VE-QFD matrix of this, top of the line hairdryer, in its current design is depicted in Table 3.13, calling for a unit price tag, as well as worth, of $135 and a system value of 1.0.

2) Example analysis

A thorough analysis of each subsystem indicated the following:

1) An improved design of the heating element could improve the efficiency of this subsystem so more heat can be utilized, improving the user experience and, therefore, the worth of the system.
2) A different motor manufacturer can supply similar-quality motors at a reduced price. This change will reduce the overall price of the system.
3) A new casing design will eliminate the use of asbestos within the system. This will slightly increase the price of a unit but substantially improve its safety.

Table 3.14 depicts the integrated VE-QFD matrix of the upgraded system. As can be seen, the cost of the updated system was reduced to $127, while the worth of the system has increased to $175 and a system value of 1.4.

Table 3.13 Hairdryer current design

<table>
<tr><th rowspan="2">Hairdryer System features</th><th colspan="16">Subsystems (Current design)</th></tr>
</table>

Hairdryer System features	Cord		On/Off switch		Fan speed switch		Motor		Fan		Heat switch		Heating element		Casing	
	Cost	Worth	Cost	Worth	Cost	Worth	Cost	Worth	Cost	Worth	Cost	Worth	Cost	Worth	Cost	Worth
Heats air													20	20		
Blows air							30	30	10	10						
Varies heat			2	2							5	5				
Varies air flow			2	2	5	5										
Portable device	5	5	5	5			5	5	5	5					5	5
Uses electricity	5	5	5	5			2	2					2	2		
Protects users			2	2	2	2	2	2	2	2	2	2	2	2	10	10
Total	10	10	16	16	7	7	39	39	17	17	7	7	24	24	15	15

Total cost [$]	135
Total worth [$]	135
System value	1.0

Table 3.14 Hair dryer updated design

Hairdryer System features	Cord		On/Off switch		Fan speed switch		Motor		Fan		Heat switch		Heating element		Casing	
	Cost	Worth	Cost	Worth	Cost	Worth	Cost	Worth	Cost	Worth	Cost	Worth	Cost	Worth	Cost	Worth
Heats air													20	30		
Blows air							20	30	10	10						
Varies heat			2	2							5	5				
Varies air flow			2	2	5	5										
Portable device	5	5	5	5			5	5	5	5					5	5
Uses electricity	5	5	5	5			2	2					2	2		
Protects users			2	2	2	2	2	2	2	2	2	2	2	2	12	40
Total	10	10	16	16	7	7	29	39	17	17	7	7	24	34	17	45

Total cost [$]	127
Total worth [$]	175
System value	1.4

3.4.4.4 Further Reading

- Kassa, 2015.
- Miles, 2015.
- Rumane, 2010.

3.4.5 Pareto Analysis

3.4.5.1 Theoretical Background

Pareto analysis is a simple technique for prioritizing and selecting a promising course of action where many alternatives approaches are competing for consideration. The *Pareto effect* is named after Vilfredo Pareto, an economist and sociologist who lived the late nineteenth and early twentieth centuries. Basically, it illustrates the asymmetry between the spent efforts and the results achieved. It states, in general terms, that for many events, roughly 80% of the effects comes from only 20% of the causes. Therefore, a Pareto analysis will direct a systems engineer encountering a problem to focus on the 20% of issues that are really important in order to deliver 80% of the results.

Pareto analysis is a creative way for identifying the most significant causes of problems for more immediate corrective actions. This includes identification and listing of problems and their causes, and then grading each problem and grouping them together by their causes. Finally, the problems associated with the highest-grade group can be targeted for resolution.

3.4.5.2 Implementation Procedure

The following procedure may implement the Pareto analysis process.

Step 1: Identify and list the problems. Review the issue at hand and identify and list all of the problems that need to be addressed.

Step 2: Score the problems. Grade each of the identified problems using a common scoring technique (e.g. Likert scale).

Step 3: Identify the root cause of each problem. Identify the fundamental cause of each problem.

Step 4: Cluster problems by root cause. Cluster the problems together by their root cause.

Step 5: Sum up each group's score. Add up the scores of each group, determining the priority of addressing each problem.

Step 6: Address the problems by score levels. Generally, the group with the highest score should be resolved first and vice versa relative to the group with the lowest score.

3.4.5.3 Example: Software Project Development Problems

1) Example background

In this synthetic example, a very large software system is being developed. However, everyone senses that the project suffers from many problems related to: (1) people involved in the project, (2) processes used by the software team, (3) the developed system itself and its supporting infrastructure, as well as (4) the technology used by the team.

2) Example analysis

A Pareto analysis was conducted and the results are depicted in Table 3.15. The individual software development problems have been clustered into four groups: (1) people-related problems, (2) process-related problems, (3) system-related problems, and (4) technology-related problems.

Table 3.15 Example software problems in a development project

#	Group	Typical problems[11]	Project losses [%][12]
1.	People	Low motivation Problem employees Unproductive work environment Inefficient project management style Lack of stakeholder interest Ineffective project sponsorship	50%
2.	Process	Unrealistic schedule Insufficient identification Unsuitable life cycle model selection Abandoning quality under pressure Unstructured software development	25%
3.	System	System scope change Research-oriented software development Ill-defined scope Fuzzy users	15%
4.	Technology	Overestimated savings from reusable software Switching tools in midway Integrating unrelated software products	10%
Total			**100%**

As can be seen in Figure 3.45, some 50% of the software project losses are attributed to people-related problems. Therefore, management should deal with these problems first. A second Pareto analysis may be carried out in order to distinguish among the six typical problems within the people-related problems in order to determine which ones contribute more toward the 50% losses and tackle them first.

---

11 Adopted from ZeePedia.com, Software Project Management (CS615), Problems in Software Projects, Process-related Problems. See: http://www.zeepedia.com/. Accessed: August 2017.
12 Based on author experience.

Project losses [%]

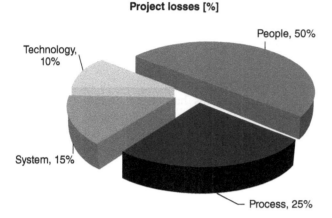

Figure 3.45 Example software project losses by category

3.4.5.4 Further Reading

• Koch, 1999.

3.5 Convergent Methods for Teams

This chapter describes a collection of convergent creative methods for teams, as encapsulated in Figure 3.46.

Figure 3.46 Convergent methods for teams

3.5.1 Delphi Method

3.5.1.1 Theoretical Background

1) General discussion

The Delphi method was developed in the early 1950s by Norman Dalkey and Olaf Helmer at the Rand Corporation, a company that carried out R&D

programs for the US Defense Department.[13] The name of the method relates to the Oracle of Delphi, a priestess at the temple of Apollo in ancient Greece, known for her ability to forecast the future. The method provides structured framework for collecting opinions and independent analysis from domain experts and then aggregate it into a unified, consensus response.

The purpose of this process is to bridge the gap between available knowledge, on the one hand, and required information on the other. Cooke (1991) provides an extensive survey and a critical examination of the literature on the use of domain expert opinion in scientific inquiry and policy making. The elicitation, representation, and use of domain expert opinion have become increasingly important because advancing technology requires more and more complex decisions. Cooke considers how expert opinion is being used today, how an expert's uncertainty is represented, how people reason with uncertainty, how the quality and usefulness of the expert opinion can be assessed, and how the views of several experts might be combined. Loveridge (2002) expanded on Cooke's seminal work and covers topics such as the selection of people for expert committees, as this is much more critical than is generally appreciated. Vose (2008) described modeling techniques based on various probability distributions (e.g. triangular, beta).

Eliciting data from domain experts can be a difficult and involved process. Keeney and von Winterfeld (1991) discuss the process of acquiring probabilities from domain experts in the complex nuclear power plant environment. Experts had to estimate failure probabilities associated with two critical valves in which a simultaneous failure could trigger a catastrophic core meltdown. The elicitation took several months to accomplish, and the uncertainties were very large, often covering several orders of magnitude in the case of probability frequencies and 50–80% of the physically feasible range in the case of some uncertain quantities.

Questionnaires are distributed to a panel of domain experts and the responses are aggregated and shared (anonymously) with the group. Experts may adjust their answers and add related feedback to further the discussion with each subsequent round. The process may be repeated several times so consensus can be reached over time as opinions are swayed. Note that the anonymity afforded each expert is crucial, as it alleviates the need to defend original positions and allows gradual shift to new opinions without incurring undo social pressure. Finally, the Delphi process assumes that decisions made by a structured group of individuals, are more likely to be accurate than those made by an individual. This is often true but should be taken with a grain of salt.

2) Delphi method under triangular distribution

---

13 The Delphi method was finally declassified in 1963.

One version of the Delphi process is to elicit data under a triangular distribution paradigm – that is, three values (minimum (a), most likely (m), and maximum (b)) are created by the experts for each required quantity. Then, in order to aggregate the raw data in such cases, one takes the following approach:

1) Postulate that the investigated phenomenon is random, bounded within a certain minimum (a) and maximum (b) range.
2) Assume also that the random variable representing the phenomenon has a most likely value (m) within the range $a \leq m \leq b$.
3) Consider a collection of n domain experts as measuring instruments with built-in errors in their measurement abilities. The most likely (m) value represents the actual measurements of the instruments, and the minimum (a) and maximum (b) values represent the respective lower and upper boundaries measured by the instruments.
4) Assume that each of the n instruments introduces an unbiased, random measuring error. Then aggregate the results using a numerical analysis – for example a Monte Carlo simulation.[14]

Based on the above, one aggregates all the responses and presents the results to the group in a *scrubbing* meeting. During that meeting, each expert has a chance to review his or her original responses in light of the groups' aggregated data. Some debates may take place as to the exact meaning of certain questions. But normally, all the experts came to a shared collective understanding of the issues. A few experts may change their original response along the line of the group. But, often, the majority will not change their opinions, even when their data are quite different from that of the group. Each expert is assumed to have a probability $p_{k,I}$ of being correct, where $p_{k,i}$ is associated with response cluster k representing individual expert i.

For a total of n experts, the aggregated response cluster is represented by a generalized discrete distribution:

$$F_k(x) = \left\{ \left(p_{k,1}, f_{k,1}(x) \right), \left(p_{k,2}, f_{k,2}(x) \right), \ldots, \left(p_{k,n}, f_{k,n}(x) \right) \right\}$$

This means that the aggregated probability density function of a value x in response cluster k for a total of n experts is

$$F_k(x) = \sum_{i=1}^{n} p_{k,i}, f_{k,i}(x)$$

This equation satisfies the mathematical and behavioral approaches discussed by Clemen and Winkler (1999). Note that the sum of several triangular distributions is not a triangular distribution, and this nonlinearity suggests that closed mathematical expressions for statistical moments of the aggregated

---

14 For a detailed description of Monte Carlo statistical methods, See Robert and Casella, 2005.

distribution are impractical. Therefore, a credible data aggregation could be accomplished by means of numerical analysis – for example, by Monte Carlo simulation[15] (Vose, 2008). Also, note that each of the n experts dealing with all the response clusters is often assumed to be correct equally likely; thus:

$$p_{k,i} = \frac{1}{n} \forall k, i$$

3.5.1.2 Implementation Procedure
The following steps can be used to implement a Delphi procedure:

Step 1: Train the participants. Recruit people to the Delphi group, explain the Delphi process, and formally describe the problem at hand.

Step 2: Elicit and collect opinions. Distribute a questionnaire containing the relevant questions, provide any needed answer from members of the Delphi group, and then collect responses from each one of them.

Step 3: Aggregate results. Collate and aggregate the responses into a single anonymous set of responses. Then give the aggregated anonymous responses to the experts for further review.

Step 4: Scrub data. Repeat this process as necessary. If one seeks consensus and there are conflicting responses, then this may require either: (1) conducting a group vote or (2) bringing the expert together (either physically or via the internet) and discussing the issue face to face.

3.5.1.3 Example: Estimating Duration of a Task

1) Example background

In this example, 10 domain experts are asked the following question: "How many person-months are required in order to complete the design of an RSA chemical reactor with half coils wrapped around it?" (Figure 3.47).

2) Example analysis

The experts use the Delphi triangular distribution method and generate 10 minimum (Min), most likely (ML), and maximum (Max) results, for a total of 30 individual values. The results of the Delphi process after

Figure 3.47 RSA Chemical reactor

15 There are several commercial tools that support the aggregation and analysis of expert data, for example, Oracle Crystal Ball (http://www.oracle.com/us/products/applications/crystalball/overview/index.html) and @RISK (http://www.palisade.com). Accessed: Oct. 2017.

two scrubbing meetings are depicted in Figure 3.48. Note the distinct outlier in the figure at: {Min = 1, ML = 2.5, Max = 5}. The three values provided by each of the 10 experts have been aggregated into a single value.

Figure 3.49 depicts the aggregated probability distribution function of the 10 experts' replies to the said question. After 10,000 Monte Carlo iterations per

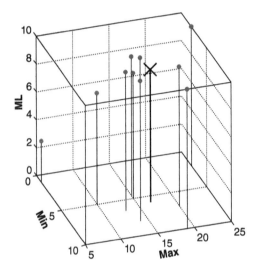

Figure 3.48 Ten experts' Delphi response

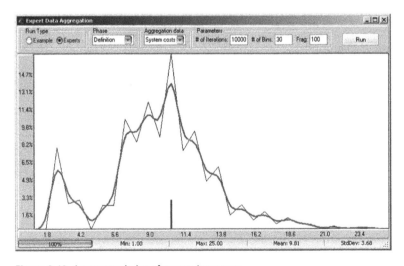

Figure 3.49 Aggregated plot of experts' response

expert, the plot indicates a minimum value of 1.0, a maximum value of 25.0, and a mean of 9.81 person-months, with standard deviation of 3.68.

3.5.1.4 Further Reading

- Cooke, 1991.
- Engel, 2010.
- Garson, 2013.
- Keeney and von Winterfeld, 1991.
- Loveridge, 2002.
- Robert and Casella, 2005.
- Vose, 2008.

3.5.2 SAST Analysis

3.5.2.1 Theoretical Background

Strategic assumptions, surfacing, and testing (SAST) is a method for approaching ill-structured problems, bringing onto the surface the assumptions that underlie people's insights and then challenging them through analysis[16]. Early SAST concepts were examined by Charles Churchman and then evolved through the collaborative efforts of Richard Mason and Ian Mitroff. The main purpose of the SAST method is to evaluate different opinions and systematically bring onto the surface hidden assumptions for explicit examination and then challenge them. In the process, people will be able to analyze and examine the relationship between underlying assumptions and the resulting engineering policies and solutions to problems. Finally, the SAST method offers mechanisms to formulate new and novel engineering visions. By and large, SAST incorporates the following key principles:

Adversarial. This principle is based on the belief that examinations of ill-structured problems are best made after consideration of opposing perspectives.

Participative. This principle is based on the premise that different groups and individuals from different position within an organization and from different professional backgrounds are best suited to resolve ill-structured problems.

Integrative. This principle is based on the premise that different views generated in an adversarial process can be brought together again into a higher-order synthesis, so that an acceptable action plan may be produced.

---

16 See: McDonald et al., 2011.

The SAST process starts with gathering individuals involved in the problem at hand. These persons should be divided into two or more small teams in a way that will minimize the conflicts within each team and maximize the differences between diverse teams. Next, each team will try to identify and adopt a preferred solution to the problem. Then, each member of each team should explore and bring to the surface the assumptions contributing to his/her opinion. Thereafter, the various teams should meet in order to discuss and argue the different points of view and, in particular, analyze and examine the validity of the assumptions raised by the team's members. Exploring peoples' assumptions seems to be the centerpiece of SAST methodology, as this process may uncover a significant set of incorrect and illogical assumptions. Some of them may be discarded and, thus, removing barriers for creative solutions. Finally, the participants are asked to hammer out a practical and integrated synthesis culminating in an agreed solution to the problem.

A useful map to ease the conduct of a SAST and similar methods is to utilize the *agreement versus certainty* diagram devised by Ralph Stacey in the 1990s.[17] This diagram can help people understand where an idea or an assumption may lie within the "agreement" versus "certainty" spectrum. Based on this knowledge, one can select the most effective resolution strategy correlated with the degree of certainty as well as the level of agreement regarding the issue at hand. Essentially, the Stacey model defines five regions (Figure 3.50):

"Rational" region. In this region, one uses information from past experience in order to predict the future. Therefore, the strategy here is to plan ones' actions in order to achieve specific outcomes and then monitor the results against the original plans.

"Political" region. This region deals with certainty regarding how outcomes could be created but high levels of disagreement regarding the specific desirable outcome. Therefore, the way forward is to negotiate a compromise and to establish an agreed solution for the good of the organization and its stakeholders.

"Judgmental" region. This region deals with issues that have a high level of agreement but not much certainty as to how one should implement them. Therefore, the goal in this region is to seek a general agreed upon future direction, even though the specific paths cannot be fully predicted.

---

17 See also: The Stacey Matrix. The basic idea. Adapted from R.D. Stacey by B.J. Zimmerman. http://adaptknowledge.com/wp-content/uploads/rapidintake/PI_CL/media/Stacey_Matrix.pdf. Accessed Nov., 2017.

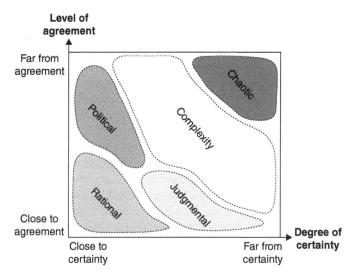

Figure 3.50 Agreement versus certainty the Stacey model

"Chaotic" region. This region deals with situations where high levels of uncertainty as well as disagreement often result in failures. Traditional methods of planning and negotiation are often insufficient in these contexts. Therefore, engineers and planners should do their best to avoid this region altogether.

"Complexity" region. There is a large area on the *agreement versus certainty* diagram that lies between the chaotic region and regions of the traditional management approaches. Unfortunately, in this area, traditional management approaches are quite limited. Therefore, engineers and managers, finding themselves in this area are advised to adopt innovative and creative strategies of operations.

3.5.2.2 Implementation Procedure

The SAST methodology may be implemented following these steps.

Step 1: Form teams. A number of individuals with diverse professional background should be formed into two or (preferably) more small teams. The aim is to maximize similarity of perspectives within each team and to maximize different perspectives among teams.

Step 2: Surface assumptions. Each team should meet separately with the intent of discovering its preferred strategy and solution to the problem at hand. Next, each team should try to identify and analyze the assumptions upon which it's preferred strategy and solution rests. It should be noted that these listed assumptions (1) should have a significant bearing on the outcome of the chosen strategy (i.e. meet the *importance* criteria) and (2) should be, as much as possible, verifiable (i.e. meet the *certainty* criteria). In addition, it is recommended that each assumption should be ranked so that for each team, the more significant assumptions will be placed at the top of the assumptions list.

Step 3: Deliberate and debate.[18] Next, all the teams will gather in a plenary session and each team will present its preferred strategy as well as the assumptions supporting it. The aim of the exercise is that each team will understand all the proposed strategies and their associated assumptions. Breaks in the plenary process should be scheduled so each team will have opportunities to review and adjust its assumptions. Finally, agreed assumptions should be identified, while problematic assumptions will be further debated and resolved.

Step 4: Generate synthesis. The aim of this step is to arrive at a compromise regarding the assumptions and their priorities so a unified and final strategy could be achieved. If the list of agreed assumptions is sufficiently long, then the disputed assumptions may be dropped and a derived strategy may be worked out. If no agreed synthesis can be achieved then points of disagreement should be noted and a decision may be attained by way of other means.

3.5.2.3 Example: Unmanned Air Vehicle Power Plant Configuration

1) Example background

The US Army Joint Project Office's (JPO) early unmanned air vehicle (UAV) acquisition effort was the short-range (SR) UAV, subsequently named the RQ-5 Hunter. A request for proposal (RFP) was issued by the US Department of Defense (DoD) in 1988, and two of five bidders were selected to compete in the development and fielding of five UAV systems. Eventually, the Israel Aerospace Industry (IAI), joined by Thompson Ramo Wooldridge (TRW), won the competition and by the end of the program in 1995 some 52 UAV systems were procured including some 416 Hunter air vehicles and other associated equipment, at an approximate cost of $2.1 billion.

2) Example analysis

At the initial bidding stage, IAI's engineers involved in the UAV SR project debated how to respond to the RFP. One engineering faction opted for a small,

---

18 Deliberation is normally considered as a collaborative effort whereas debate is normally considered as an adversarial process.

single-engine UAV, similar to the RQ-2 Pioneer, which has been developed at IAI and used extensively by the US Navy, Marine Corps, and Army. A second engineering faction opted for a large, dual engine UAV that some IAI's engineers were toying with for sometimes. IAI management ordered the UAV preliminary design department to develop two alternative designs for the two configurations and present it to management for a final decision on the desired biding configuration (Table 3.16).

Table 3.16 General characteristics: single and dual engine configuration

General characteristics	Single engine	Dual engine
Payload (Kg)	45	90
Length (Meter)	4	7
Wingspan (Meter)	5.2	11
Height (Meter)	1	1.9
Gross weight (Kg)	200	727
Speed (Km/hour)	200	90–160
Range (Km)	200	125
Endurance (Hours)	10	21
Ceiling (Meter)	4,600	5,500
Fuel capacity (Liters)	50	300

The massive difference characterizing the two design configurations reflect the sharply divided sets of assumptions held by the two engineering factions vis-à-vis the US Army JPO's desired capabilities of the UAV-SR program (Table 3.17).

Table 3.17 Assumptions: Single and dual engine configuration

Single engine assumptions	Dual engine assumptions
Significantly lighter and smaller UAV	Significantly increased UAV propulsion reliability
Low fuel consumption	Double the UAV payload weight
Smaller radar cross-section	Double the UAV mission endurance
Land and small shipboard deployment via net capturing (Rear mounted engine)	Improved reliability due to dual engine configuration
Simpler Integrated Logistics Support (ILS)	Land and large aircraft carrier deployment
Significantly less expensive air vehicle	
Low single UAV unit cost	

However, in this particular case, not much SAST deliberation and debate was actually undertaken. As it happened and, unbeknown to IAI's management, the UAV preliminary design department put very few engineers to design the single-engine UAV variant in accordance with the RFP requirements. All other engineers in the department were assigned to the dual-engine UAV variant. Not surprisingly, as decision time arrived, only the dual-engine variant design was sufficiently advanced for bidding. The single-engine design was not sufficiently mature and could not be selected as a bidding option. IAI (together with TRW) proceeded to win the UAV-SR competition with the dual-engine RQ-5 Hunter.

Figure 3.51 depicts a second-generation UAV, the MQ-5B Hunter. Most importantly, this vehicle uses two Mercedes engines, requiring HFE diesel fuel, commonly utilized by the US armed services, thus, relieving the services from carrying light fuel (regular automobile gasoline) for their UAV fleet.

Figure 3.51 Unmanned air vehicle MQ-5B Hunter

3.5.2.4 Further Reading

• Flood and Jackson, 1991.	• Midgley, 2000.
• Mason and Mitroff, 1981.	• Rodrigues, 1997.
• McDonald et al., 2011.	• Stacey, 2012

3.5.3 Cause-and-Effect Diagram

3.5.3.1 Theoretical Background

Cause-and-effect diagrams (also called fishbone diagrams as well as Ishikawa diagrams), were originally developed as a quality control tool by Kaoru Ishikawa in the late 1960s. However, the technique may be used to discover the root causes of a problems as well as uncover bottlenecks in a given processes and so forth. Causes may be arranged by chronological order or according to their level of importance, depicting relationships and hierarchies of events. This can help engineers in examining root causes, of problems, and weighing the relative effects of different causes. Construction of a cause-and-effect diagram starts by identifying the problem at hand. A brainstorming process is advisable, so once the team agrees on the statement of the problem, it will be possible to move to the next phase. Again, the team should identify the major causal categories or the "main bones" in the fishbone diagram. This may also be done by way of a brainstorming session or other creative group process. Finally, the group should identify the specific possible causes and attach them to the appropriate branches.

3.5.3.2 Implementation Procedure
Cause-and-effect diagrams could be created using the following procedure:

Step 1: Identify the problem. Discuss within the group and agree on an exact and succinct problem statement.

Step 2: Identify major causal categories. Brainstorm the categories of causes of the problem. Often, such categories may include methods, equipment, people, materials, measurement, as well as the environment.

Step 3: Identify specific possible causes. Next, for each of the above causal categories, identify specific possible causes contributing to the problem.

Step 4: Draw cause and effect diagram. Draw the cause-and-effect diagram and analyze it for correctness and completeness. If the diagram is too crowded, consider splitting up relevant branches.

3.5.3.3 Example: The *Titanic* disaster

1) Example background

The RMS *Titanic* was a British passenger ship that was on her maiden voyage from the United Kingdom to New York City. On the night of April 14, 1912, the *Titanic* struck an iceberg in the North Atlantic and sank in less than three hours. Over 1,500 people, two-thirds of the *Titanic's* passengers and crew perished, mostly because there were not enough lifeboats to rescue everyone on board (Figure 3.52).

2) Example analysis

A cause-and-effect diagram describing the causes leading to the *Titanic* disaster is depicted in Figure 3.53.

Figure 3.52 Sinking of the *Titanic* (Engraving by Willy Stower)

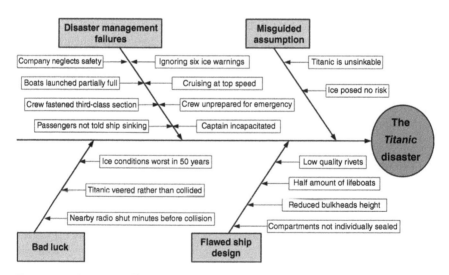

Figure 3.53 Cause-and-effect analysis: The *Titanic* disaster

3.5.3.4 Further Reading

- Ishikawa, 1990.

3.5.4 Kano Model Analysis

3.5.4.1 Theoretical Background

Sometimes, designers of new systems receive detailed specifications regarding the system they are expected to design and build. Sometimes however, such specifications do not exist or they are sketchy or, invariably, they are going to change quite substantially. In such circumstances, designers face a fundamental dilemma: which features should be included in the design and which ones should not? If important features are not included, customers will reject the system. If the system is loaded with good features but they render the system expensive, customers will also reject the system.

1) Qualitative Kano model

One way to mitigate this problem is to utilize the Kano model, developed by Noriaki Kano in the 1980s. In order to better understand customer preferences and satisfaction, one may utilize the Kano model, which identifies five emotional responses to either features' availability or its absence within a given system. These emotional responses may be visualized as curves on a graph, where the x-axis is the degree of feature's implementation within a given system and the y-axis is the satisfaction level (Figure 3.54).

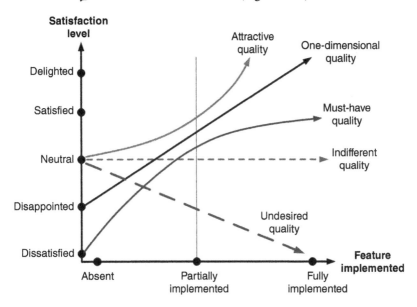

Figure 3.54 Kano model: emotional responses to features' implementation

Attractive quality. Customers do not normally expect such an attribute, but its presence provides satisfaction and delight. An example of an attractive quality would be sunglasses with a nice case. In general, attractive features fulfill previously unmet but often unrecognized needs.

One-dimensional quality. Such an attribute results in satisfaction when fulfilled and dissatisfaction when not fulfilled. That is, there is a linear relationship between this product's feature and the level of satisfaction. An example of one-dimensional quality would be the cost of a given product. The lower the cost, the more satisfied the customer or user will be, and vice versa (well, up to a point).

Must-have quality. Such an attribute is one that customers expect the product to contain. However, if not included, customers and users will be dissatisfied. An example of must-have quality would be an ordinary mirror. When a mirror is smooth, customers accept it as normal. However, when the mirror is coarse, distorting the image, customers are dissatisfied.

Indifferent quality. Such an attribute refers to product's features that customers and users do not care about either way. Therefore, they do not affect customer satisfaction or dissatisfaction. An example of indifferent quality would be the internal design of a cell phone. In fact, most people have no clue about this rather complex issue.

Undesired quality. Such an attribute refers to product's features, which are unneeded and even annoy most customers and users. An example of this would be a handheld television remote control that, for some reasons, contains much too many buttons.

2) Quantitative Kano model

Several scholars extended the Kano model to include quantitative features. This extension is especially effective in a group setting where the opinion of each member should contribute to the overall design of a system (i.e. which features should be designed and built and which ones should not).

One may use the following quantitative model.[19]

$$SI = \frac{A+O}{A+O+M+I+U}$$

$$DI = \frac{-(A+M)}{A+O+M+I+U}$$

$$ASC = \frac{|SI| + |DI|}{2}$$

where:

SI = Satisfaction index
DI = Dissatisfaction index
ASC = Average satisfaction coefficient

---

19 See: Mkpojiogu and Hashim, 2016.

A = Attractive quality
O = One-dimensional quality
M = Must-have quality
I = Indifferent quality
U = Undesired quality

3.5.4.2 Implementation Procedure

The following steps may be used in order to implement a quantitative Kano model analysis:

Step 1: Identify expected features. Identify the expected features of the system or product under consideration. A group setting is recommended.
Step 2: Survey desired system features. Conduct a survey in which each person in the group will classify each of the expected system's feature into one of the following classes: attractive, one-dimensional, must-have, indifferent, and undesired.
Step 3: Generate Kano table. Collate the survey responses and generate a Kano feature categorization table.
Step 4: Assign values to quality attributes. Assign value to each of the quality attributes according to the following evaluation rules: Must-have > One-dimensional > Attractive > Indifferent > Undesired. For example, by Likert-type rating scale: Must-have = 5, One-dimensional = 4, Attractive = 3, Indifferent = 2 and Undesired = 1.
Step 5: Create satisfaction table. Generate a User/Customer Satisfaction table, culminating in the satisfaction index (SI) and dissatisfaction index (DI) for each expected system's feature.
Step 6: Prioritize system's features. Prioritize each expected system's feature in accordance with its average satisfaction coefficient (ASC).

3.5.4.3 Example: Preferred Features in a New Car

1) Example background

A passenger cars manufacturer is planning to develop and manufacture a new passenger car. A 10-member Kano team is assembled to rate the five top-level feature-groups of the expected car (Table 3.18).

2) Example analysis

Table 3.19 depicts the resultant Kano feature-group categorization table. Here, each column depicts the number of team members selecting a given attribute for each system's feature.

Table 3.20 depicts the resultant user/customer satisfaction table. As can be seen in the right-most column, the Chassis electronics feature-group was assigned the highest ASC score and the Infotainment feature-group was assigned the lowest ASC score.

Table 3.18 Passenger cars' feature-group and individual features[20]

Feature group	Individual features
Chassis electronics	• ABS – Anti-lock braking system • TCS – Traction control system • EBD – Electronic brake distribution • ESP – Electronic stability program
Passive safety	• Air bags • Hill descent control • Emergency brake assist system
Driver assistance	• Lane assists system • Speed assist system • Blind spot detection • Park assists system • Adaptive cruise control system • Pre-collision assist
Passenger comfort	• Automatic climate control • Electronic seat adjustment with memory • Automatic wipers • Automatic headlamps; adjusts beam automatically • Automatic cooling temperature adjustment
Infotainment	• Navigation system • Vehicle audio • Information access

Table 3.19 Kano feature-group categorization table

Feature	Must-have (M)	One-dimensional (O)	Attractive (A)	Indifferent (I)	Undesired (U)	Total
Chassis electronics	3	2	5			10
Passive safety	4	2	2	2		10
Driver assistance	5	4		1		10
Passenger comfort		3	2	5		10
Infotainment	1	3	1	3	2	10

---

20 To simplify the Kano example, passenger cars' features-groups are used instead of individual features. See: Automotive electronics, Wikipedia, https://en.wikipedia.org/wiki/Automotive_ electronics. Accessed Dec. 2017.

Table 3.20 User/Customer satisfaction table

Feature	Must-have (M)	One-dimensional (O)	Attractive (A)	Indifferent (I)	Undesired (U)	Satisfaction index (SI)	Dissatisfaction index (DI)	Average satisfaction coefficient (ASC)
Chassis electronics	15	8	15	0	0	0.61	-0.79	0.70
Passive safety	20	8	6	4	0	0.37	-0.68	0.53
Driver assistance	25	16	0	2	0	0.37	-0.58	0.48
Passenger comfort	0	12	6	10	0	0.64	-0.21	0.43
Infotainment	5	12	3	6	2	0.54	-0.29	0.41

3.5.4.4 Further Reading

- Berger et al., 1993.
- Mkpojiogu and Hashim, 2016.
- Moorman, 2012.

3.5.5 Group Decisions: Theoretical Background

The purpose of this section and the next one is to acquaint engineers with the theory and practice of group decision process.[21] This is an important issue when a group of people must choose one course of action out of multiple options.

3.5.5.1 Theoretical Background

Group decisions are often made by technical experts, convened to resolve specific engineering and management issues. Such groups may be partially active throughout the entire system life cycle and may be scheduled to act whenever needed. For example, a group may be selected in order to decide an optimal system's design, appropriate test and qualification strategy, production strategy, and the like. Typically, technical reviews are often conducted by means of such decision groups. They provide managers, system designers, builders, test engineers, and production engineers with valuable insights into the state of the system with which they are involved. Evaluation and decision processes carried out within such groups have distinct advantages over similar processes performed by individuals due to the following:

- Research shows that the effectiveness of groups as decision makers is generally superior to each member of the group individually. Groups can discuss issues and process information and are more likely to identify errors in logic and facts as well as reject incorrect solutions.
- By nature, groups bring to the table a broad representation of personalities and opinions so that more ideas are generated and the space for evaluation increases. In addition, a group represents greater informational resources and possesses more accurate memory of facts and events than does each individual member.
- Groups generally set standards for conducting evaluations and making decisions. Usually, following formal procedures solidifies the process and ensures that all aspects of a problem have been addressed. Well-defined

---

21 (a) Making group decisions. See: https://www.uvm.edu/crs/resources/citizens/decision.htm. Accessed: August 2017.
(b) Groupthink. See: http://www.blendedbody.com/GroupThink/Groupthink-PatternOfThought CharacterizedbySelf-Deception.htm. Accessed: August 2017.

decision rules (e.g. majority rule, unanimous decision, quantitative decision procedures) ensure, at least to some extent, that all group members will have the opportunity to air their views so open issues are settled in a wise manner.

- By and large, people are more likely to follow through if decisions have been made through an accepted group process. This increased commitment for implementation fosters diligence and expedience, as well as better cooperation among the members of the group.

This section provides relevant theoretical background information related to: (1) making group decisions, (2) factors affecting group decisions, (3) leadership styles and group decisions, and (4) risks in group decisions (Figure 3.55).

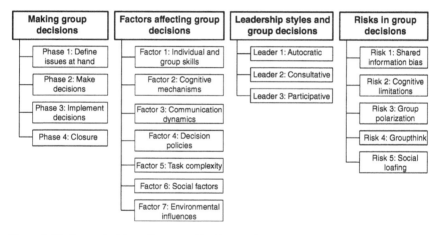

Figure 3.55 Group decisions: Theoretical background

3.5.5.2 Making Group Decisions

The basic assumption for this discussion is that the members of a decision group are professionally and intellectually suitable to perform their assignments. For instance, if the task involves reviewing a technical issue, all group members should have some expertise and knowledge that will apply to the technology involved. Based on this assumption, there are four basic phases involved in a typical group decision process.

Phase 1: Define issues at hand

The first phase of the group decision process starts with a group orientation and development of shared mental model of the issue. More specifically, the group tries to arrive at an accurate understanding of the system at hand. This may be achieved by means of discussion as well as exchanging and sharing relevant information.

If initial evaluation of the data available to the group identifies a problem, then the nature of the problem, the extent and seriousness of the problem, as well as the likely cause of the problem and the possible consequences of not dealing effectively with it are analyzed. Based on this analysis, the group will normally generate a number of appropriate and feasible alternative lines of action among which an acceptable choice of one or more actions should exist.

Phase 2: Make decisions
During this phase, the group uses one of several decision schemes to select a single alternative line of action from the various alternatives originally proposed by the group. Typical decision schemes may be: (1) an individual (usually a manager) who makes the decision for the group, (2) voting using a majority rule or consensus rule (where all members of the group must agree to a specified decision), and so on.

Phase 3: Implement decisions
During this phase, the group reviews the implementation of the adopted solution and evaluates the consequences of this decision.

Phase 4: Closure
The group needs to be fully cognizant of the relative merits and disadvantages of each available alternative in order to learn how the group can be more effective in the future. More specifically, postmortem (i.e. after the issue had been dealt with) discussions provide valuable learning lessons to the group, facilitating a retrospective look at past decisions and the decision-making process itself.

3.5.5.3 Factors Affecting Group Decisions
Research in several disciplines (e.g. economics, business, engineering, psychology) indicates that both individual and group characteristics influence group dynamics and decision-making processes. Current research shows that group process effectiveness in terms of decision-making speed, correctness, or accuracy often depends on the following factors.

Factor 1: Individual and group skills
Individual and group skills, communication skills, and problem-solving skills among group members are important components of effective groups decision process. Similarly, group skills such as conflict resolution, group goal setting, or egalitarian leadership foster effective group performance.

Factor 2: Cognitive mechanisms
Cognitive mechanisms include the mental activities involved in processing information and their related dynamic mental models. Cognitive strategies are the formal mechanism controlling the mental processing of information, whereas heuristics are informal mechanisms controlling the mental processing of information.

Factor 3: Communication dynamics
Beyond the communications skills of individuals within the group, the characteristics of the communication process itself is significant to group dynamics and decision making. Communication patterns among group members expose information power relationships and the social status of individual group members.

Factor 4: Decision policies
Decision policies are the agreed-upon rules that cement the required discipline for group decision making. Such decision policies may be formal – for example, Delphi technique or majority vote or nominal group methods. Conversely, decision policies may be informal – for example, broad group processes. The aim of informal processes is to deliberate openly and democratically in order to obtain reasoned agreement among equally qualified group participants.

Factor 5: Task complexity
Task complexity significantly affects the behavior and dynamics of the group. Complexity can be measured in many ways, including the amount of information that must be absorbed and processed, the number of possible decision options available to the group, and the number of steps required to perform each individual task (e.g. evaluating the behavior of a system's performance).

Factor 6: Social factors
Social factors determine the nature and dynamics of interpersonal relationships within the group. They often include interpersonal influence and power as well as group network cohesiveness and individual role assumed by each group member.

Factor 7: Environmental influences
Environmental factors affect group decision making. Organizational characteristics such as size, formal structure, and culture influence the decision-making processes. In addition, factors such as working environment and financial or time pressure can produce stress, which also affect group behavior.

3.5.5.4 Leadership Styles and Group Decisions
Typically, leaders of decision groups are critical components of the decision process. In a rather broad and general way, leaders may be categorized into the following three leadership styles:

Leader 1: Autocratic

Under autocratic leadership style, leaders tend to solve problems on their own based on information available to them at the time. The information or advice provided by group members is utilized only when it coincides with their own ideas or when a leader encounters irrefutable proof that he/she is wrong. Otherwise, they seldom seek information or advice from group members.

Leader 2: Consultative

Consultative leaders tend to share the problem-solving process with members of the group. However, they still rely heavily on their own knowledge, experience, and opinions.

Leader 3: Participative

Participative leaders discuss the issue with the members of the group and together the leader and members devise an appropriate response. Under this management style, the leader acts as a chairperson of a committee and, by and large, accepts a group decision, which typically is arrived at on the basis of decision by majority rule or by consensus.

3.5.5.5 Risks in Group Decisions

Group evaluation and decision processes are not always successful. First, all such group processes are time consuming. If derived solutions and appropriate mitigating solutions are not timely, the group process may be a failure. In addition, sometimes the group makes a bad decision. Among causes that may be to blame for a bad decision are biases in sharing information, cognitive limitations, group polarization and, most notoriously, groupthink phenomena as well as plain old social loafing. The following describes these pitfalls, often found in bad decisions made by groups:

Risk 1: Shared information bias

Shared information bias is the tendency for groups to discuss issues familiar to all members and avoid examining information that only a few members know. This leads to poor decisions making due to ignorance of important facts by the group. For example, evaluating system test information where certain failures are known to some members but are not exposed to the rest of the group may cause judgment errors and heuristic biases.

Risk 2: Cognitive limitations

Poor communication skills as well as biases in an individual's cognition and motivation can often lead to judgment errors on the part of individuals in the group. Another cognitive limitation on the part of individuals is the tendency to seek out information that confirms their inferences rather than disconfirms them. Again, this may lead to errors in judgment and a failed decision process. In addition, individuals tend to overestimate their judgmental accuracy because they remember mostly the times their decisions were confirmed. Finally, some group participants lack inquiry and problem-solving skills or their information processing is limited relative to other persons, affecting their cognitive abilities.

Risk 3: Group polarization

Research in social comparison theory identifies the phenomenon of group polarization, the tendency to respond in a more extreme way when making a choice as part of a group. Under this condition a group has difficulty assessing the facts rationally and often fail to reach a decision acceptable to all (Figure 3.56).

There are a number of possible explanations to group polarization incidents: First, it is likely that extreme majority alternatives get more group discussion time. Second, often, extreme individuals become more extreme in the heat of an argument. More often than not, group polarization manifests itself when the group (1) lacks maturity and heterogeneity, (2) contains persons tending to egocentrism, or (3) most commonly is managed by a person lacking conflict resolution skills.

Figure 3.56 Polarization: Not an effective group strategy

Risk 4: Groupthink

Irving Janis's (1972) groupthink theory states that decision-making groups will sometimes succumb to a groupthink phenomenon. This occurs when group members become so focused on achieving concurrence that the search for consensus overrides any realistic assessment of other views. Groups affected by groupthink ignore alternatives and tend to take irrational actions. A group is especially vulnerable to groupthink when the group is insulated from outside opinions and is highly cohesive. Symptoms of groupthink are group pressures toward uniformity, invariably expressed in either overt or covert criticism of any dissenting views.

Typically, the group tends to overestimate its power and invulnerability and manifest close-mindedness and stereotype views about the world outside the group. Other typical causes for groupthink are structural failures in the makeup of the group, entrapment in sunk costs, control by an autocratic leader or a domineering member in the group and finally, plainly defective decision-making processes. Groupthink is a particularly vicious phenomenon resulting in a system that either does not meet requirements or contains problems that were not properly addressed.

Groupthink can be prevented or its effect can be greatly mitigated by taking the following steps:

1) **Enhance group decision process.** This entails assigning the role of devil's advocate to one or a few members of the group. Given this title publicly, a person would more readily voice different or contradictory views in the group discussions. In addition, the enhanced group process should mandate the obligation to always create multiple alternatives for an eventual selection and adoption of a preferred approach. It will also require reexamining advantages, weaknesses and potential risks of each alternative discussed by the group. Finally, the enhanced group process should require that a contingency plan be established in case something goes wrong with the current approach.

2) **Obtain outside expert.** The group should attempt to obtain expert or outside advice. This is important in order to correct group misperceptions and biases. It should be noted that such outside advice should be received from persons of different background, persuasions and beliefs who will not succumb themselves to existing group dynamics and pressure.

3) **Adopt effective decision-making technique.** The group should adopt an effective decision-making technique that will eliminate the tendency of the group to get trapped in stereotyped views. One technique that may be effective is to divide the decision group into two or more smaller groups, which would discuss the issues separately and then present their findings in a joint session.

4) **Select participative leader.** Finally, autocratic leaders should adopt a more open style of leadership. In addition, domineering members of the group must be persuaded to make their suggestions later, after others members have had their say.

One should also be aware that groupthink phenomenon is rarely recognized by members of these groups. As a result, the group will not usually take steps to remedy this tendency. Unfortunately, only after a particularly disastrous error in judgment on the part of the group will it be open to corrective action.[22]

Risk 5: Social loafing

Research shows that, sometimes, people do not work as hard in a group setting as they work alone. This is especially true on easy tasks in which individual contributions are blended and become indistinguishable. For example, in rope-tugging experiments, Maximilien Ringelmann[23] (1861–1931) showed that the larger the group, the less effort individual expand (i.e. one person pulled a rope at 100 units, two people at 186, three people at 255, and eight people at 392 units). Researchers suggest the following reasons for social loafing:

1) **Diffusion of responsibility.** Naturally, in a group setup the responsibility for the final outcome is diffused among members of the group. More specifically, often, members of the group are less exposed to individual responsibility and this may lead to a reduction of efforts.

2) **Free-rider effect.** Sometimes members of a group sense the benefit of belonging to a group in terms of prestige and power and yet feel that their individual contribution is not appreciated. As a result, they are likely to offer little in return and often practice decisional avoidance tendencies (e.g. avoiding responsibility, ignoring alternatives, procrastination, etc.).

3) **Sucker effect.** In a group situation, everyone is benefiting and getting credit. Often individual members do not want to be the ones who do all the work without specific recognition. As a result, sometimes members are willing to do what they conceive as their fair share but not more than that. In other words, contribute as little as possible.

Based on this phenomenon, it is fair to conclude that quite often some of the participants in a decision group do not contribute to the full extent of their capabilities. However, research shows that people contribute their best when they think their efforts will help them achieve outcomes they personally value.

---

22 For example, after the Bay of Pigs invasion fiasco (1961), US President John Kennedy sought to avoid groupthink in his cabinet meetings. He encouraged cabinet members to discuss possible solutions within their own departments and invited outside experts to share their viewpoints. Occasionally, he divided his cabinet into subgroups to break the group cohesion and sometimes he deliberately left the cabinet room for a while in order to avoid pressing his own opinion. Later, in September 1962, the Soviet government placed offensive nuclear missiles in Cuba, precipitating a crisis that came closest to a strategic nuclear war. The same group that blundered into the Bay of Pigs tackled this political and military challenge with distinguished wisdom and ingenuity.
23 See: The Ringelmann effect, https://en.wikipedia.org/wiki/Ringelmann_effect. Accessed: Aug. 2017.

Therefore, it is possible to identify several social factors that may eliminate or at least reduce social loafing tendencies.

From a positive standpoint, group work should include public acknowledgment of each individual's personal efforts and contributions. Social research shows that people rise to the occasion when the task is challenging and appealing. Therefore, group leaders and relevant outsiders should instill within the group the notion that participation in such a group is a meaningful and important task. Another factor affecting social loafing is group size as well as familiarity among the group members and cohesiveness within the group. In general, people prefer to work with buddies rather than strangers, within a smaller and neatly tied group where people can speak their minds freely.

From a negative standpoint, individuals within a group tend to work hard and contribute to the limit of their abilities if they expect they or the entire group to be punished for poor performance.

3.5.5.6 Further Reading

- Best, 2001.
- Gallagher, 2008.
- Hirokawa and Poole, 1996.
- Ishizaka and Nemery, 2013.
- Jahan et al., 2016.
- Janis, 1972.
- Lu et al., 2007.
- Torrence, 1991.
- Vroom and Yetton, 1976.

3.5.6 Group Decisions: Practical Methods

This section describes three group evaluation and decision approaches: (1) informal approach, (2) formal approach, and (3) quantitative approach (Figure 3.57).

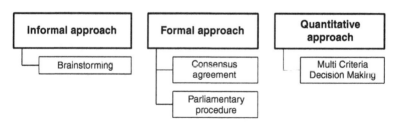

Figure 3.57 Group decisions: Practical methods

3.5.6.1 Informal Approach: Brainstorming

Brainstorming is an informal but useful method for making decisions in a group setting. Although brainstorming is usually considered a divergent creative method for teams, the unique nature of this evaluation and decision brainstorming form is that the group is involved in convergent thinking, seeking to reduce the solution space into a single, most appropriate outcome. Researchers indicate that when dealing with relatively simple problems, members of a brainstorming group often utilize logic, established standards, probabilities, and knowledge to sway their fellow members one way or the other. General knowledge plays a crucial role in this process. It provides pathways to solutions as well as criteria for selecting attractive results. However, when a problem becomes complex, members of a brainstorming group resort to abstract means like gut feeling, instincts, or intuition.

3.5.6.2 Formal Approach: Consensus Agreement

Formal group decision process is diametrically opposed to obtaining ideas and reaching conclusions by way of a traditional brainstorming. Often, a formal approach is advantageous since evaluating complex technical problems is extremely difficult. First, such difficulties stem from the complexity of the technical issues associated with modern systems. Second, the diversity of agendas and people who are involved in the design, building, evaluations, reviews, and decisions make the entire process that much more difficult.

Consensus agreement is a process of coming to an agreement on a particular technical issue. A group decision meeting conducted by consensus is usually less formal and the team leader must be willing to share control and allow more leeway in the group discussions.

As a rule, an issue brought up for discussion will be debated until the group reaches an agreement that all sides can accept. In other words, the group cannot take action that is not agreeable to each and every member in the group. Consensus does not necessarily mean unanimity, nor does it mean that all sides are satisfied with the solution but, at least, everyone must agree that they can "live with" and support the decision since it is the best solution acceptable to the group. Depending on national culture, personalities and the specific issues, reaching consensus takes considerable amount of time, but the outcome is often worth it.[24]

First, consensus agreement fosters open communication. People talk with one another regarding the technical issues at hand and their ideas about possible solution. This exchange provides the basis for designing workable and acceptable alternative.

Second, consensus agreement encourages more informed decisions. It is based on diverse opinions delivered in an open atmosphere and encourages greater creativity and a larger number of options leading to more satisfactory decisions.

---

24 For example, on the North American continent societies have existed for time immemorial. Before 1600, five nations – the Mohawks, Oneidas, Onondagas, Cayugas, and Senecas – formed the Haudenosaunee Confederation, which makes decisions on a consensual basis to this day. See: http://www.haudenosauneeconfederacy.com/index.html. Accessed: Nov. 2017.

Third, people who interact together to understand the issues and who have selected an optimal solution using consensus will see the reasoning behind a specific decision and, once consensus is reached, members tend to accept it. As a result, all members of the group tend to cooperate in the implementation and give the proposed decision ample opportunity to succeed.

There are situations where consensus agreement does not seem to be the most prudent way to conduct group decision. For example, sometimes the issues are simply not so important or the alternative solutions are not significantly different in their effect on the problem. A one-sided management decision can be taken with minimal risk. Sometimes, the extreme opposite occurs where the group is so polarized and emotionally charged that productive face-to-face discussions are not possible. Another example presents itself occasionally when an immediate decision must be made, as a wrong decision is better than a late decision.

3.5.6.3 Formal Approach: Parliamentary Procedure

Parliamentary procedure is also a process of coming to an agreement on a particular technical issue, and its purpose is also to help a group evaluate technical subjects efficiently while preserving a spirit of harmony. It is based on democratic principles of voting as practiced at national levels. Namely, the decisions of the majority are upheld, but voices of dissenting opinions are heard. Parliamentary procedure is simple to implement. First every member of the decision group has equal rights (this precludes the team leader from having unilateral decision power). Second, each issue presented to the group is entitled to discussion time.

When using parliamentary procedure, the dynamic within the decision group is usually quite accommodating and informal. Sometimes, however, this is not the case. For instance, when the technical issues are complex or when they are controversial and disagreements can cause an impasse. Another example is when the decision group is rather large or representing different organizations subscribing to different agendas. In such situations, the conflict resolution skills of the team leader and the careful management of the decision process are paramount.

The key difference between consensus agreement and parliamentary procedure is that in parliamentary procedure voting results tend to create a "win-lose situation." As a result, the losers often are unwilling to support the winning position, which hampers implementation of the decision. In contrast, under consensus agreement, usually synthesis of values and ideas manifest themselves rather than one side wins and the other loses.

3.5.6.4 Quantitative Approach: Multiple Criteria Decision-Making

Multiple criteria decision making (MCDM), a branch of operational research, is a formal quantitative method for making a quantitative decision based on the opinion of different people. Often, one is mostly interested in the aggregation of multiple opinions within a group containing individuals who often may have quite different or opposing opinions.

In a group, every person has individual preferences, so he or she may choose between a given set of alternatives. More precisely, each individual may choose his or her favorite alternative from each pair of alternatives. For example, given three alternatives: a, b, and c, each person could choose between each pair of these alternatives, for instance the combination {a > b, a > c, and c > b} could be a valid preference set of an individual in the group.

Social choice or, more appropriate for our domain, engineering choice (EC), is the collection of all possibilities in conjunction with their respective choice sets, and the aggregation of individual preferences. That is, given that each individual has a certain profile of preferences, the engineering choice is a function that transforms the aggregate set into the level of the collective.

The philosophy and theory of various MCDM methods is a subject of intense research in several fields (e.g. operational research and economics) and is beyond the scope of this book. The interested reader may refer to the further literature section, at the end of this section and to the Bibliography chapter at the end of Part III. Nevertheless, the following example presents a simple appetizer for readers wishing to acquaint themselves with one MCDM method.

3.5.6.5 Simple Example: Multiple Criteria Decision-Making

There are many mathematical ways to obtain data from individuals in a group and then aggregate it into a unified group decision. Let us visualize one simple method of making a group decision by the following example: A technical committee is convened to decide how to deal with a serious budget overrun and a significant schedule delay in an engineering development project. The committee comprises 13 members. It must rank four alternative actions:

Action A. Replace the main contractor.
Action B. Redesign and rebuild one problematic subsystem.
Action C. Develop and produce the system in two builds, postponing problematic capabilities by a year.
Action D. Terminate the entire project.

Each committee member has equal voting weight within the decision group. He or she ranks the four alternatives (A, B, C, D) in order of importance. This is done by assigning four points to the most attractive action, three points to the next alternative, and so forth. The result of the committee members' voting is depicted in Table 3.21.

As can be seen, alternative A is the most valued choice. Nevertheless, it is quite puzzling to see these results (i.e. four members selected one ranking set,

Table 3.21 First example: Committee member vote

Group	Member	A	B	C	D
		Alternatives/Points			
Mostly A supporters	1	4	2	1	3
	2	4	2	1	3
	3	4	2	1	3
	4	4	2	1	3
Mostly C supporters	5	3	1	4	2
	6	3	1	4	2
	7	3	1	4	2
Mostly B supporters	8	2	4	3	1
	9	2	4	3	1
	10	2	4	3	1
	11	2	4	3	1
	12	2	4	3	1
	13	2	4	3	1
Total		37	35	34	24

three members selected a second ranking set, and six members selected a third ranking set). Typically, one would expect that independent individuals with integrity would exhibit much greater variance in their alternative action rankings.

Let us examine the results. First, one might ask, what is the likelihood that such results would have occurred if each ranking set had equal probability? (This is an unrealistic but still an interesting yardstick.) We start by noting that each committee member has a total of 4! = 24 possible ranking combinations. So 13 members have a total of $S = 24^{13}$ ranking set combinations.

We select 3 combinations out of 24 and then further select 1 combination out of the 3 and assign it to the first group of 4 out of 13 individuals. We then select 1 combination out of the remaining 2 and assign it to the second group of 3 individuals out of the remaining 9. Last, we select 1 combination out of the remaining 1 and assign it to the last 6 committee members:

$$N_1 = \binom{24}{3}\binom{3}{1}\binom{13}{4}\binom{2}{1}\binom{9}{3}\binom{1}{1}\binom{6}{6} = 2024 \times 3 \times 715 \times 2 \times 84 \times 1 \times 1 = 729,368,640$$

As can be seen, the probability of this result (based on our yardstick as our sampling space) is extremely low:

$$p_1 = \frac{N_1}{S} = \frac{729,368,640}{24^{13}} = 8.32 \times 10^{-10}$$

The above result may be contrasted with a hypothetical case where each committee member selects a unique ranking solution. In this case we select 13 combinations out of 24 and assign it to 13 committee members:

$$N_2 = \binom{24}{13} \times 13! = 2,496,144 \times 6,227,020,800 = 1.554 \times 10^{16}$$

As can be seen, the probability of this result seems "within an expectable range":

$$p_2 = \frac{N_2}{S} = \frac{1.554 \times 10^{16}}{24^{13}} = 0.0177$$

So we observe that P_1 is about seven or eight orders of magnitude smaller than P_2, a very significant difference. One way to explain this puzzling situation is to speculate that the committee members did not vote as free agents with total dedication to the interest of the project but, possibly, were aware of what decision would be acceptable to their respective bosses.[25]

Further analysis of the voting patterns brings another possible "deceptive" strategy common in group decision making. That is, adding a nonrealistic alternative in order to distort the voting results[26]. Let us look at the voting patterns if we eliminate the fourth (the deceptive) alternative. Now, each committee member will assign three points to the most attractive alternative, two points to the next alternative, and so forth. The result of the committee members' voting is depicted in Table 3.22. Now, alternative B scored the highest and, remarkably, alternative A got the lowest score.

---

25 Some readers may disagree with the validity of this example. Is it reasonable to use the above yardstick? Is the resulting speculation valid? Nineteenth century British Prime Minister Benjamin Disraeli characterized three kinds of lies: "Lies, damned lies, and statistics." We are aware that mathematicians may exercise professional caution about the applicability of statistical inference, knowing that sometimes reality may not conform to assumptions on which these inferential models are constructed. Nevertheless, we think that within engineering this example is telling. As observed by Laplace (Théorie analytique des probabilités, 1820), "The theory of probabilities is at bottom nothing but common sense reduced to calculus."
26 Kenneth Joseph Arrow was a joint winner of the Nobel Prize in Economics in 1972. He is mostly known for contributions to social choice theory, notably, *Arrow's impossibility theorem*. The condition of *Independence of Irrelevant Alternatives* (IIA) was first proposed by Arrow in 1951.

Table 3.22 Second example: Committee member vote

Group	Member	Alternatives/points		
		A	B	C
Mostly A supporters	1	3	2	1
	2	3	2	1
	3	3	2	1
	4	3	2	1
Mostly C supporters	5	2	1	3
	6	2	1	3
	7	2	1	3
Mostly B supporters	8	1	3	2
	9	1	3	2
	10	1	3	2
	11	1	3	2
	12	1	3	2
	13	1	3	2
Total		**24**	**29**	**25**

3.5.6.6 Further Reading

- Al-Shammari and Masri, 2015.
- Arrow et al., 2002.
- Kaliszewski et al., 2016.

3.6 Other Creative Methods

This chapter describes other creative methods that are not traditionally considered "creative" but, the author believes, are important tools in the creative engineer's arsenal (Figure 3.58).

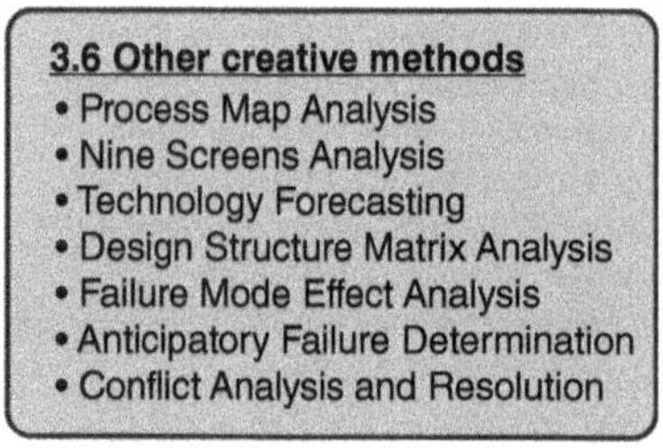

3.6 Other creative methods
- Process Map Analysis
- Nine Screens Analysis
- Technology Forecasting
- Design Structure Matrix Analysis
- Failure Mode Effect Analysis
- Anticipatory Failure Determination
- Conflict Analysis and Resolution

Figure 3.58 Other creative methods

3.6.1 Process Map Analysis

3.6.1.1 Theoretical Background

A *process* is a structured set of activities that transform inputs into outputs.[27] A *process flow* is a network of activities and flows (i.e. material, information, or energy) that make up the entire process, and a *process map* describes graphically this network. Process maps are created in order to gain better understanding of a process itself, build stronger communication, clarify process boundaries, and assist organizations in becoming more efficient. Finally, process maps break down the complexity of processes, providing an excellent platform to plan forthcoming projects.

Virtually, all process maps follow the supplier, input, process, output, customer (SIPOC) construct. Here: (1) supplier is the entity that provides input to a process, (2) input is all that is used to produce outputs from a process, (3) process is a set of steps or activities carried out in order to convert inputs into outputs, (4) output is the outcome or physical products emerging from a process, and (5) customer is the entity that uses the output of the said process. Process maps may be divided into the following families.

Top-level process maps
The purpose of top-level process maps is to illustrate, in a simplified and generic way, the (1) supplier, (2) input, (3) process, (4) output, and (5) customer of a given process. Typically, the inner structure of top-level process maps is intentionally concealed in order to distinguish their broad essence.

Detailed-level process maps
Detailed-level process maps describe the process flow at a high-level but at substantially higher resolution. For example, detailed-level process maps will show beginning and end of a sub-processes, activities that are done, and points where decisions are made, together with appropriate labels, back flows, parallel processes, etc. In addition, interfaces and detailed internal and external flow information is normally presented in appropriate granularity.

Cross-functional (swim lanes) maps
Cross-functional (*swim lanes*) maps are often used when a process requires contribution from two or more persons, departments, or sub-processes. Cross-functional maps help in clarifying who is responsible for producing each flow element and who is expected to receive it.

---

27 According to the Merriam-Webster dictionary, A "process" is defined as "a series of actions or operations conduced to an end; especially: a continuous operation or treatment especially in manufacture."

In cross-functional mapping, processes are grouped visually into individual (often) horizontal "lanes," Each lane is identified by a label (e.g. person's name, department name, sub-processes ID, etc.). Individual flow components are identified on the chart by drawing vertical line between the producer's and the user's lanes.

3.6.1.2 Implementation Procedure

Over the years, process mapping has become straightforward because of the availability of many software packages that provide multitude of capabilities beyond mere drawings. Usually, constructing a process map involves the following steps

Step 1: Define process boundary. That is, define where each process begins and ends and what is the process's environment, from which it receives inputs and to which it transmits output.

Step 2: Define map parameters. Decide what category of map is desired and what is the required level of details.

Step 3: Define process steps. Determine the specific sequence of process steps.

Step 4: Draw the process map. Based on the above steps, draw the process map.

3.6.1.3 Example: Research Project Operations

1) Example background

In this example a research project is undertaken by a consortium composed of universities, large industrial firms, and small and medium enterprises (SMEs). The project is expected to generate a software package as well as research papers and reports.

2) Example analysis: top level

A top-level process map is depicted in Figure 3.59. The project receives its scientific and management objectives from the funding agency as well as

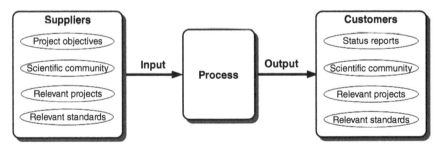

Figure 3.59 Research project: Top-level process map

other customers. It uses state of the art information from the scientific community, relevant other projects, and relevant standards (i.e. the suppliers of this project). The project generates ongoing status reports for the funding agency and provides new scientific knowledge to the scientific community, relevant other projects, and relevant standards (i.e. the customers of this project).

3) Example analysis: Detailed level

There are numerous formats to draw a detailed process map. One example of a detailed-level process map is depicted in Figure 3.60. As can be seen, the individual subprocesses as well as the individual flow components between these subprocesses are specifically designated. For example, subprocess WP2 (methodology development) produces, among others, flow component D2.2, which is used by subprocess WP3 (tool development) and WP4 (pilot projects).

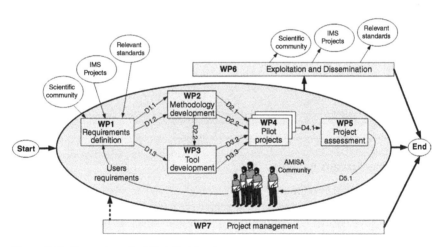

Figure 3.60 Research project: Detailed-level process map

4) Example analysis: Cross-functional (swim lanes) map

A typical cross-functional process map is depicted in Figure 3.61. Here, each one of the sub-processes as well as the suppliers and customers are assigned a horizontal "lane" and individual flow components are identified by vertical line between a producer's and a user's lanes. Here again, subprocess WP2 (methodology development) produces flow component D2.2, which is used by subprocess WP3 (tool development) and WP4 (pilot projects).

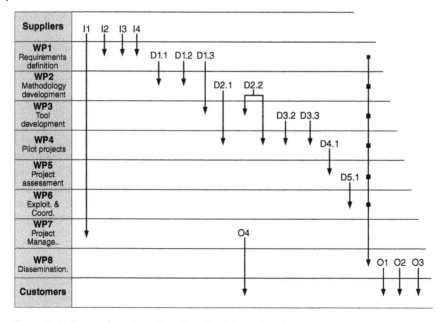

Figure 3.61 Research project: Cross-functional (swim lanes) process map

3.6.1.4 Further Reading

- Damelio, 2011.

3.6.2 Nine-Screens Analysis

3.6.2.1 Theoretical Background

When engineers develop a system or seek to resolve certain system problems, they naturally tend to concentrate on the system itself at present. The nine screens analysis proposed by G. Altshuller in the 1980s provides a way to examine the system from nine different vantage points or more.[28] This includes viewing the system as it is as well as viewing the lower and higher system levels. In addition, this approach depicts the temporal aspects of the system namely, its past, present, and future, creating a 3 × 3 matrix with nine screens (Figure 3.62).

The nine-screens concept encourages engineers to examine the system from both system's levels and time perspectives, stimulating new ideas by looking beyond current design requirements. More specifically, by examining how the system increases its ideality as it has evolved from the past to the present, engineers could envision how the system may continue its evolution into the future. This process could expand their horizon, inspiring creativity vis-à-vis the current system.

---

28 The number of screens may be expanded by considering larger temporal space (e.g. distant past, past, present, future, distant future, etc.) and/or larger system stages (e.g. super-supersystem, supersystem, system, subsystems, components, etc.).

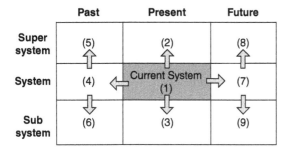

Figure 3.62 Nine-screens matrix

3.6.2.2 Implementation Procedure

A procedure for implementing the nine-screens approach may be composed of the following steps.

Step 1: Identify the system. The system or problem under investigation is identified and this information is placed in the center of the 3 × 3 matrix (i.e. present-system).

Step 2: Identify the supersystem (environment). The environment of the system is identified and this information is placed in the top-middle cell of the matrix (i.e. present-supersystem).

Step 3: Identify the subsystems. The main subsystems are identified and this information is placed in the bottom-middle cell of the matrix (i.e. present-subsystems).

Step 4: Repeat for the past system. Steps 1, 2, and 3 should be repeated vis-à-vis the past, precursor to the current system. The results should be placed in the leftmost three (past) cells of the matrix.

Step 5: Analyzed the system ideality. The increase (or decrease) in the ideality of the system should be analyzed. Specifically, both the useful functions (UF) and the nonuseful functions (NUF) should be estimated.

Step 6: Project for the future system. Based on the results from step 5, a projection is made as to the future system, its environment, and its main subsystems should be done. Thereafter, the results should be placed in the rightmost three (future) cells of the matrix.

Step 7: Seek insights. Further understanding regarding the system or the problem under investigation may be extracted based on the data accumulated in the nine screens.

3.6.2.3 Example: Rocket Engine Design

1) Example background

Advanced rocket engine is to be designed. The design process should be supported by a nine-screens analysis.

2) Example analysis

A nine-screens analysis of an advanced rocket engine design is described in Table 3.23 and the following text.

Table 3.23 Nine-screens analysis for a rocket engine design

	Past	Present	Future
Super system	Aircraft in Earth's atmosphere	Rocket in near Earth's space	Spacecraft in interstellar space
System	Jet engine	Rocket engine design	Interstellar ion propulsion engine
Subsystems	• Intake fan • Compressor • Combustor • Turbine • Nozzle	• Fuel/Oxidizer pump • Turbine • Gas generator • Thrust chamber • Nozzle extension	• Interstellar gas scooper • Plasma accelerator • Fusion reactor

A rocket engine[29] design is used here to illustrate a nine-screens analysis. A rocket engine is a specialized type of jet engine. However, in contrast with jet engines, rocket engines use on-board stored fuel (e.g. liquid hydrogen) and oxidizer (e.g. liquid oxygen) so the rocket can travel in vacuum, propelling payloads into space. Rocket engines deliver thrust in accordance with Newton's third law. That is, they produce thrust by the expulsion of hot gases, which have been accelerated to a high speed through a propelling nozzle. Compared to jet engines, rocket engines are relatively light and have high thrust but have low propellant efficiency. The three vertical cells in the Nine-screens matrix exhibit the environment and the subsystems of a typical rocket engine (Figure 3.63).

Figure 3.63 A typical rocket engine

---

29 See also: Rocket engine, Wikipedia, the free encyclopedia. https://en.wikipedia.org/wiki/Rocket_engine#Throttling. Accessed: November 2017.

Jet engines[30] are internal combustion, air-breathing power plants, burning petroleum-based fuels. This engine discharges a fast-moving jet that generates thrust that pushes an aircraft forward. An engine is composed of an engine core and a central shaft on which three key elements (i.e. intake fan, compressor and turbine) rotate at high speed. The intake fan or a turbofan sucks large quantities of air from the environment. Some air flows through the engine core and the rest bypasses the core, producing added thrust to the engine as well as cools the engine (Figure 3.64). The compressor pushes the incoming air into smaller and smaller areas, resulting in an increase in the air pressure and energy potential.

Figure 3.64 A jet engine

The hot and pressurized air is forced into the combustion chamber, where it is mixed with fuel. The fuel combines with the oxygen in the compressed air, producing high-intensity blaze and expanding hot gases. These gases move through the turbine blades, causing the central shaft to rotate and, in turn, causing the compressor and the intake fan at the front, to rotate with it. Finally, the nozzle, the engine part that produces the thrust, combines the hot gases expelled from the turbine and the bypassed cold air, producing the thrust that propel the aircraft forward.

---

30 See more: Jet engine, Wikipedia, the free encyclopedia. https://en.wikipedia.org/wiki/Jet_engine. Accessed: November 2017.

The advantage of the rocket engine is that it carries its fuel and oxidizer on board the rocket so it is independent of its supersystem (as far as propulsion source). This is contrary to a jet engine, which is an air-breathing machine. Is this a hallmark of an improved ideality? Well, not necessarily, because each engine is perfectly adapt to function within its own supersystem. Nevertheless, this capability is crucial in the case of a future engine, discussed next.

A proposed theoretical interstellar ion propulsion engine[31] is based on a ramjet engine using a giant interstellar gas scooper (ranging in diameters from kilometers to thousands of kilometers). Hydrogen from the interstellar space, used as the sole fuel source, is gathered and, under the combination of the high velocity of the spacecraft and an appropriately constructed magnetic field, is forced into a progressively constricted space until a thermonuclear fusion occurs. The magnetic field then directs the energy as spacecraft exhaust, thereby accelerating the spacecraft in the interstellar space. Fundamentally, if the thrust of the engine is significantly greater than the drag created by the scooped ions, then this engine will be capable of accelerating the spacecraft to an incredible velocity (Figure 3.65).

Figure 3.65 An interstellar ion propulsion engine

31 See more: Interstellar travel, Wikipedia, the free encyclopedia. https://en.wikipedia.org/wiki/Interstellar_travel. Accessed: Nov. 2017.

Assuming an ion propulsion engine is feasible and further supposing that other interstellar travel problems could be solved (e.g. building an appropriate spacecraft, resolving navigation issues, finding way to decelerate the spacecraft, landing safely on a desired planet and, of course, resolving the long-term human interactions dilemma), then this approach will enable mankind the luxury of colonizing other planets. In this respect the ideality of this system is improved by orders of magnitude relative to a chemically propelled rocket. But, of course, existing technology is far from reaching any practical application of such an engine.

3.6.2.4 Further Reading

- Berk, 2013.
- Hunecke, 2010.
- Long, 2012.

3.6.3 Technology Forecasting

3.6.3.1 Theoretical Background

In general, the purpose of technology forecasting is to help evaluate the likelihood as well as the significance of potential future developments so that engineers and managers could make effective and efficient planning. A variety of techniques have been developed for technology forecasting. For example (1) surveillance, i.e. examining the technological state of the art and finding source of innovation, (2) expert opinions, i.e. seeking experts to provide advice on the subject, (3) trend analysis, i.e. following time-series data and identifying evolutionary patterns and driving forces behind similar systems, (4) modeling and simulation, i.e. using simulation forecasting tools to estimate how systems could evolve, and (5) scenario analysis, i.e. examining possible future events and evolution patterns.

Here, a practical forecasting methodology is provided, starting with existing systems, which evolve over specified amount of time, in according with TRIZ's laws of technological system evolution.

3.6.3.2 Implementation Procedure

A procedure for implementing technology forecasting using TRIZ methodology may be composed of the following steps:

Step 1: Study technology and management environment. One should start by studying the technological state of the art. This includes identifying and analyzing the existing firm's products, markets, customers' habits and changing economic trends. In addition, one should analyze the internal organization's vision and currently defined missions.

Step 2: Identify system to be analyzed. In this step, one should identify the specific system that will be the focus of the technology forecasting.

Step 3: Identify top-level system and environment. In this step, the top-level system, its boundary, and its environment (i.e. super system) should be defined.

Step 4: Identify detailed system and interfaces. Next, the subsystems and their interfaces (i.e. internal interfaces) as well as the interfaces between the system and its environment (i.e. external interfaces) should be defined.

Step 5: Identify system evolution stage. Define the system evolution stage using the S-curve model (Section 4.2.1 Modeling Systems Evolution – S-Curve) appropriate to each subsystem. For example, one may use a generic S-curve model depicted in Table 3.24.

Table 3.24 Generic S-curve stages

Stage	Name	Meaning
1	Initial concept	Basic scientific concepts are observed and reported
2	First implementation	First commercial implementation and use of technology
3	Societal recognition	Technology is recognized by society at large
4	Resources decline	Resources supporting the technology start to decline
5	Technology maximized	Technology maximizes its potentials
6	Technology declines	Usage of technology declines
7	Emerging technology	New and improved technology emerges

Step 6: Define relevant time horizon. The relevant time horizon for system evolution should be determined. Typically, representatives from management, finance, marketing, sales, and engineering will discuss the issues or, preferably, utilize an established method like a Delphi procedure to agree on it.

Step 7: Use laws of technical system evolution. Examine each TRIZ, "Basic laws of technical systems evolution" (Table 3.25) in order to identify relevant technical and/or business parameters likely to evolve and affect the future value of each subsystem during the relevant time frame (Alternatively, one may utilize Appendix B: Extended laws of technical systems evolution).

Step 8: Ascertain subsystems' current value. Estimate the current value of each subsystem based on either the cost of the relevant subsystem in the open market or the cost associated with developing, producing, maintaining, and disposing of it.

Step 9: Define current and future parameters. Evaluate each subsystem's technical and business parameters in terms of their Initial (I) and Future (F) stages on the S-curve model.

Step 10: Assign weight to parameters. For each subsystem, estimate the weight of each parameter relative to the other ones, ensuring a sum weight equal to 1.0 per subsystem.

Table 3.25 Basic laws of technical system evolution

#	Basic laws of evolution
1	Law of increasing degree of ideality
2	Law of uneven evolution of subsystems
3	Law of transition to higher-level system
4	Law of increasing dynamism of systems
5	Law of transition from macro to micro-levels
6	Law of completeness
7	Law of shortening the energy flow path
8	Law of increasing controllability
9	Law of harmonizing rhythms of systems' parts

Step 11: Compute initial and final weighted factors. For each subsystem, compute the initial and final weighted factors for each parameter as well as their corresponding totals.

Step 12: Compute future worth and value gain. Based on the subsystem's current value (S), compute its expected future worth (S′) using the arithmetic rule of three as well as its expected value gain (S′ S).

3.6.3.3 Example: Genomic Personalized Medicine

1) Example background

A large and well-established pharmaceutical company is interested in expanding its operations into the personalized healthcare business. This involves the development of several new genomic drugs for treating complex diseases[32] (diseases that are influenced by a combination of individuals' genomes and environmental factors). The CEO establishes a scientific and engineering taskforce with a mandate to explore this area and generate a technology forecast regarding the economic potential of such undertaking. The taskforce provides the following information.

Human genome

Every person has a unique genome. A genome is an organism's complete set of DNA,[33] including all of its genes. Each genome contains all of the information needed to build and maintain that organism. The human genome, for

---

32 Typical complex diseases include various types of cancer, asthma, diabetes, epilepsy, hypertension, manic depression, and schizophrenia.

33 DNA: Deoxyribonucleic acid is an extremely long macromolecule that is the main component of chromosomes and is the material that transfers genetic characteristics in all life forms (Dictionary.com).

example, is made up of 23 chromosome pairs with a total of about 3 billion DNA base pairs and estimated 20,000–25,000 human protein-coding genes. Personalized medicine uses a patient's genome in order to determine the most appropriate drug therapy to a given disease.

Development in human genome research

The Human Genome Project (HGP) was an international project aiming at identifying the sequence of nucleotide base pairs and mapping all the genes of the human genome. The project, costing over $3 billion, started in 1990, and was officially complete in 2003.

Fundamentally, a key for applying affordable personalized medicine is the ability to perform a genome sequencing of a single individual at a reasonable price, and indeed, this price has dropped dramatically from the $3 billion spent by the original HGP to as little as $1,000 in 2016, and is expected to be further reduced to about $100 per single genome sequencing within a decade or so[34] (Figure 3.66).

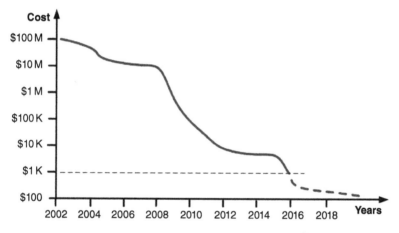

Figure 3.66 Cost of sequencing a genome of a single individual[35]

Healthcare paradigm shift

Under classical disease treatment, physicians provide nearly uniform healthcare to all patients suffering from the same disease, following the paradigm: Observation => Treatment => Uncertain response. This reactive medical care

34 M. Herper, "Illumina Promises to Sequence Human Genome For $100 – But Not Quite Yet," *Forbes*, Jan. 9, 2017. https://www.forbes.com/sites/matthewherper/2017/01/09/illumina-promises-to-sequence-human-genome-for-100-but-not-quite-yet/#266dc8a1386d. Last accessed March 2018.
35 Derived from data computed by the National Human Genome Research Institute (NHGRI) and other sources.

strategy attempts to maximize the ratio of positive therapeutic results versus risks of therapeutic failures or patients' undesirable side effects.

Personalized medicine, sometimes called *Stratified medicine, precision medicine* or *P4 medicine*, refers to the use of an individual's genetic characteristics in order to guide medical decisions with regard to the prevention, diagnosis, and treatment of each disease. So, under this new paradigm, physicians make more effective medical decisions depending on a patient's genomic characteristics. That is, patients are offered specific interventions strategies tailored to their predicted medical response or risk associated with a given disease.

The use of information and data from a patient's genotype and phenotype (level of gene expression and/or clinical information) helps in initiating a preventative measure that is particularly suited to a particular patient. More specifically, individuals' biomarkers are used as diagnostic tools, which could predict the therapeutic response to specific drugs (Figure 3.67).

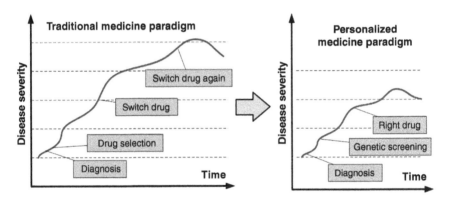

Figure 3.67 Healthcare paradigm shift

The net result of this healthcare paradigm shift is a more efficient healthcare service at reduced public cost and significant reduction of side effects and suffering for patients who cannot benefit from certain drug treatments. According to Illumina,[36] during the year 2014, an estimated quarter of a million human genomes have been completely sequenced by researchers around the globe. The company estimates that this number will double about every 12 months, reaching 1.6 million yearly individual genomes by 2017.

Along with this development, many genomic drugs become available on the market. For example, according to data published in 2012 by the US Food and Drug Administration (FDA), more than 110 marketed drugs have

---

36 Illumina, Inc. is a US company that develops, manufactures, and markets integrated systems for the analysis of genetic variation and biological function.

pharmacogenetic biomarkers on their label. Personalized medicine's greatest impact has been in cancer therapy. In particular, melanoma, thyroid cancer, colorectal cancer, lung and pancreatic cancers as well as breast cancer contain genetic mutations that could be targeted by genetic drugs.

Data from the field

In order to appreciate the economic benefits of personalized medicine, one can look at a research work reported by Robert Langreth in 2008. He compared Erbitux treatment and cost comparisons with and without KRAS testing to direct personalized treatment (DPT).

KRAS, known by its formal name KRAS Mutation Analysis, detects the presence of the most common KRAS gene mutations in the DNA of cells in tumor tissue in order to help guide cancer treatment. KRAS mutation analysis is ordered primarily to determine if a person with metastatic colon cancer or non–small-cell lung cancer is likely to respond to standard therapy, an anti-EGFR drug therapy. Tumors with the KRAS mutation do not respond to anti-EGFR therapy. The results of the analysis are quite impressive: (1) 60% reduction in cost per success, (2) 40% of patients spared side effects from ineffective treatments and (3) overall success is unchanged at 25% (Figure 3.68).

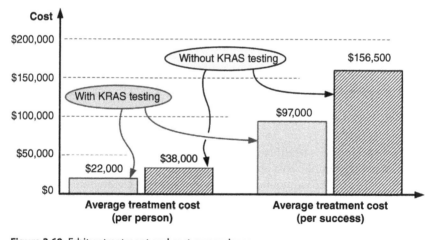

Figure 3.68 Erbitux treatment and cost comparisons

2) Example analysis

1) Personalized medicine's S-curve stage for the society at large is three (technology is recognized by society at large). However, for this pharmaceutical company, it is two (first commercial implementation and use of technology).

2) The relevant time-horizon for this analysis is 10 years.

3) Table 3.26 depicts the technology forecasting analysis of the proposed personalized medicine. Accordingly, and according to Figure 3.69, the advantages of personalized medicine for this pharmaceutical company are significant.

Table 3.26 Technology forecasting: Personalized medicine

Evolution laws			Parameter	Initial	Future	Max	Weight	Calculation	
L1	L2	L3	p	I	F	M	W	W*I	F*I
1	3	5	Increase range of drug selection	3	5	7	0.15	0.45	0.75
8			Decrease drug dosage	3	4	7	0.05	0.15	0.20
1	2	5	Increase drug efficacy	2	4	7	0.10	0.20	0.40
4	6		Decrease recurrence risk	3	3	7	0.15	0.45	0.45
1	9		Reduce cost per success	1	4	7	0.25	0.25	1.00
1	5	8	Spare patients from side effects	1	4	7	0.15	0.15	0.60
1	2	3	Increase overall healthcare success	3	5	7	0.15	0.45	0.75
Total							**1.00**	**2.10**	**4.15**

Current price [$M] (S) =	100
Future worth [$M] (S′) =	205
Gain [$M] (S′−S) =	105

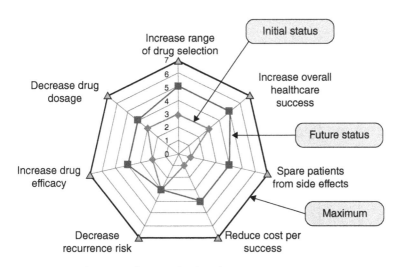

Figure 3.69 Advantage of personalized medicine

3.6.3.4 Further Reading

- Bejan and Lorente, 2011.
- Christensen, 1992.
- Cooke, 1991.
- Hyndman and Athanasopoulos, 2013.
- Langreth, 2008.
- Loveridge, 2002.

3.6.4 Design Structure Matrix Analysis

3.6.4.1 Theoretical Background

Design Structure Matrix (DSM), also referred to as dependency structure matrix, is a modeling technique that can be used for developing and managing complex systems. DSM captures the elements of a system and their interactions within a system, as well as between the system and its environment. Therefore, DSM is an excellent tool for defining systems' architecture as well as processes and organizations architectures. DSM's key advantage is derived from its intuitive representation, visual nature, and compact format.

A DSM is a square matrix in which the system components are identified in the rows to the left of the matrix as well as the columns above the matrix. The off-diagonal cells indicate relationships between the system's elements. There are several conventions for interpretation DSMs, but, in general, a marking of a given cell indicates some linkage between two components (i.e. identified in the relevant raw and column of the matrix).

A common convention identifies output produced by a component along the raw of that component and input consumed by a component is identified along the column of that component.

For example, the system described in Figure 3.70 is composed of three components A, B, and C. Component A produces output (x), which is consumed by component B. Similarly, component A also produces an output (y), which is consumed by component C. Finally, component C produces an output (z), which is consumed by component A.

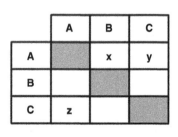
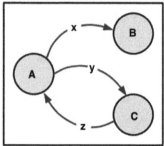

Figure 3.70 Three-component system

A multiple domain matrix (MDM) is an extension of the original DSM model. It depicts an integrated view of multiple domains.

Two or more DSMs, each representing a specific domain, may be placed in tandem in order to model relations between components of different systems or domains. For example, the two systems depicted in Figure 3.71 represent two domains (e.g. parts of a machine A, B, C and teams of engineers 1, 2. The relationships between specific machine parts and engineering teams are depicted by link α and link μ.

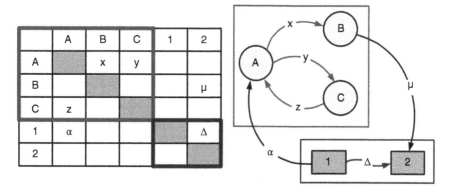

Figure 3.71 Multiple domain matrix (MDM)

3.6.4.2 Implementation Procedure

Implementation of a DSM entails the following steps:

Step 1: Define system of interest. Define the system of interest, including its elements at the desired level of granularity.

Step 2: Define environment. Define the environment of the system at the desired level of granularity.

Step 3: Determine relationships. Determine the nature of the relationships linking the different elements.

Step 4: Create matrix. Create a square matrix, identifying the top raw and left column with the system's elements as well as the environment to the system.

Step 5: Identify relations among elements. For each relevant cell, identify the relationship linking each of the elements in the system

3.6.4.3 Example: Handheld Hairdryer

1) Example background

A handheld hairdryer is an electromechanical device designed to blow hot air over wet hair in order to dry it. This system is described in Section 3.4.4.3, Example: Hairdryer Design.

2) Example analysis

The purpose of this example is to analyze the types of interfaces among components as well as between components and the system's environment. For this end, the following types of interfaces are defined:[37] (1) spatial, (2) material, (3) energy, and (4) information.

The resulting DSM, modeling the hairdryer's components and their various types of interfaces, is depicted in Figure 3.72. By way of example, the fan speed switch (C) interface provides energy (electricity) to the motor (D). Along the same line, the environment (I) provides material (air) to the fan (E).

Figure 3.73 depicts the energy, material, and spatial interfaces in the hairdryer. Energy interfaces define flow of energy from/to the environment as well

		A	B	C	D	E	F	G	H	I
Cord	A		3						1	
On/off switch	B			3			3		1	
Fan speed switch	C				3				1	
Motor	D					1			1	
Fan	E				1			1, 2	1	
Heat switch	F							3	1	
Heating element	G					1			1	2
Casing	H	1	1	1	1	1	1	1		
Environment	I	3				2				

Figure 3.72 DSM modeling the components and interfaces of a hairdryer

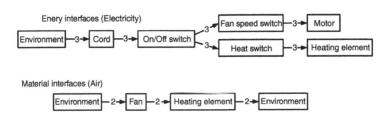

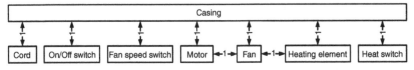

Figure 3.73 Spatial, material, and energy interfaces in a hairdryer

---

37 See: Pimmler and Eppinger, 1994.

as between a set of two components. Material interfaces define the flow of material between two components. Finally, spatial interfaces define the relative position and orientation between two components.

3.6.4.4 Further Reading

• Engel and Reich, 2015.	• Maurer and Lindemann, 2007.
• Eppinger and Browning, 2012.	• Pimmler and Eppinger, 1994.

3.6.5 Failure Mode Effect Analysis

3.6.5.1 Theoretical Background

1) General concepts

Failure mode effect analysis (FMEA)[38] is a bottom-up procedure for: (1) analysis of potential failure modes within a system or a process and (2) determining how to eliminate such problems. This is accomplished by identifying the potential types of problems that may occur, their causes, and the potential frequency with which they may impact the system or the process at hand. The analysis proceeds with estimating the effects of such failures should they occur. Next a determination is made as to how such events may be detected and/or prevented (e.g. modifying the system design, improving the manufacturing process, etc.). Finally, under the FMEA procedure, the actual handling of these corrective actions takes place (Figure 3.74). FMEA is widely used in various phases of products' life cycles, especially during the design and manufacturing of systems and their corresponding processes.

The ultimate purpose of FMEA is to take actions to eliminate or reduce potential future failures. Therefore, a key FMEA practice is to prioritize the handling of these potential failures according to: (1) how serious their consequences are, (2) how frequently they occur, and (3) how easily they can be detected and fixed.

2) Basic FMEA terms

Some of the basic FMEA terms are:

Failure cause. The underlying cause of the failure or the cause, which may initiate a process leading to failure (e.g. defects in design, deficient manufacturing process, poor quality etc.).

---

38 See also: Failure mode and effects analysis, Wikipedia, the free encyclopedia. https://en.wikipedia.org/wiki/Failure_mode_and_effects_analysis. Accessed: Nov. 2017.

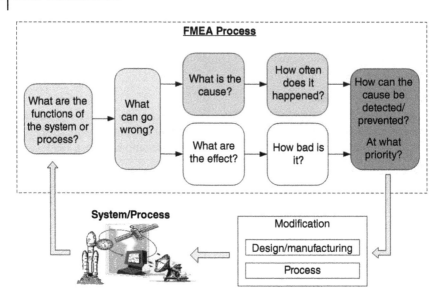

Figure 3.74 Typical FMEA process

Failure mode. The characteristics a system or process may fail. It refers to a complete description under which the failure may occur, how the system is being used, and the final results of the failure.

Failure effect. The immediate consequences of a failure on the operation, functionality, or status of the system at hand.

Failure severity. The consequences of a failure mode – that is, the worst potential consequence of that type of failure, determined by the degree of injury, property damage, or system damage that could ensue.

3) Basic Types of FMEAs

There are four basic types of FMEA processes, although most practitioners tend to match and mix them as they see fit.

Design FMEA. This process is performed on a system or service during the system design phase. Under FMEA, systems are analyzed in order to determine how failure modes affect the system operation. This leads to better understanding of design deficiencies, which can then be corrected so impact of failure modes is avoided or reduced.

Functional FMEA. This FMEA process focuses on the intended functionality, or use, of a system. For example, the FMEA on an automobile design would investigate the behavior of an automobile of that design without paying much attention to its detailed structure. The FMEA could (1) analyze the potential problem stemming from each potential loss of functionality, (2) estimate the statistical probability of such problem, and (3) estimate the potential damage to the automobile, its occupants or the environment of the car. Finally, the functional FMEA would attempt to offer remedy to such problems and a priority for implementing each solution.

System FMEA. This *white-box* FMEA can be used to analyze a system at any level, from the lowest component up to the system level. At the lowest level, it looks at each component in the system to determine the ways in which the component may fail and how these failures affect the system. In this procedure, the detailed make-up and structure of the system takes central stage. The focus shifts from mere system functionality to clear understanding of potential failures and mutual interactions of each individual part of the entire complex system. In the automobile example above, this could mean attention would be given to the intricacies and failure modes of, say, the steering mechanism, the tires, and the gas tank, as well as every other essential part of the vehicle.

Process FMEA. This process is mostly focused on the manufacturing processes, although other engineering processes (e.g. system development, systems verification, and validation, systems operations, etc.) may be undertaken. The procedure identifies possible failure modes in the process, limitations of resources, equipment, tooling, gauges, operator training, or potential sources of error. As in the other FMEA types, this information is used to determine the corrective actions that need to be taken.

4) FMEA standards

There exist several FMEA standards. Virtually all provide sample inspection forms and instruction documents. They also identify criteria for the quantification of risk associated with potential failures and offer general guidelines on the mechanics of completing FMEA procedures. In addition, most standards describe FMEA procedures encompassing functional, interface, and detailed FMEAs, as well as certain pre-analysis activities (FMEA planning and functional requirement analysis), post-analysis activities (failure latency analysis, FMEA verification, and documentation) and applications to hardware, software, and process design. Many FMEA software tools support these standards. The following are a few examples of available FMEA standards:

MIL-STD-1629A (1980). This FMEA standard describes a method used mostly by government, military and commercial organizations worldwide. This standard was formally canceled without replacement in 1998; however, it remains in wide use for military and space applications today, providing formulas for determining criticality and allows rating of failure modes by severity class. A variant of this standard, the failure modes effects and criticality analysis (FMECA) was issued in 1993.

SAE J1739 (2002). This FMEA standard is based on a procedure defined by major international automobile companies and their suppliers. It has been adopted and recommended by the Society of Automotive Engineers (SAE).

ARP5580 (2012). The SAE recommends this FMEA standard for nonautomobile applications. It is intended for use by organizations whose product or system development processes use FMEA as a tool for assessing the safety and reliability of system elements within their product improvement processes.

Many organizations use a combination of different standards, modifying them to suit their needs for their particular applications.

3.6.5.2 Implementation Procedure

Typical FMEA process may be divided into the following steps.

Step 1: Prepare for a FMEA process

Before starting with a FMEA process, it is important to undertake some preliminary work to confirm that robustness and past history are considered in the analysis. FMEA is initiated by describing the system and its functions or the process that must undergo FMEA evaluation. A good understanding of the FMEA object simplifies the further analysis. This way, an engineer can observe which uses of the system are desirable and which are not. It is important to consider both intended and unintended uses of the system, where unintended use includes improper operation, unexpected environmental effects on the system or perhaps malicious use by a hostile user.

Next, a system block diagram is created depicting an overview of the major components or process steps and how they are related. These are the logical relations around which the FMEA can be developed. Finally, a well-defined set of procedures, forms, and worksheets must be created that define important information about the system (e.g. revision dates, names of the components, etc.). In addition, all the items or functions of any corresponding element should be listed in a logical manner.

FMEA activities should be supported by appropriate database tools, as the procedure tends to be tedious and time consuming. Several techniques can be used to reduce the tedium, time, and thus cost of performing a FMEA. For example, failure mode distribution standards can be used to assign common failure modes. Standard reports and input formats may be created to streamline the failure data collection and reporting process. Custom failure mode libraries can also be created and reused for future projects. Several software tools supporting efficient FMEA procedures and standards are available commercially. Such tools can reduce the overall cost of performing and improve the robustness of the FMEA process.

Step 2: Determine FMEA severity level

In this step, one determines all potential failure modes based on the functional requirements of the system and their effects. Examples of failure modes are loss of braking ability in a car and malfunction of a lathe machine in an assembly line. As one failure can lead to another failure, it is critical to analyze all the ramifications of each failure type that can occur. A failure effect is defined as the result of a failure mode on the function of the system as perceived by the user, operator, or other affected individuals.

Examples of failure effects are degraded performance, noisy operation, or discomfort by or even injury to a user. Customarily, each potential failure effect is assigned a severity rating (S) from 1 to 10. For example, Table 3.27 depicts typical FMEA parameters. These systems-oriented FAME elements are derived

Table 3.27 Design FMEA severity evaluation criteria

Effect	Severity of effect	Rating
Hazardous, without warning	Very high severity rating when a potential failure mode affects safe system operation or involves noncompliance with regulation without warning.	10
Hazardous, with warning	Very high severity rating when a potential failure mode affects safe system operation or involves noncompliance with regulation with warning.	9
Very high	System inoperable (loss of primary function).	8
High	System operable, but at a reduced level of performance. Customer very dissatisfied.	7
Moderate	System operable but comfort/convenience item(s) inoperable. Customer dissatisfied.	6
Low	System operable but comfort / convenience item(s) operable at a reduced level of performance. Customer somewhat dissatisfied.	5
Very low	Fit & finish / squeak & rattle item do not conform. Defect noticed by most customers (greater than 75%).	4
Minor	Fit & finish / squeak & rattle item do not conform. Defect noticed by 50% of customers.	3
Very minor	Fit & finish / squeak & rattle item do not conform. Defect noticed by discriminating customers (less than 25%).	2
None	No discernible effect.	1

from Military Standard MIL-STD-1629A and the Society of Automotive Engineers, SAE-J1739. These rating numbers help an engineer to prioritize the failure modes and their effects. A severity rating of 9 or 10 is generally associated with those effects that would cause injury to a user or otherwise result in litigation. In such a case, actions must be taken to change the system by either eliminating the failure mode or protecting the user from its effect.

Step 3: Determine FMEA failure rate
In this step, it is necessary to look at the cause of a failure and the frequency with which it may occur. Looking at similar products or processes and the failures that have been documented for them can help in this task. A failure cause may be a design weakness, manufacturing flaws, operation error, and the like. All potential causes for a failure mode should be identified, analyzed, and documented. An occurrence rating (O), customarily in the range of 1–10 (Table 3.28), should be assigned to each failure mode.

Table 3.28 Design FMEA occurrence evaluation criteria

Probability of failure	Likely failure rates over design life	Rating
Very high: Persistent failures	≥100 per thousand items	10
	50 per thousand items	9
High: Frequent failures	20 per thousand items	8
	10 per thousand items	7
Moderate: Occasional failures	5 per thousand items	6
	2 per thousand items	5
	1 per thousand items	4
Low: Relatively few failures	0.5 per thousand items	3
	0.1 per thousand items	2
Remote: Failure is unlikely	≤ 0.01 per thousand items	1

Step 4: Determine FMEA detection rate
A detection rating (D) represents the general ability to detect a system defect or a failure mode by means of a planned set of tests and inspections. In this step, engineers look at the system mechanisms that are responsible for detecting potential failures, thus preventing actual failures from occurring. For example, the oil pressure indicator in a car is a mechanism that detects low oil pressure and warns the driver about a potential engine seizure. Engineers then carry out testing, analysis, monitoring, and other actions in order to detect or

prevent failures. From these design control efforts, an engineer can learn how likely it is for a failure to be identified or detected. Typical detection ratings are depicted in Table 3.29.

Table 3.29 Design FMEA detection evaluation criteria

Detection	Likelihood of detection by design control	Rating
Absolute uncertainty	Design control will not or cannot detect a potential cause or mechanism and subsequent failure mode, or there is no design control.	10
Very remote	Very remote chance the design control will detect a potential cause or mechanism and subsequent failure mode.	9
Remote	Remote chance the design control will detect a potential cause or mechanism and subsequent failure mode.	8
Very low	Very low chance the design control will detect a potential cause or mechanism and subsequent failure mode.	7
Low	Low chance the design control will detect a potential cause or mechanism and subsequent failure mode.	6
Moderate	Moderate chance the design control will detect a potential cause or mechanism and subsequent failure mode.	5
Moderately high	Moderately high chance the design control will detect a potential cause or mechanism and subsequent failure mode.	4
High	High chance the design control will detect a potential cause or mechanism and subsequent failure mode.	3
Very high	Very high chance the design control will detect a potential cause or mechanism and subsequent failure mode.	2
Almost certain	Design control will almost certainly detect a potential cause or mechanism and subsequent failure mode.	1

Step 5: Compute risk priority value

A risk priority number (RPN) is a quantitative determination of risk based on multiple factors. Traditionally, RPN is defined as the product of the severity rating (S), occurrence rating (O), and detection rating (D) values of each failure mode:

$$RPN = S \times O \times D$$

The failure modes that have the highest RPN should be given the highest priority for corrective action. While the above traditional RPN computation is widely used, every project has a unique set of circumstances, and a one-size-fits-all approach to RPN calculation may not produce the most effective results for an analysis. For example, in some situations, such as where human safety is at risk, the RPN_S could be more meaningful where the severity rating (S) is weighted much more heavily:

$$RPN_S = S^2 \times O \times D$$

3.6.5.3 Example: Cardiac Pacemaker

1) Example background

An implanted cardiac pacemaker device is a small, battery-powered medical device that performs biventricular pacing in patients with congestive heart failure. It delivers a sequence of electrical impulses to the heart muscles by electrodes in order to regulate the beating of the heart (Figure 3.75 and Figure 3.76).

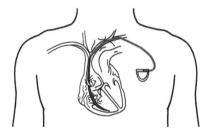

Figure 3.75 Pacemaker embedded in the body

Figure 3.76 Pacemaker impulse shape

2) Example analysis

Figure 3.77 depicts a block diagram of a pacemaker system in its environment. The system is composed of six subsystems (1) a pulse generator, (2) sense leads, (3) impulse leads, (4) a receiver, (5) a transmitter, and (6) a battery to power the system. The system delivers appropriate electrical impulses to the patient heart, based on analysis of its continual behavior. A heart surgeon "installs" the

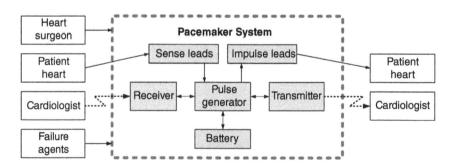

Figure 3.77 Pacemaker block diagram: system and environment

Table 3.30 FMEA analysis for pacemaker failure modes

#	Failures modes	Failure effects	Failure severity (1–10)	Potential failure causes	Failure occurrence (1–10)	Failure prevention	Failure detection (1–10)	RPN (1–1,000)	RPN$_5$ (1–10,000)
1	Undersensing and Oversensing	Patient's heart is subjected to untimely impulses.	8	• PM setup • Sense-leads	3	Verify design, manufacturing and setup of PM and its sense-leads.	6	144	1152
2	Loss or failure of output	Patient heart does not receive impulses.	6	• Battery • Impulse-leads	6	Ensure battery meets design specs.	5	180	1080
3	Inappropriate pulse rate, shape or intensity	Patient's heart is subjected to improper impulses.	8	Pulse generator	2	Ensure pulse generator meets design specs.	3	48	384
4	Inappropriate lead position	Heart pacing is not effective.	6	PM setup	7	Verify heart surgeon's implantation procedure.	3	126	756
5	Extra cardiac stimulation	Patient's heart is subjected to unneeded impulses.	8	• PM setup • Pulse generator	3	Ensure pulse generator meets design specs.	3	72	576
6	True pulse generator failure	Patient's heart receives improper impulses.	6	Pulse generator	6	Ensure pulse generator meets design specs.	5	180	1080
7	Pacemaker-mediated tachycardia[39]	Patient's heart starts to race dangerously.	10	Pulse generator	2	Ensure pulse generator meets design specs.	3	60	600

---

39 Tachycardia comes from the Greek words tachys (accelerated) and cardia (of the heart). A heart rate of over 100 P/M at rest, is generally considered as tachycardia.

system in the patient body and a cardiologist provides lifelong maintenance service to the system.

Yarlagadda (2013) classified failure modes and causes of pacemakers' malfunctions. Table 3.30 uses a subset of Yarlagadda's listing, depicting an example of a FMEA analysis. Given limited corrective resources, the RPN column identifies items 2 and 6 as the ones to be attended first. However, according to the RPN_S column, item 1 should be attended first.

3.6.5.4 Further Reading

- ARP5580, 2012.
- Dyadem Press, 2003.
- Modarres et al., 1999.
- MIL STD 1629A, 2001.
- SAE J1739, 2002.
- Stamatis, 2003.
- Yarlagadda, 2011.

3.6.6 Anticipatory Failure Determination

3.6.6.1 Theoretical Background

Anticipatory Failure Determination (AFD)™, sometimes called *subversion analysis,* was developed by Genrich Altshuller and fellow TRIZ researchers during the late 1980s. Basically, it is a systematic method for analyzing systems' failures and predicting potential system failures. This section, therefore, concentrates on a method for early identification of potential systems failures.

AFD methodology offers several strategies to analyze failure scenarios. In particular, we are interested in identifying possible failure initiation events and then investigate the resulting failure propagation emanating from each of them. Initiating events are defined as failures of individual subsystems or components of the system, as well as unexpected external events inducing systems' failures. Thus, in a given system, one would work through each system element, asking, "What would happen if this part failed?" or "What kind of external event can cause this part to behave in an unplanned manner?" This process is viable because identification of initiating events and, consequently, failure scenario trees can be carried out at various levels of details and thoroughness.

Unfortunately, there is a fundamental problem with failure analysis using such questions. Engineers are, first and foremost, builders, creators, and solvers of problems. Intrinsically, they proceed with their psychological inertia, i.e. unable to generate a wide spectrum of failure questions and mishap scenarios. The reason for this is that engineers (and human beings in general) are subject to a psychological denial under which their brains resist contemplating unpleasant events.

AFD proposes an ingenious doctrine to combat this psychological phenomenon. Engineers should implement *inverted logic* rather than concentrate on risks pertinent to the system's specifications. More specifically, under inverted logic, engineers are asked to transform themselves into subversive agents, deliberately trying to cause systems' failures. So, inverted logic leads engineers to change their attitude toward failures and generate inverted questions. Such questions are useful in counteracting the tendency of humans to deny unpleasant events. For example, one may ask an inverted question: "How can I sabotage the system?" When one applies his/her engineering skills, then his mind opens up to the full spectrum of failure possibilities. In addition, AFD stresses the concept of resources. For any system failure to occur, all the necessary failure components must be present, within the system or its environment.

3.6.6.2 Implementation Procedure

The following is a general procedure for implementing an anticipatory failure determination analysis:

Step 1: Define system and its resources. In this step, the system under consideration, its subsystems and components are defined to the desired level of granularity. In addition, resources associated with the environment of the system are also defined.

Step 2: Define mission of system. In this step, the mission of the system, assuming successful performance of the mission (success scenario, S_0) is defined. The definition describes each phase, the processes within each phase, and the results achieved at the end of each phase.

Step 3: Formulate inverted mission. In this step, the mission defined in step 2 is inverted. That is, within each phase, each process produces system results, which differ or contradict the original mission results.

Step 4: Force system to fail. In this step, all the possible initiating events ($IE_{i,j}$) causing scenarios trees (S_i) that lead to harmful end states ($ES_{i,j,k}$), are defined. One may search for failure scenarios by employing a commercially available AFD software package.

Step 5: Identify failure resources. In this step, all the resources (i.e. conditions) available in or around the system that might be instrumental in contributing to a failure are identified.

Step 6: Invent new solutions. In this step, one can use the AFD principle that all the resources (conditions) necessary for an initiating event must be present in a situation in order for the event to actually occur. Conversely, if at least one of the necessary resources is not present, then the failure event will not occur. This principle is most valuable in guiding the search for system's failure elimination – namely, remove from the system one of these necessary resources.

3.6.6.3 Example: Anticipatory UAV System Failures

The following is an anticipatory failure determination example of an unmanned air vehicle (UAV) system.

1) <u>UAV system resources</u>

Under AFD methodology, system "resources" are defined as all the substances, components, configurations or other factors presented in a situation. A simplified set of resources in the UAV system example is the six subsystems described below and depicted in Figure 3.78.

Ground control system (GCS). The GCS is a shelter, often mounted on a truck, in which a UAV pilot and other UAV personnel are located. The UAV team flies the unmanned aircraft, observe video and infrared image stream acquired by the UAV and control the entire UAV system.

Ground data terminal (GDT). The GDT is a ground unit containing a powerful transmitter and receiver. It receives commands from the GCS and transmits them to the UAV and similarly, it receives UAV telemetry status, as well as video and infrared stream, and sends them to the GCS.

Air vehicle (AV). The AV is an unmanned craft designed to take off, fly and land automatically or manually and carry various payloads and support systems to a desired altitude and location and transmit live video and infrared pictures from that location.

Air data terminal (ADT). The ADT is the airborne counterpart of the GDT, performing quite similar activities.

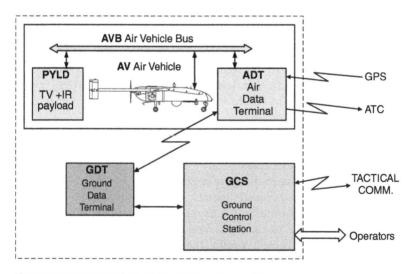

Figure 3.78 Unmanned air vehicle (UAV) system architecture

Payload (PYLD). The PYLD is a unit containing specialized cameras, mounted on a gimbaled platform attached to the AV. It is capable of viewing the external world, in visible as well as infrared frequencies and sends the data to the ADT for transmission to the ground.

Air vehicle bus (AVB). The AVB is a data-bus connecting the ADT, AV and PYLD and allowing the transfer of command, status and other data among these subsystems.

2) UAV system mission

Prior to performing an anticipatory failure determination for this system, one should define clearly the exact mission of the system. In other words, for a failure scenario to be understood, the "success" (or as-planned) scenario (denoted S_0), must be clearly specified. In this example, an Unmanned Air Vehicle (UAV) system is designed to take off automatically from an air strip, cruise to a given altitude and location, perform its visual surveillance mission and then cruise back home and land automatically at the home base. A successful UAV mission is accomplished by way of five phases (see below and in Figure 3.79.

Phase 1: Take off automatically. The UAV performs an automatic take-off from an airstrip.

Phase 2: Cruise to target. The UAV flies along a designated route to a designated altitude and location.

Phase 3: Perform the mission. The UAV flies in a predefined flight path and direct its cameras to certain sets of locations.

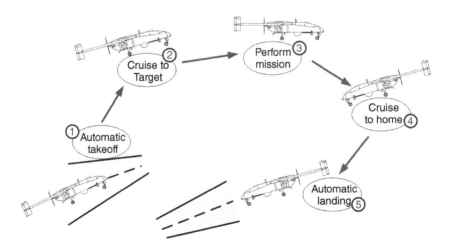

Figure 3.79 A planned unmanned air vehicle (UAV) operational scenario (S0)

Phase 4: Cruise to home. The UAV flies along a designated route back to the original airstrip.

Phase 5: Land automatically. The UAV performs an automatic landing on the airstrip and comes to a standstill at a designated place on the airstrip.

AFD considers S_0 as a trajectory in the state space of the system; depicting general relations between system's mission phases and time (Figure 3.80).

3) Formulating the inverted mission

Since S_0 is the planned scenario, any failure scenario (S_i) that departs from this plan must have a point of departure from normal system operations. The initiating event ($IE_{i,j}$) of S_i may be generated due to internal system failure or due to an unanticipated external disturbance. Two such initiating events are depicted in Figure 3.81. From each initiating event, an outgrowth of related failure scenarios emerges, which is referred to as a *failure scenario tree*. Each path through the tree represents a particular scenario, depending on what happens after the initiating event. Each branch of the tree continues until it reaches some system end state ($ES_{i,j,k}$).

For example, we, as subversive agents, create two events producing two failure scenario trees (Figure 3.81). The first failure tree, occurs during the mission phase "Cruise to target," emanates from event $IE_{0,0}$ and ends at one of four system end states $\{ES_{0,0,A}, ..., ES_{0,0,D}\}$ and a second failure scenario, occurring during mission state "Cruise to home," emanates from event $IE_{0,1}$ and end at one of three system end states $\{ES_{0,1,A}, ..., ES_{0,1,C}\}$.

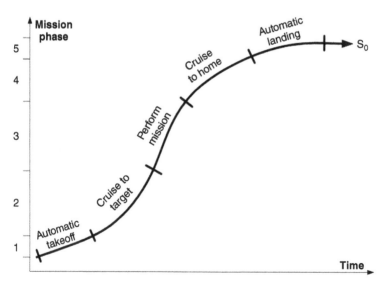

Figure 3.80 UAV system state (system's mission phases versus time)

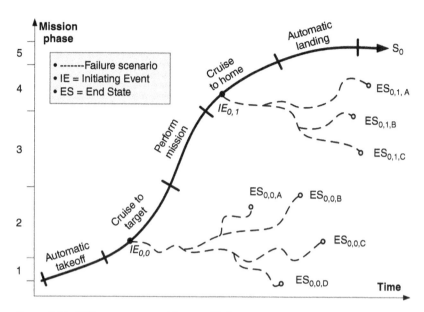

Figure 3.81 UAV system states with several failure scenarios

4) Making the system to fail

Figure 3.82 depicts the six UAV subsystems along the vertical axis.

As mentioned, the particular UAV mission S_0, having distinct phases of operation, are represented along the horizontal axis, forming a time-like axis. For each combination of UAV subsystem and mission phase, one can identify any number of initiating events ($IE_{i,j}$). Next, one draws outgoing failure tree (S_i, $i \neq 0$) from each of these initiating events. This is done so that the set of paths in each tree represents a complete set of scenarios emerging from that event and leading to multiple end states ($ES_{i,j,k}$). For a given resolution of system structure and mission phases, the combination of components and phases are finite, therefore, a "complete" set of system failure scenarios may be created.

For example (as seen in Figure 3.81 and Figure 3.82), one can impose several individual failures by the initiating event $IE_{0,0}$, "Loss of communication between the ground data terminal (GDT) and the UAV," which occurs during the "Cruise to target" phase of the UAV mission. This initiating event occurred because we as saboteurs, intentionally disconnected the antenna cable from the GDT transmitter. This situation means that the UAV operator at the ground control center (GCS) is unable to control the UAV or receive any data from it. Four end states have been identified:

$ES_{0,0,A}$ – The UAV is out of control. It flies until it runs out of fuel, at which time, it crashes to the ground.

UAV mission phases:

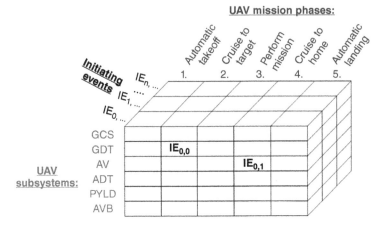

Figure 3.82 Three-dimensional space of initiating failure events in a UAV system

$ES_{0,0,B}$ – The UAV recognizes the loss of transmission condition and initiates its automatic "return to home" procedure. The UAV is then returns and automatically lands safely at home base.

$ES_{0,0,C}$ – Similar to $ES_{0,0,B}$ but, unfortunately, the global home coordinate address provided to the UAV was pointing to the southern hemisphere instead of the northern hemisphere. The UAV proceed to fly away from home base, runs out of fuel and crashes to the ground.

$ES_{0,0,D}$ – The UAV operators initiate a GDT emergency procedure, reestablishing the proper operation of the GDT. The communication between the GDT and the UAV is restored however the UAV mission is aborted and the UAV is returned home.

Let us now consider the second initiating event $IE_{0,1}$ – "UAV fuel runs out," which occurs during the "Cruise to home" phase of the UAV mission. This initiating event occurred because we as saboteurs, intentionally punctured the fuel tank of the UAV. This situation means that the UAV engine will stop running within a short time. One can impose three scenarios or end states for this initiated event:

$ES_{0,1,A}$ – The engine of the UAV stops. Without propulsion the UAV lose its ability to remain airborne. The air vehicle exits its flight envelope and crashes to the ground.

$ES_{0,1,B}$ – The UAV operators recognize the problem and direct the UAV to glide without propulsion and then land at a secondary landing strip located in the vicinity of the stricken UAV. This procedure is successful.

$ES_{0,1,C}$ – Similar to $ES_{0,1,B}$ but the procedure is unsuccessful due to a lack of automatic landing facilities at the secondary landing strip. The UAV hits the landing strip toward the end of the runway and crashes against the landing strip perimeter.

5) Identifying available failure scenario resources

In this example, the initiating event $IE_{0,0}$, "Loss of communication between the ground data terminal (GDT) and the UAV," created the first failure. The resources that were probably instrumental in contributing to this failure are the GDT (e.g. transmitter problem) and the ADT (e.g. receiver problem) within the UAV. The initiating event $IE_{0,1}$, "UAV fuel runs out," created the second failure. The resources that were probably instrumental in contributing to this failure are the UAV (e.g. fuel tank leakage) and the UAV operations support team (e.g. improper refueling process).

6) Inventing new solutions

In this example, the first failure could have been avoided by performing a timely verification, validation, and testing (VVT) of the design, manufacturing, operations, and maintenance of the GDT and ADT subsystems in order to eliminate the source of the failure. Similar procedure is required in dealing with the second failure as well as possibly; improving the UAV operations support team training and management control.

3.6.6.4 Further Reading

```
• Brue, 2003.          • Kaplan et al., 1999.
• Engel, 2010.         • Middleton and Sutton, 2005.
• Haimes, 2009.
```

3.6.7 Conflict Analysis and Resolution

3.6.7.1 Theoretical Background

Conflicts, or differences of viewpoints, inevitably arise whenever human beings interact with one another.[40] For example, consider disputes that may take place among different units within a company when it must decide upon the development, manufacturing and marketing of a new product. Different entities within the company will have to cooperate as much as possible within some type of group decision-making process, in order to arrive at an agreeable plan of action.

---

40 This section was inspired by the Hipel and Obeidi (2005) and Pinto and Kharbanda (1995-A) papers.

Conflicts and their resolutions have been subjects to many studies and researches. For example, Ruble and Thomas (1976) proposed a two-dimensional model of conflicts and resolutions strategies. They suggest that parties to a conflict (henceforth, decision makers, or DMs) make implicit trade-offs in their desire to seek their own gains versus their willingness to satisfy other DMs relevant to the conflict. On the one hand, DMs seek to maximize their own advantages and on the other hand to cooperate with the other DMs in order to maintain satisfactory relationships. This two-dimensional model has five distinct and recognizable types of conflict-handling styles (Figure 3.83).

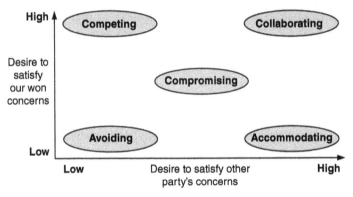

Source: Ruble and Thomas (1976).

Figure 3.83 Model of conflicts and resolutions strategies

These conflict-handling styles are described as follows:

Competing behavior
This type of behavior emanates from a somewhat stubborn and uncooperative attitude of the relevant DM. By and large, such DM, be it an individual person or a group, have little regard for satisfying the other DM's needs, viewing conflicts as hard win-lose propositions. Competing behavior is often utilized by insecure DMs who will use any device in order to get their way.

Accommodating behavior
This type of behavior emanates from a spirit of passivity and amiability, usually in an effort to become a "team player." Accommodators usually approach conflicts seeking to find ways to either defuse the situation or to allow other DMs interests to be fulfilled. This type of behavior can be welcomed in situations where it is more critical to resolve a conflict than to allow it to simmer and poison the operating dynamics among the DMs involved in the conflict. On the other hand, conflicts, within reason, tend to promote different opinions and viewpoints, enriching the overall process.

Avoiding behavior

This type of behavior expresses disregard and, sometimes, actual disdain to either satisfy the DM's own objectives as well as the aims of other DMs to the conflict. In this respect, avoiding behavior is one that is at the same time unassertive and uncooperative. Avoiding behavior tends to be an effective method for sidestepping conflicts the DM does not seek. However, if such a DM is in a position of responsibility, this shirking of obligations could end badly for the individual or the group one leads or the organization as a whole.

Compromising behavior

This type of behavior combines some assertiveness and cooperativeness. It represents a desire by one DM to satisfy some of its objectives but also a willingness to concede to some of the demands made by other DMs. Generally, compromisers consider conflicts as a soft win–lose situation, realizing that obtaining some of their objectives necessitates accepting other DMs' requirements. In other words, compromising behavior stems from the realization that, under most circumstances in life, making concessions is a necessary strategy in order to achieve certain objectives, especially, in conflict situations.

Collaborating behavior

This type of behavior is present where DMs exhibit strong tendencies for both assertiveness and cooperativeness. Collaborators often seek win-win solutions by conducting negotiations among all the DMs, seeking to find agreements that, as much as possible, fully satisfy all DMs. By and large, collaborating among several DMs under conflict situation requires a great deal of flexibility, creativity, and extensive channels of communication amongst all concerned.

The graph model for conflict resolution (GMCR) constitutes a formal (mathematical) and systematic approach to studying and resolving real-world disputes and conflicts. Under most circumstances, it is aligned with the compromising strategy for conflict resolutions mentioned above.

The GMCR approach to conflict resolution is particularly suited to be implemented as a flexible tool to be used within competitive environments. The model is based on ideas from both graph and game theories, extending the realm of multiple objectives–multiple participants (MOMP) decision-making processes. In theory, any finite number of DMs may be defined and each DM may control any finite number of conflict options. In addition, a particular DM can represent an individual person or a small group of people or even a big organization or a nation. Finally, the graph model can describe and distinguish between reversible and irreversible real-life actions.

3.6.7.2 Implementation Procedure

Implementation of a basic GMCR methodology within an actual conflict entails the following steps.

Step 1: Identify the decision makers (DMs). Identify all the individuals or parties that are involved in the conflict.

Step 2: Identify options available to the DMs. Identify all relevant options that are controlled by each DM.

Step 3: Determine relative or specific preference of options. For each DM, determine the relative or specific preference of each option under the said DM.

Step 4: Compute set of possible conflict states. Since an option can be taken or not by the DM that controls it, there exist a set of 2^n possible conflict states in the conflict space (where n represent the total number of options).

Step 5: Define option contradictions. Define option contradictions. These may manifest themselves within the domain of a specific DM or amongst several domains of different DMs.

Step 6: Compute set of feasible conflict states. Use the option contradictions data in order to identify infeasible conflict states and then remove them from the result obtained in step 4, thus creating a set of feasible conflict states.

Step 7: Compute the preference ranking of states per each DM. For each DM, use its relative or specific preference of options in order to compute the preference ranking of the conflict states.

Step 8: Compute the preference ranking of states for all DMs. Use the relative or specific preference of options in order to compute the combined preference ranking of conflict states for all DMs involved in the conflict.

Step 9: Negotiate with all DMs on a reasonable compromise. Based on the combined preference ranking of conflict states, negotiate with all DMs for implementing a reasonable resolution of the conflict by way of mutual compromises.

3.6.7.3   UAV Design Conflict Example

1) <u>Example background</u>

The purpose of this example is to describe a way to resolve individual's or group's conflicts by using the graph model for conflict resolution (GMCR) methodology through compromises made by decision makers (DMs). This example is based on the unmanned air vehicle (UAV) system described in Section 3.5.2.3, Example: Unmanned Air Vehicle Power Plant Configuration.

2) <u>Example analysis</u>

The UAV design conflict example contains two DMs: the single-engine faction and the dual-engine faction. Each one of them has three options. Table 3.31 lists each of the two DMs followed by a list of design options under each DM control. The option priorities of each of the two DMs could be defined either in relative terms (e.g. U-cost > D-cost > Size) or in specific terms. However, in this particular example, the options priorities are defined in specific terms (1, 2, 3).

Table 3.31 Decision makers, options priorities and status quo conflict state

DMs	#	Names	Descriptions	Priorities	Status quo state
			Options		
Single engine	1.	U-cost	Minimal UAV unit cost	3	0
	2.	D-cost	Minimal development cost	2	0
	3.	Size	Minimal air vehicle size	1	0
Dual engine	4.	Payload	Large payload	3	1
	5.	Endurance	Long endurance	2	1
	6.	Ceiling	High flight ceiling	1	1

In general, each DM defines its design strategy by assigning values to its options. A "1" indicates the option is selected by the DM controlling it, whereas a "0" corresponds to a rejected option. However, in this particular example, management selected all the design options favored by the dual-engine faction, while none of the options favored by the Single engine faction were selected. This status quo state represents the existing design strategy selected at the time of conducting the GMCR analysis.

The reader should note that although an individual option represents a binary choice, the DM can break a given option to a set of several mini-options in order to expand his/her control granularity. For example, the single engine faction DM can break the "minimal UAV unit cost" option, defined in Table 3.31, into several, more specific UAV unit costs options (e.g. $1 million, $2 million, etc.).

Since an option can be taken or not, by each DM there exists a set of $2^6 = 64$ possible conflict states in the UAV design conflict example. However, the set of possible conflict states may contain some states, which, in reality, are unlikely to happen. In general, there are two main reasons for such infeasibilities. First, for a given DM, some options may be mutually exclusive and cannot be selected at the same time. Second, if one DM chooses one option then, sometimes, another DM may not be able to select an option, which directly contradicts the first selection. In this UAV design conflict example, all the contradictions belong to the second category and are described in Table 3.32.

Table 3.32 Option contradictions

#	Single-engine options	Dual-engine options
1	Minimal UAV unit cost	Long endurance
2	Minimal air vehicle size	Large payload

All the infeasible conflict states are identified and removed from the set of possible conflict states, thus creating a new set of feasible conflict states. Table 3.33 illustrates the remaining 36 feasible conflict states in the UAV design conflict example. Note that conflict state 57 corresponds to the status quo state.

The ordinal rankings of the conflict states for the single-engine and the dual-engine DMs are individually computed and are depicted in Table 3.34. The data are presented from the most preferred conflict states (at the top) to the least preferred conflict states (at the bottom). Also, a group of conflict states enclosed in parenthesis are equally preferred for the given DM.

For example, the single-engine DM equally prefers the conflict states in the first set: (8, 40), and this set is more preferred to the conflict states in the second set: (4, 12, 36, 44).

Notice that the status quo (conflict state 57) is, by far, not the preferred solution for the single-engine DM. In fact, conflict state 44 exhibits the most optimal systemwide design solution (considering both DMs). Here, the single-engine DM fulfills its minimal UAV unit cost option as well as its minimal development cost option. Likewise, the dual-engine DM fulfills its large payload option as well as its high flight ceiling option. In addition, for both DMs, conflict states (12, 59) followed by conflict states (27, 42) are more preferred than the status quo state.

Normally, conflict evolution is conducted along a compromising procedure, where each DM, in turn, is asked to make a small concession to the other DM. However, in the case of the UAV design conflict example, the status quo (conflict state 57) leaves no concession option for the single-engine DM. Therefore, the two DMs must pursue the collaborating procedure. First, the two DMs agree that the project should adhere to a minimal development cost target. This agreement pushed the UAV design conflict example to conflict state 59. Next, the two DMs agree that the UAV design team should pursue a design leading to a minimal UAV unit cost. However, the upshot of this last decision means that the long endurance, originally pursued by the dual-engine DM is not attainable. Together, these collaborating agreements pushed the UAV design conflict to conflict state 44, the most preferred solution for both DMs. The evolution of the UAV design conflict example is depicted in Figure 3.84.

To sum it up, engineering evaluation of the options commensurate with conflict state 44 favors a UAV design powered by two engines, which realizes the following requirements: (1) minimal UAV unit cost, (2) minimal development cost, (3) carrying large payload, and (4) achieving high flight altitude.

Table 3.33 Feasible conflict states in the UAV design conflict example

DM	#	Options Names	\multicolumn{18}{c}{Conflict states}																	
			1	2	3	4	5	6	7	8	9	10	11	12	17	19	21	23	25	27
Single engine	1	U-cost	0	1	0	1	0	1	0	1	0	1	0	1	0	0	0	0	0	0
	2	D-cost	0	0	1	1	0	0	1	1	0	0	1	1	0	1	0	1	0	1
	3	Size	0	0	0	0	1	1	1	1	0	0	0	0	0	0	1	1	0	0
Dual engine	4	Payload	0	0	0	0	0	0	0	0	1	1	1	1	0	0	0	0	1	1
	5	Endurance	0	0	0	0	0	0	0	0	0	0	0	0	1	1	1	1	1	1
	6	Ceiling	0	0	0	0	0	0	0	0	0	0	0	0	0	0	0	0	0	0

DM	#	Options Names	\multicolumn{18}{c}{Conflict states}																	
			33	34	35	36	37	38	39	40	41	42	43	44	49	51	53	55	57	59
Single engine	1	U-cost	0	1	0	1	0	1	0	1	0	1	0	1	0	0	0	0	0	0
	2	D-cost	0	0	1	1	0	0	1	1	0	0	1	1	0	1	1	1	0	1
	3	Size	0	0	0	0	1	1	1	1	0	0	0	0	0	0	1	1	0	0
Dual engine	4	Payload	0	0	0	0	0	0	0	0	1	1	1	1	0	0	0	0	1	1
	5	Endurance	0	0	0	0	0	0	0	0	0	0	0	0	1	1	1	1	1	1
	6	Ceiling	1	1	1	1	1	1	1	1	1	1	1	1	1	1	1	1	1	1

Table 3.34 Preference ranking of conflict states in the UAV design conflict example

Single-engine DM	Dual-engine DM
(8, 40)	(57, 59)
(4, 12, 36, 44)	(25, 27)
(6, 38)	(41, 42, 43, 44)
(2, 10, 23, 34, 42, 35)	(9, 10, 11, 12, 49, 51, 53, 55)
(3, 11, 19, 27, 43, 51, 59)	(17, 19, 21, 23)
(5, 21, 37, 53)	(33, 34, 35, 36, 37, 38, 39, 40)
(9, 17, 25, 33, 41, 49, 57)	(1, 2, 3, 4, 5. 6, 7, 8)

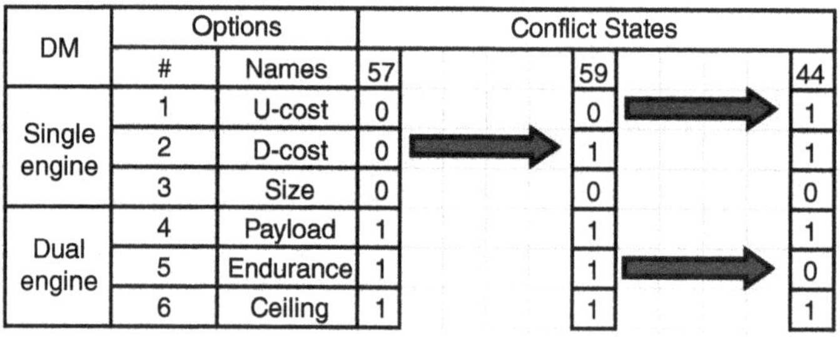

DM	Options		Conflict States			
	#	Names	57	59		44
Single engine	1	U-cost	0	0	→	1
	2	D-cost	0	1		1
	3	Size	0	0		0
Dual engine	4	Payload	1	1		1
	5	Endurance	1	1	→	0
	6	Ceiling	1	1		1

Figure 3.84 Evolution of the UAV design conflict example

3.6.7.4 Further Reading

- Hipel and Obeidi, 2005.
- Fang et al., 1993.
- Fisher et al., 2011.
- Obeidi, 2006.
- Pinto and Kharbanda, 1995-A.
- Pinto and Kharbanda, 1995-B.
- Ruble and Thomas, 1976.
- Shell, 2006.

3.7 Bibliography

Al-Shammari M. and Masri H. eds. (2015). *Multiple Criteria Decision Making in Finance, Insurance and Investment.* Switzerland: Springer.

Altshuller G., (1996). *And Suddenly the Inventor Appeared: TRIZ, the Theory of Inventive Problem Solving,* 2nd ed. Technical Innovation Center, Inc.

ARP5580 (2012). *Recommended Failure Modes and Effects Analysis (FMEA) Practices for Non-Automobile Applications.* SAE International, May.

Arrow, J. K., Sen, K. A. K., and Suzumura, K. eds. (2002). *Handbook of Social Choice and Welfare,* Vol. 1. North Holland.

Ball L. et al. (2014). *TRIZ Power Tools, The Skill that Will Give You the Confidence to Do the Rest,* Skill #1 Resolving Contradictions, November.

Bar-Cohen Y. (2005). *Biomimetics: Biologically Inspired Technologies.* Boca Raton, FL: CRC Press.

Bejan A., and Lorente S. (2011). The constructal law origin of the logistics S curve. *Journal of Applied Physics* 110, 024901.

Bensoussan B.E., and Fleisher C.S. (2015). *Analysis Without Paralysis: 12 Tools to Make Better Strategic Decisions,* 2nd ed. FT Press.

Benyus J.M. (2002). *Biomimicry: Innovation Inspired by Nature.* New York: Harper Perennial.

Berger C., Blauth R., Boger D., et al. (1993). Kano's method for understanding customer-defined quality. *Journal of Japanese Social Quality Control* 2 (4): 3–35.

Berk J. (2013). *Unleashing Engineering Creativity.* CreateSpace Independent Publishing Platform.

Best, J. (2001). *Damned Lies and Statistics: Untangling Numbers from the Media, Politicians, and Activists.* University of California Press.

Brostow A.A. (2015). *Become an Inventor: Idea-Generating and Problem-Solving Techniques with Element of TRIZ, SIT, SCAMPER, and More.* CreateSpace Independent Publishing Platform.

Brue G., and Launsby R. (2003). *Design for Six Sigma.* New York: McGraw Hill Professional.

Buzan T. (2006). *Ultimate Book of Mind Maps.* London: Thorsons Publishers.

Carlson C. (2012). *Effective FMEAs: Achieving Safe, Reliable, and Economical Products and Processes using Failure Mode and Effects Analysi.* Hoboken, NJ: John Wiley & Sons.

Christensen C.M. (1992). Exploring the limits of the technology S-curve. *Production and Operations Management* 1 (4): 334–366.

Cooke M.R. (1991). *Experts in Uncertainty: Opinion and Subjective Probability in Science.* Oxford University Press.

Damelio R. (2011). *The Basics of Process Mapping,* 2nd ed. Boca Raton, FL: Productivity Press.

de Bono, E. (2015). *Lateral Thinking: Creativity Step by Step.* Harper Colophon; Reissue edition.

de Bono, E. (1993). *Serious Creativity: Using the Power of Lateral Thinking to Create New Ideas.* New York: Harperbusiness.

de Bono, E. (1999). *Six Thinking Hats.* Back Bay Books.

de Bono, E. (2017). *Parallel Thinking.* Random House UK.

de Bono, E. Thinking Systems homepage: http://www.debonothinking systems.com.

Dyadem Press. (2013). *Guidelines for Failure Mode and Effects Analysis (FMEA), for Automotive, Aerospace, and General Manufacturing Industries.* Boca Raton, FL: CRC Press, 2003.

Eberle B. (2008). *Scamper: Creative Games and Activities for Imagination Development.* Prufrock Press, Inc.

Eisner H. (2005). *Managing Complex Systems: Thinking Outside the Box.* Hoboken, NJ: Wiley-Interscience.

Engel, A., and Reich, Y. (2015). Advancing architecture options theory: Six industrial case studies. *Systems Engineering* 18: 396–414.

Engel, A. (2010). *Verification, Validation and Testing of Engineered Systems (Wiley Series in Systems Engineering and Management).* Hoboken, NJ: John Wiley & Sons.

Eppinger, S.D., and Browning T.R. (2012). *Design Structure Matrix Methods and Applications.* Cambridge, MA: MIT Press.

Fang, L., Hipel, K.W., and Kilgour, D.M. (1993). *Interactive Decision Making: The Graph Model for Conflict Resolution.* New York: Wiley-Interscience.

Fey, V., and Rivin E. (2005). *Innovation on Demand: New Product Development Using TRIZ.* Cambridge University Press.

Fine, L.G. (2009). *The SWOT Analysis: Using Your Strength to Overcome Weaknesses, Using Opportunities to Overcome Threats.* CreateSpace Independent Publishing Platform.

Fisher, R., Ury, W.L., and Patton, B. (2011). *Getting to Yes: Negotiating Agreement Without Giving In.* Penguin Books; Updated, Revised edition.

Flood, R.L., and Jackson, M.C. (1991). *Creative Problem Solving: Total Systems Intervention,* New York: John Wiley & Sons.

Gadd, K. (2011). *TRIZ for Engineers: Enabling Inventive Problem Solving.* Hoboken, NJ: John Wiley & Sons.

Gallagher, S. (2008). *Brainstorming: Views and Interviews on the Mind.* New York: Academic.

Garson, G.D. (2013). *The Delphi Method in Quantitative Research.* Statistical Associates Publishers.

Haimes, Y.Y., (2009). *Risk Modeling, Assessment, and Management,* 3rd ed. Hoboken, NJ: Wiley Blackwell.

Hipel, K.W., and Obeidi, A. (2005). Trade versus the environment: strategic settlement from a systems engineering perspective, *Systems Engineering* 8 (3).

Hirokawa, Y. R., and Poole, S. M. eds. (1996). *Communication and Group Decision Making,* 2nd ed. Thousand Oaks, CA: Sage Publications.

Hunecke, K. (2010). *Jet Engines: Fundamentals of Theory, Design and Operation.* United Kingdom: The Crowood Press.

Hyndman, R.J., and Athanasopoulos G. (2013). *Forecasting: Principles and Practice.* OTexts, https://www.otexts.org/fpp.

Ishikawa, K. (1990). *Introduction to Quality Control.* Boca Raton, FL: Productivity Press.

Janis, I.L. (1972). *Victims of Groupthink: A Psychological Study of Foreign Policy Decisions and Fiascoes.* Boston: Houghton Mifflin Company.

Kaliszewski, I., Miroforidis, J., and Podkopaev, D. (2016). *Multiple Criteria Decision Making by Multiobjective Optimization: A Toolbox.* New York: Springer.

Kaplan, S., Visnepolshi, S., Zlotin, B., and Zusman, A. (1999). *Tools for Failure & Risk Analysis: Anticipatory Failure Determination (AFD) & the Theory of Scenario Structuring.* Ideation International.

Karl, T.R., Melillo, J.M., and Peterson, T.C. ed. (2009). *U.S. Global Change Research Program, Global climate change impacts in the United States.* Cambridge University Press.

Kassa, A.O. (2015). *Value Analysis and Engineering Reengineered: The Blueprint for Achieving Operational Excellence and Developing Problem Solvers and Innovators.* Boca Raton, FL: Productivity Press.

Keeney, L.R., and von Winterfeld, D. (1991). Eliciting probabilities from experts in complex technical problems. *IEEE Transactions on Engineering Management* 38 (3), 191–201.

Kim, K., Seo, E., Chang, S.K., Park, T.J., and Lee, S.J. (2016, February). Novel water filtration of saline water in the outermost layer of mangrove roots. *Scientific Reports* 6, Article number: 20426. Accessed Aug. 21, 2017.

Koch, R. (1999). *The 80/20 Principle: The Secret to Achieving More with Less.* New York: Crown Business; Reprint edition.

Lakhtakia, A., and Martín-Palma, R.J. eds. (2013). *Engineered Biomimicry.* Amsterdam: Elsevier.

Langreth, R. (2008, May). Imclone's gene test battle. Forbes.com.

Long, K.F. (2012). *Deep Space Propulsion: A Roadmap to Interstellar Flight.* New York: Springer.

Loveridge, D. (2002, June). *Experts and Foresight: Review and Experience*, Paper 02–09, PRES. Manchester, UK: The University of Manchester.

Loveridge, D. (2002). *Experts and Foresight: Review and Experience*, Paper 02-09, PRES. Manchester, UK: University of Manchester.

Lu, J., Zhang, G., and Ruan, D. (2007). *Multi Objective Group Decision Making: Methods, Software and Applications with Fuzzy Set Techniques.* Imperial College Press.

Mason, R.O., and Mitroff, I.I. (1981). *Challenging Strategic Planning Assumptions.* New York: John Wiley & Sons.

Maurer, M., and Lindemann, U. (2007, October). Structural awareness in complex product design – the multiple-domain matrix, 9th international Design Structure Matrix conference, DSM'07, Munich, Germany.

McDonald, D., Bammer G., and Deane P. (2011). *Research Integration Using Dialogue Methods.* ANU E Press.

Merton, R.K. (1990). *Focused Interview*, 2nd ed. New York: Free Press.

Michalko, M. (2006). *Thinkertoys: A Handbook of Creative-Thinking Techniques*, 2nd ed. New York: Ten Speed Press.

Middleton, P., and Sutton J. (2005). *Lean Software Strategies: Proven Techniques for Managers and Developers*. Productivity.

Midgley, G. (2000). *Systemic Intervention: Philosophy, Methodology, and Practice*. New York: Springer.

Miles, L.D. (2015). *Techniques of Value Analysis and Engineering*. Lawrence D. Miles Value Foundation.

MIL-STD-1629A. (1980, November). *Military Standard Procedures for Performing a Failure Mode, Effects and Criticality Analysis*. Washington, DC: U.S. Department of Defense.

Mkpojiogu, E.O.C., and Hashim, N.L. (2016, February). Understanding the relationship between Kano model's customer satisfaction scores and self-stated requirements importance *SpringerPlus* 5: 197.

Modarres, M., Kaminskiy, M., and Krivtsov, V. (1999). *Reliability Engineering and Risk Analysis: A Practical Guide*. Boca Raton, FL: CRC Press.

Moorman, J. (2012, October). Leveraging the Kano Model for optimal results, *UX Magazine* 882, https://uxmag.com/articles/leveraging-the-kano-model-for-optimal-results. Accessed August 2017.

Novak, J.D., and Musonda D. (1991). A twelve-year longitudinal study of science concept learning. *American Educational Research Journal* 28 (1): 117–153.

Novak, J.D. (2009). *Learning, Creating, and Using Knowledge: Concept Maps as Facilitative Tools in Schools and Corporations*, 2nd ed. Routledge.

Obeidi, A. (2006). Emotion, Perception and Strategy in Conflict Analysis and Resolution. Ph.D. thesis, University of Waterloo, Waterloo, Ontario, Canada.

Osborn, A.F., and Bristol L.H. (1979). *Applied Imagination: Principles and Procedures of Creative Thinking*, 3rd ed. Charles Scribner's Sons.

Park, Y.T., Jang H., and Song H. (2012). Determining the importance values of quality attributes using asc., *Journal of Korean Society for Quality Management* 40 (4): 589–598.

Parsch A. (2016). Northrop Grumman (TRW/IAI) BQM-155/RQ-5/MQ-5 Hunter, http://www.designation-systems.net/dusrm/m-155.html. Accessed Nov. 28, 2016.

Passino K.M. (2004). *Biomimicry for Optimization, Control, and Automation*. London: Springer; 2005 edition.

Pimmler U.T., and Eppinger, S.D. (1994). Integration analysis of product decompositions. Working Paper #3690-94-MS. Cambridge, MA: MIT Sloan School of Management.

Pinto, J.K., and Kharbanda, O.P. (1995-A). Project management and conflict resolution. *Project Management Journal* 26 (4): 45–54.

Pinto, J.K., and Kharbanda, O.P. (1995-B). *Successful Project Managers: Leading Your Team to Success.* Van Nostrand Reinhold.

Proctor, T. (2013). *Creative Problem Solving for Managers: Developing Skills for Decision Making and Innovation,* 4th ed. New York: Routledge.

Quinlan, J.R. (1987). Simplifying decision trees. *International Journal of Man-Machine Studies* 27 (3): 221.

Richardson, D. (2016). *Transparent: How to see Through the Powerful Assumptions That Control You.* Clovercroft Publishing.

Robert, P.C., and Casella G. (2005). *Monte Carlo Statistical Methods,* 2nd ed., New York: Springer.

Rodrigues, L.J. (1997). *Unmanned Aerial Vehicles DoD's Acquisition Efforts.* GAO/T-NSIAD-97-138.

Ruble, T.L., and Thomas, K.W. (1976). Support for a two-dimensional model of conflict behavior. *Organizational Behavior and Human Performance* 16, 221–237.

Rumane, A.R. (2010). *Quality Management in Construction Projects.* Boca Raton, FL: CRC Press.

SAE J1739 (2002). *Potential Failure Mode and Effects Analysis in Design (Design FMEA) and Potential Failure Mode and Effects Analysis in Manufacturing and Assembly Processes (Process FMEA) and Effects Analysis for Machinery (Machinery FMEA).* Society for Automotive Engineers.

Salamatov, Y. (1999). *TRIZ: The Right Solution at the Right Time.* The Netherlands: Insytec.

San, Y.T. (2014). *TRIZ Systematic Innovation in Business and Management.* FirstFruits Sdn Bhd.

Savransky, S.D. (2000). *Engineering of Creativity: Introduction to TRIZ Methodology of Inventive Problem Solving.* Boca Raton, FL: CRC Press.

Shell, G.R. (2006). *Bargaining for Advantage: Negotiation Strategies for Reasonable People,* 2nd ed. New York: Penguin Books.

Skinner, D.C. (2009). *Introduction to Decision Analysis,* 3rd ed. Probabilistic Publishing.

Sloane, P. (2006). *The Leader's Guide to Lateral Thinking Skills: Unlocking the Creativity and Innovation in You and Your Team Paperback,* 2nd ed. Kogan Page.

Stacey, R.D. (2012). *Tools and Techniques of Leadership and Management: Meeting the Challenge of Complexity.* New York: Routledge.

Stamatis, H.D. (2003). *Failure Mode and Effect Analysis: FMEA from Theory to Execution,* 2nd rev. ed. Milwaukee: Quality Press.

Sternberg, R.J. (1998). *Handbook of Creativity.* Cambridge University Press.

Sternberg, R.J. (1999). A propulsion model of types of creative contribution. *Review of General Psychology* 3: 83–100.

Torrence, R. S. (1991). *How to Run Scientific and Technical Meetings.* New York: Van Nostrand Reinhold.

Vose, D., (2008). *Risk Analysis: A Quantitative Guide*, 3rd ed. Hoboken, NJ: John Wiley & Sons.

Vroom, H.V., and Yetton, W. P. (1976). *Leadership and Decision Making.* Pittsburgh, PA: University of Pittsburgh Press.

Yarlagadda, C. (2011). Pacemaker Malfunction, Clinical Presentation, Medscape, http://emedicine.medscape.com/article/156583. Accessed: August 2017.

Zwicky, F. (1969). *Discovery, Invention, Research Through the Morphological Approach.* Toronto: The Macmillan Company.

Part IV

Promoting Innovative Culture

"The chief enemy of creativity is good sense."
Pablo Picasso (1881–1973)

4.1 Introduction to Part IV

As mentioned earlier, innovation is a process of transforming creative ideas or inventions into new products, services, business processes, organizational processes, or marketing processes that generate value for relevant stakeholders. Part IV: Promoting Innovative Culture is composed of 10 chapters, as follows (Figure 4.1).

Chapter 4.1 Introduction to Part IV. This chapter describes the contents and structure of Part IV.
Chapter 4.2 Systems evolution. This chapter describes the intrinsic way systems are evolving. The chapter covers the topics of modeling systems evolution by way of an S-curve and the laws of systems evolution from TRIZs[1] perspectives.
Chapter 4.3 Modeling the Innovation process. This chapter describes classes and types of innovative processes as well as various technological innovative processes. In addition, this chapter characterizes the funding of the innovation process.
Chapter 4.4 Measuring creativity and innovation. This chapter describes how to measure creativity and innovation. The chapter points out that, firstly, the organization's management should define its innovation objectives. Then, it describes the actual measuring of the innovation process. In addition, the chapter describes a framework by which an organization can evaluate its innovative status using the innovation capability maturity model (ICMM).

---

1 TRIZ stands for the Russian acronym "Theory of Inventive Problem Solving."

Practical Creativity and Innovation in Systems Engineering, First Edition. Avner Engel.
© 2018 John Wiley & Sons, Inc. Published 2018 by John Wiley & Sons, Inc.

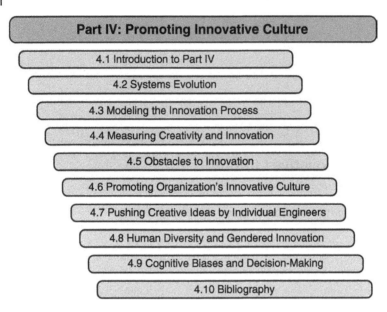

Figure 4.1 Structure and contents of Part IV

Chapter 4.5 Obstacles to innovation. This chapter describes typical organization obstacles to innovation. These include factors like: human habits, cost consideration, institutional reactivity, lake of knowledge and markets apprehension. Finally, it lists innovation obstacles and their relevance to each one of four classes of innovations.

Chapter 4.6 Promoting organization's innovative culture. This chapter describes various means to promote an organization's innovative culture. These include issues related to leadership, organization, people, assets, culture, values, processes, and tools. Finally, it provides a set of practical steps that may be adopted in order to advance the innovation process within organizations.

Chapter 4.7 Pushing creative ideas by individual engineers. This chapter attempts to provide insight and advice to many frustrated engineers who come up with creative ideas but are unable to advance them through multiple layers of resisting bureaucracy, not truly amenable to innovation. The chapter analyzes why large organizations seldom innovate and provides helpful innovation advice to creative engineers seeking to overcoming these obstacles.

Chapter 4.8 Human diversity and gendered innovation. This chapter describes the linkage between human diversity and the innovation process. It discusses the recent shift in gender paradigm as well as gender disparity in terms of innovation implications. Finally, it provides a set of six specific strategies designed to advance gendered innovation.

Chapter 4.9 Cognitive biases and decision-making. This chapter describes various types of cognitive biases and how they relate to strategic decisions made by engineers and manager affecting, among others, innovation processes.

Chapter 4.10 Bibliography. This chapter provides bibliography related to Part IV topics.

4.2 Systems Evolution

4.2.1 Modeling Systems Evolution – S-Curve

Utilizing emerging technology in new systems is often delayed by special interest groups, which have vested interest in old systems. Nevertheless, system and especially technological systems, evolve continuously due to changes in technological developments, stakeholders' desires, economic conditions, political circumstances, psychological factors, and so forth. For example, the automotive-electronics application domain is steadily growing and, so far, with no sign of resource decline (Figure 4.2). In fact, current, leading-edge luxury vehicles use as many as 100 electronic control units (ECUs), connected by some five or more dedicated buses.

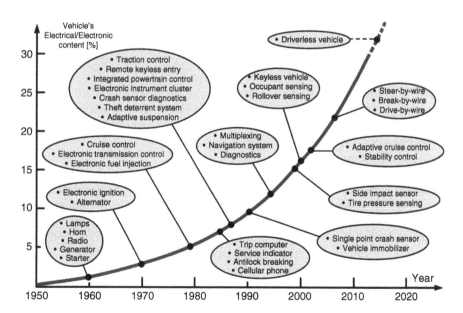

Figure 4.2 Vehicle electrical and electronic content[2]

---

2 Inspired by Hellestrand (2005) and Chong (2010).

The evolution of technological systems may be modeled by an S-curve[3] identifying typical stages of development, reflecting changes of the system's benefit-to-cost ratio over time. The length and slope of each segment on the S-curve depends on technical as well as on economics, human psychology, and other factors (Figure 4.3).

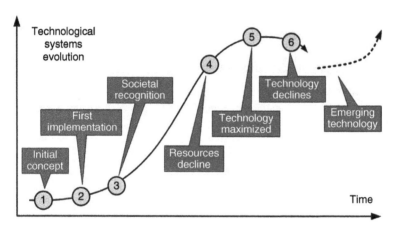

Figure 4.3 Technological systems evolution (S-curve)

Technological systems evolution along an S-curve typically follows these stages:

Stage 1: Initial concept. In this stage, basic scientific concepts are observed and reported, leading to the emergence of a new technological system. The curve at this point, exhibits an incubation period, when substantial creative investment yields slow tangible results.

Stage 2: First implementation. In this stage, the first commercial implementation and use of the new technological system is achieved / demonstrated.

Stage 3: Societal recognition. In this stage, the new technological system reaches an inflection point on the S-curve, leading to a period of rapid growth, becoming recognized and used by society at large.

Stage 4: Resources decline. In this stage, the system continues to develop in terms of technical capacities while its cost goes down. However, the resources supporting the system start to decline.

Stage 5: Technology maximized. In this stage, the system reaches its maximum potential in terms of technical capabilities and commercial possibilities.

---

3 S-curve is a sigmoid function depicting a typical shape of engineered systems technology life cycle (Stewart, 1981).

Stage 6: Technology declines. In this stage, the growth of the system starts to stagnate and often declines as its supporting technology reaches its limits. **Emerging technology.** New and improved technology emerges and a new next-generation technological system (described by another S-curve) appears to gradually replace the existing one.

4.2.2 Laws of Systems Evolution

As mentioned before, TRIZ was developed by Genrikh Altshuller and his colleagues during the second half of the twentieth century in the former Soviet Union. TRIZ is a rich methodology consisting of a wide variety of tools and techniques, some of which are discussed in various parts of this book.

The laws of systems evolution are useful TRIZ tools because they provide strong indication as to how technological systems may evolve in the future. Under TRIZ methodology, the evolution of technological systems follows a set of *evolution laws*. These laws define a general direction of technological system's evolution. More specifically, technological systems evolve along repeatable interactions among systems components, as well as between systems and their environments.

This process continues until a system exhausts its available technological or commercial resources. This, eventually, leads to the replacement of a given system by a more advanced system that performs its function in a superior way. The reader should note however, that unlike the laws of nature, TRIZ laws are elastic and applicable in a rather general way. In addition, the reader should note that the set of laws has been evolving over the years and different TRIZ researchers identify slightly different laws under slightly different names. The laws of technological systems evolution are described[4] in Figure 4.4 and in the following text.

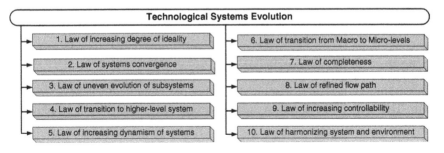

Figure 4.4 Laws of technological systems evolution

---

4 This book follows a set of rules and naming conventions scheme, integrating elements of several TRIZ researchers.

4.2.2.1   Law of Increasing Degree of Ideality

Under TRIZ, an ideal system is described as one that delivers all functions at the required time, needing no physical space, consuming no energy, material, or information, all the above at zero cost. So, a true ideal system must deliver an infinite number of positive effects without incurring any negative effects at the required time. A model describing the degree of ideality (in slightly simplified form) was originally proposed by Boris Goldovsky in 1974.

$$Ideality = \frac{\sum_{i=1}^{\infty} Useful_i}{\sum_{j=1}^{\infty} Harmful_j} \rightarrow \infty$$

Where:

$Useful_i$ - Positive effect (i);
$Harmful_j$ - Negative effect (j);
i - Useful variables (i);
j - Harmful variables (j).

Of course, such a system could not be realized, but the concept is important because engineers always strive to increase the degree of a system's ideality. Figure 4.5 depicts the concept of system's ideality and evolutionary potential.

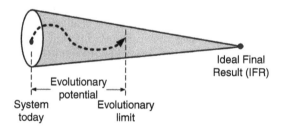

Figure 4.5 Visualizing system's ideality and evolutionary potential

As can be seen, an ideal system is defined as the ideal final result (IFR), which can never be achieved. In fact, under real-life condition one can only strive to improve the system up to a point, defined as the evolutionary limit of the system. This may be achieved by implementing one or more of the following strategies:

1) Increase the number or magnitude of the useful functions.
2) Decrease the number or magnitude of the harmful functions.
3) A combination of the above two strategies:
    3.1) Decrease slightly the number or magnitude of the useful functions but decrease significantly the number or magnitude of the harmful functions.

3.2) Increase slightly the number or magnitude of the harmful functions but increase significantly the number or magnitude of the useful functions.

The law of increasing degree of ideality states that all technological systems evolve in the direction of increased degree of ideality. That is, the system tends to exhibit improvement in the ratio between its overall useful effects and its overall harmful effects. This, in practical terms, means that during system's evolution it may perform additional functions, use fewer resources, improve its yield at a reduced overall cost and so forth.

4.2.2.2 Law of Systems Convergence

The law of system convergence represents a pattern of systems' evolution in which the number of elements from which the system is built tends to decrease over time, without deterioration in the performance of the system itself. This reduction is customarily accompanied by a cost reduction and performance improvement leading, by definition, to enhancement in systems' ideality. Sometimes, the functional capability of the system is maintained by redistribution of useful functions to the remaining elements of the system (this process is called *system merging*). In other cases, some inner system capability may disappear altogether without affecting the external capabilities of the system.

Many researches show that design of various systems tends to evolve and improve again and again over many years due to market pressure combined with technology advances and engineering ingenuity. For example, Ehrlenspiel et al. (2007) describe the evolution of a transmission system with hydraulic torque converters (Figure 4.6). Over the years, this system was redesigned

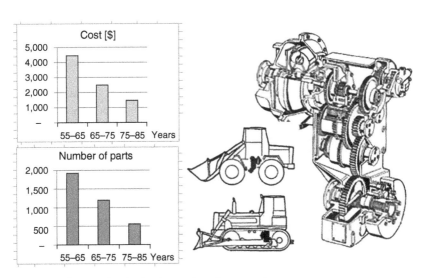

Figure 4.6 Continual reduction in number of parts and costs

numerous times, reducing the number of parts by about 70%. In the process, the cost of the system was reduced by 70% (inflation considered).

4.2.2.3 Law of Uneven Evolution of Subsystems

According to TRIZ, different components of technological systems evolve at different pace. Some progress rapidly while others stay stable for a long time. Accordingly, the law of nonuniform evolution of technological systems states that the rate of evolution of various components within a system is not uniform. And as a general rule, more complex systems tend to exhibit more nonuniform evolution of their components. A system undergoing a nonuniform evolution exhibits one or more contradictions among its components. This, in turn, limits the ideality of the system as a whole.

This phenomenon stems from changing design priorities in the course of a system's evolution. In the early stages of the system's development, the design priorities concentrate on the system's performance and reliability. Later, design priorities shift to optimize performance characteristics. Finally, design priorities expand into issues of environmental, ecological and other considerations. At the same time, design priorities affect the evolution pace and direction of individual components and technological systems. More specifically, as technological system evolve, different parts hinder further improvement of the overall system's performance. Recognizing such barriers, scientists and engineers shift their priorities, and concentrate on improving these technologically legging parts (Karasik, 2011).

A good example of this phenomenon is the nonuniform evolution of the mainframe computers from the early 1950s to the early 1970s. The central processing unit (CPU) has evolved rapidly from electromagnetic relays technology to vacuum tubes technology to transistors technology and into small, integrated circuits (IC). These early ICs included small-scale integration (SSI) containing about 10 electronic gates on a chip to medium-scale integration (MSI) containing up to 100 electronic gates on a chip.

At the same time, computer memory relied on magnetic core technology, which was slow, energy gobbling, and extremely expensive. Magnetic core memory stored data on arrays of small rings of magnetized ferrite material. Each ring stored one bit of data that may be switched from "0" to "1" by changing the polarity of its magnetic field. For example, in the early 1950s, the most advanced and reliable computer memory was a 32 × 32 cm core memory plane containing 1,024 bits (128 bytes) of data (Figure 4.7).

In 1954, the International Business Machine (IBM) corporation packaged 144 of these planes and created the IBM 737, Magnetic core storage unit. The 737 was an auxiliary element attached to IBM main-frame computers, offering 4,096 36-bit words (Figure 4.8). The mainframe computers transitioned into an all integrated circuit (IC)-based memory technology almost 20 years later, in 1972, with the introduction of the IBM 370/3145. The 3145 had a memory capacity of up to 512 K bytes, running up to five times faster than earlier models.

Figure 4.7 1,024 bit core memory

Figure 4.8 IBM 737 Magnetic core storage

The technology associated with the CPU portion of the early mainframe computers was evolving rapidly because the design priorities stressed the computational capabilities and reliability of the CPU subsystem. In addition, software size was relatively small and could be swapped between disk memory and core memory, providing sufficient computational bandwidth. Size and related issues (e.g. electrical wiring, electricity consumption, cooling, work environment etc.) were resolved by establishing massive computer centers. Systems' cost issues were partially resolved by leasing the machines rather than buying them outright.

Over the years, needs were changing. Software packages grow exponentially in the financial, manufacturing, and government sectors. Consequently, the need for overall systems speed increased by leaps and bounds. Finally, cost factor became significantly more important. Accordingly, design priorities shifted, leading eventually to the evolution of faster, smaller, and more energy-efficient as well as cheaper IC-based memory. This, in turn, closed the components' nonuniformity gap within the computer industry.

4.2.2.4 Law of Transition to Higher-Level System
When technological systems evolve, they reach a point where they exhaust their local resources and their further evolution likelihood diminishes. At such times, these technological systems tend to transition into higher-level systems. This is often manifested in one of two flavors.

One flavor is that existing systems expand from monosystems to bisystems and then to a polisystems and finally to adjustable systems (Figure 4.9). These expanded systems have either identical components (by duplicating component of earlier stages) or similar components that are different in some aspects or sometimes, additional components.

Figure 4.9 Transition to higher-level systems

For example, Figure 4.10 depicts a wrench transition to higher-level systems.

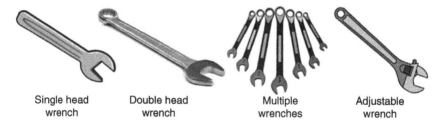

| Single head wrench | Double head wrench | Multiple wrenches | Adjustable wrench |

Figure 4.10 Example: Wrench transition to higher-level systems

Another flavor of this type of transition to higher-level systems (sometimes called super-system) arises when systems integrate with other systems and continue to evolve within that environment. For example, a self-resourced inertial navigation system (INS) in an aircraft uses gyroscopes in order to determine the position, orientation and location of the aircraft. Later, the system is integrated with a terrestrial radio navigation system (e.g. LORAN – LOng RAnge Navigation) in order to improve the accuracy of the navigation system. Then, at a later stage, this combined system is further integrated with a satellite-based global positioning system (GPS), again, in order to further improve the navigation accuracy of the aircraft. This includes position and orientation in space, speed, and acceleration in all three axes as well as 3D location in space (Figure 4.11).

Inertial Navigation System

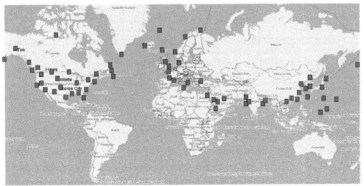

LORAN-C navigation system

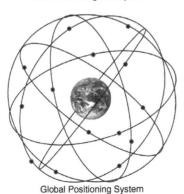

Global Positioning System

Figure 4.11 System integrates with super-system

4.2.2.5   Law of Increasing Dynamism of Systems

Technological systems are created in order to perform specific functions in response to stated needs. Their design reflects prevailing environments, available technologies as well as original stakeholders' expectations. Over time however, environments, technologies as well as stakeholders' expectations evolve and so do technological systems. This act of setting free the dynamic powers of systems is called Dynamization. Generally speaking, dynamization (flexibility) leads systems to become more adaptable to the changing environments and needs. This is often manifested in expended functional capabilities. Accordingly, the law of increasing dynamism, states that technological systems evolve in the direction of more flexible systems and multi functionality. Several directions of increasing dynamism have been identified:

Design dynamization. Design dynamization deals with the evolution of systems from the point of view of their design. The design dynamization is composed of two types. The first type is substance (mostly hardware) dynamization, which is depicted in Figure 4.12.

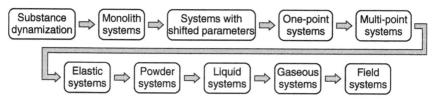

Figure 4.12 Substance dynamization

Field dynamization. In TRIZ, the concept of "field" covers a verity of phenomena: electromagnetic field (radiowaves, microwaves, infrared light, visible light, ultraviolet-rays, X-rays, gamma rays, etc.), electrostatic field, magnetic field, force (mechanical, gravity, centrifugal, inertial, friction, adhesion, coriolis, nuclear, etc.), and many more. The evolutionary process of field dynamization is depicted in Figure 4.13.

Figure 4.13 Field dynamization

Composition dynamization. Composition dynamization deals with the evolution of systems from the point of view of their composition. This evolutionary process is depicted in Figure 4.14.

Figure 4.14 Composition dynamization

Internal structure dynamization. Internal structure dynamization deals with the evolution of systems from the point of view of their internal structure. This evolutionary process is depicted in Figure 4.15.

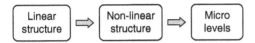

Figure 4.15 Internal structure dynamization

Function dynamization. Function dynamization deals with the evolution of systems from the point of view of their functionalities. By and large, systems' evolutionary process is characterized by the growing number of functions these systems can perform. For example, Figure 4.16 compares the key capabilities of an older cell phone (Nokia 8210 circa 1999) and an Apple iPhone 6.

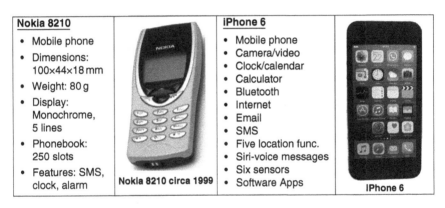

Figure 4.16 Older cellphone versus iPhone 6

4.2.2.6 Law of Transition from Macro- to Micro-levels

Technological systems are composed of substances that could be viewed hierarchically as incorporating the following physical structures (Figure 4.17).

Figure 4.17 Physical structures of technological systems

According to this view, a physical structure in a given level within the hierarchy constitutes a micro-level structure for the structures that occupy higher levels in the hierarchy. The law of transition to Micro-level asserts that technological systems evolve toward an increasing use of micro-level structures. More specifically, the evolution of a given system begins at the macro-level and moves toward the micro-level. By and large, such evolution entails a transition from a technology based on a given physical principle, into higher technology based on a different physical principle. Also, such transitions stems from the progress of science and the desire to exploit the advantages ingrained in properties of dispersed materials and particles physics. Fey and Rivin (2005) provide a good illustration of the law of transition from macro- to micro-level by considering metal machining tools as a system that evolves over time.

Crystal lattice example: Milling

Traditional machining (milling) technologies operate at the crystal lattice level. This is a process of using rotary cutters to remove material from a workpiece. It covers a wide variety of different operations involving a wide range of machine tools (Figure 4.18).

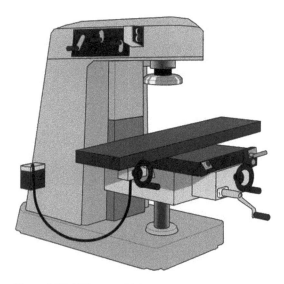

Figure 4.18 Milling machining

Molecules example: Electrochemical machining

Electrochemical machining technologies operate at the molecular level. It is a method of removing material from a workpiece by an electrochemical process. This process is mostly used to form complex shapes made of extremely hard materials (e.g. titanium nickel, cobalt, rare-earth alloys, etc.). However the method is limited to products made of electrically conductive materials (Figure 4.19).

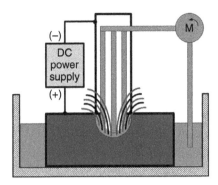

Figure 4.19 Electrochemical machining

Atoms and ions example: Plasma arc cutting

Plasma arc machining technologies operate at the atom and ions level. It is a process whereby high-energy electrical discharges cuts through materials by means of an accelerated jet of high temperature plasma. The hot plasma blasts, melts and removes a thin slice from the workpiece, which must be made of electrically conductive material such as steel, aluminum, brass, or copper (Figure 4.20).

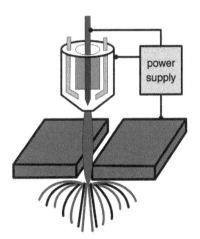

Figure 4.20 Plasma arc cutting

Elementary particles example: Laser welding

Laser machining technologies operate at the elementary particles level. Laser welding is a fabrication process that joins materials by causing fusion at the edges of two or more workpieces (e.g. steel, aluminum, titanium, etc.). The laser generates a concentrated, high power, light beam, affecting small heat zones, which enables narrow and deep welds at high welding rates (Figure 4.21).

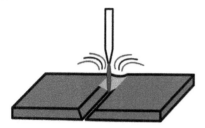

Figure 4.21 Laser welding

4.2.2.7 Law of Completeness

Technological systems refer to all types of manmade artifacts, including technical products as well as technical methods, techniques, and organizations. Under TRIZ methodology, the law of completeness states that any viable autonomous technological system should consist of four principal components[5] (Figure 4.22):

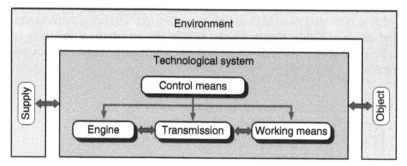

Figure 4.22 Basic structure of autonomous technological system

Working means directly perform the primary function of the system (i.e. affects the object).

An engine provides the necessary energy needed to produce the expected function.

A transmission channel[6] transmits the energy produced by the engine into the working means.

Control means govern one or more of the previous components. As a result, the system as a whole becomes adaptable and flexible for the user.

The environment of the system consists of:

A supply provides specific inputs into the system.

An object is acted upon by the system.

---

5 Some TRIZ scholars claim that the law of completeness lacks generality because there are viable systems that violate the law as stated. For example, a house is a viable autonomous technological system that includes none of the above principle components (Karasik, 2008).

6 Some TRIZ scholars suggest that, the concept of transmission channel could be expanded to consider additional types of flows like Material and Information (Cascini et al., 2009).

At an early stage of technological systems evolution, some functionalities are performed by humans. However, as these systems evolve, some control means are added and human participation is gradually phased out. Thereafter, functions that used to be performed by human are delegated to technological components. As they evolve, systems reach a point where they can govern and control themselves.

That is, systems begin to make their own decisions and humans are relegated to assume general supervision positions. Consequently, the trend to replace human by systems' components will, most likely, lead to increase in systems' efficiency and robustness.

For example, an electric kettle is a system, used for boiling water. The kettle's "working means" include a base for electrical interface, water container, lid, spout, and handle. The kettle's "engine" is an electric heating element. The kettle's "transmission channel" includes a collection of electric cables. Finally, the kettle's "control means" includes a manual On-Off switch, and a thermostat that is triggered by the steam rising from the boiling water, turning off the electric current. This thermostat, by the way, is a good example of system's evolution whereby a human action has been delegated to a technological component (Figure 4.23).

Figure 4.23 An electric kettle

4.2.2.8 Law of Refined Flow Path

The law of refined flow path asserts that technological systems evolve in the direction of optimizing the flow path between sources and their working means (i.e. flow includes energy, matter and/or information). For example, according to the law of completeness, the energy from an engine passes via an intermediate transmission channel to the working means, which creates the required system's output.

Now, the law of refined flow path states that during the evolution of systems, the positive effect of helpful energy, matter, or information becomes more efficient, whereas the negative effect of harmful energy, matter or information is significantly reduced.

Example 1: Increasing the positive effect of helpful energy flow
In many systems energy flow path is long and inefficient. According to the law of refined flow path, such flows are improved as systems evolve. A diesel generator system is composed of a diesel engine coupled mechanically with an electric generator. In this system the engine burns fuel (i.e. chemical energy) to create heat. The heat energy is converted into mechanical energy that rotates a

shaft coupling the engine with an electric generator. Finally, the rotation of the rotor inside the stator of the generator creates a magnetic field which, in turn, produces electricity (i.e. electrical energy).

However, a fuel cell is an electrochemical device that produces electricity directly. This is done by combining hydrogen and oxygen to produces electricity. So here, energy flows directly from chemical energy into electrical energy with no moving parts and minimal creation of heat. Energywise, fuel cells are approximately twice as efficient as an internal combustion engine and, in addition, the conversion process is environmentally benign. Only heat and water are emitted as byproducts of this process (Figure 4.24).

(a)   (b)

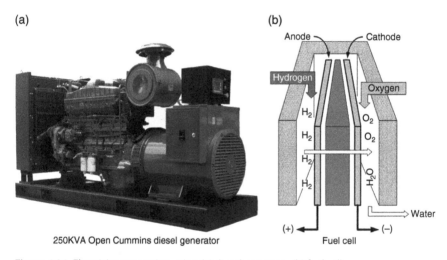

250KVA Open Cummins diesel generator   Fuel cell

Figure 4.24 Electricity generation using (a) diesel generator (b) fuel cell

Example 2: Reducing the negative effect of harmful matter flow
In many systems, harmful flow path are not desired. According to the law of refined flow path, such flows are reduced as systems evolve. For example, nowadays virtually all modern automobiles use catalytic converters. This device is used to reduce pollution by converting toxic exhaust emissions emanating from an internal combustion engine into nontoxic substances.

The catalyst[7] is made from platinum or a similar metal such as palladium or rhodium. The input of a catalytic converter is connected to the engine, receiving hot, polluted fumes. As the gases pass over the catalyst, chemical reactions take place on the surface of the catalyst, breaking apart the pollutant gases and converting them into relatively safe gases to be discharged through the automobile tailpipe (Figure 4.25).

---

7 A catalyst is a substance that causes or accelerates a chemical reaction without itself being affected.

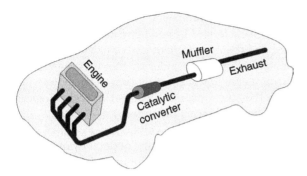

Figure 4.25 Reducing automobile pollution

4.2.2.9 Law of Increasing Controllability

Over time, technological systems evolve to become more controllable. As a result, these systems become more adaptive, enabling them to resolve contradictory requirements more effectively as well as withstand changes in the systems' environment. Accordingly, the law of increasing controllability, states that, as systems evolve, the level of control interactions among each of the system elements increases.

Fundamentally, there are two distinct flavors of increasing controllability trends. Under the first one, systems increase the number of controllable states: (1) from a single state to (2) multiple discrete states to (3) multiple variable states. Under the second one, systems increase the level of controllable states: (1) from uncontrolled system to (2) uncontrolled and human intervention system to (3) automated state to (4) self-controlled system (Figure 4.26).

| Uncontrolled system | Uncontrolled & human intervention system | Automated system | Self-controlled system |

Figure 4.26 Evolving railroad crossing protection system

4.2.2.10 Law of Harmonizing Systems and Environment

Viable technological systems are characterized by good physical and dynamical harmonization among systems and their constituent parts as well as between systems and their environments. For example, under rhythm harmonization, each part of the system performs its mission at the right time in harmony with other parts. If this is not the case, then the constitutive parts of the technological system may interfere with each other, rendering the system less effective.

Along this line, the law states that the necessary condition for optimal performance of technological systems is coordination of the periodicity of actions of their parts. More specifically, this law refers to the vibrations frequencies of individual parts within the system as well as the sequence of actions performed by each part and the synchronization among them. In this interpretation of the law, there are two categories of systems' evolutions: Under one category, systems evolve to perform more functions within the same amount of time. Under the second category, systems evolve to perform the same set of functions within a shorter timeframe. Finally, sometimes systems evolve as a combination of the above two categories.

For example, a hard disk drive (HDD), first introduced in late 1950s, stores and retrieves digital information using rotating disks, coated with magnetic material together with a magnetic head, which writes data onto and reads data from the disks surfaces.

One of the earliest hard disk drives was the IBM 350 disk storage unit. The system was configured with 50 magnetic disks providing a capacity of 50 Mbytes at average seek time of about 600 milliseconds (Figure 4.27). Much smaller HDDs, abundantly used within computer systems, have been evolving to improve both their functional capabilities (e.g. memory capacity, physical volume, weight)

Figure 4.27 IBM 350 Disk Storage Unit

as well as their dynamic capabilities (e.g. Access time) in accordance with the law of harmonizing the rhythms of parts of the system (Table 4.1).

Another interpretation of this law deals with harmonization between a system and its environment. For example, a train system must operate within a harmonized environment including railroad tracks, railroad stations, bridges, power distribution network, and the like (Figure 4.28).

Photo: An electrically hauled container-freight train. On the West Coast Main Line near Nuneaton in Warwickshire, England.

Table 4.1 Evolving HDD characteristics over time[8]

Parameter	1956	2017
Capacity [Bytes]	3.75×10^6	14×10^{12}
Physical volume	$1.9\,\text{m}^3$	$34\,\text{cm}^3$
Weight	910 Kg	62 g
Average access time [mSec]	600	2.5–10
Price / Megabyte	$9,200	$0.032
Data density/square inch [Bits]	2×10^3	1.3×10^{12}
Average MTBF [Hours]	2×10^3	2.5×10^6

Figure 4.28 Container-freight train

---

8 Adapted from: Hard disk drive, Wikipedia, https://en.wikipedia.org/wiki/Hard_disk_drive. Accessed: December 2017.

4.2.3   Further Reading

- Altshuller, 1996.
- Cascini et al., 2009.
- Chong, 2010.
- Ehrlenspiel et al., 2007.
- Fey and Rivin, 2005.
- Hellestrand, 2005.
- Karasik, 2008.

- Karasik, 2011.
- Petrov, 2002.
- Salamatov, 1999.
- San, 2014.
- Savransky, 2000.
- Stewart, 1981.

4.3   Modeling the Innovation Process

4.3.1   Classes and Types of Innovations

4.3.1.1   Classes of Innovations

Innovation may be defined as "the act of introducing something new."[9] This "something new" may belong to one of the following classes:

Systems or services innovations. Such systems or services should be new or substantially improved relative to existing, state of the art ones. These significant improvements may consist of technical attributes, components, materials, software or other functional characteristics. Most of the examples in Part III of this book relate to the systems' innovation class.

Business process innovations. Business process innovation include application of new or significantly improved business processes or core production facilities or delivery methods. These applications may consist of new or significantly improved manufacturing techniques and/or production equipment or distribution strategies. For example, FedEx, UPS, and Amazon moved aggressively into the supply chain and logistics, enabling e-commerce for millions of people and firms throughout the world. They also have developed and adopted new technologies for managing advanced distribution centers, package tracking and vehicles routings as well as integration of their logistic data with customers through easy-to-use application program interfaces (APIs).

Organizational innovations. These innovations include implementation of new organizational and/or financial business models within firms and their interactions with suppliers, employees and customers. For example, Google organizational structure is quite flat. This means that Google's employees, teams, or groups can bypass middle management and report directly to their top management. Similarly, employees can also meet and share information

---

9 American Heritage Dictionary of the English Language, Houghton Mifflin Harcourt; 5th edition, 2016.

across teams' boundaries. In addition, Google's organizational culture is particularly open and innovative, emphasizing excellence and hands-on experimentation.

Marketing innovations. Marketing innovations include the introduction of new and holistic marketing methods involving products' design, packaging, distribution, promotion, pricing and the like. For example, Amazon, starting as a humble internet bookstore, has gone on to dominate many segments of retailing, to the extent that it has expedited the demise of many brick-and-mortar chains.

4.3.1.2 Types of Innovations

Along an orthogonal axis, one can identify four types of innovative efforts. The first one, which may be designated as Type I innovation, is one where the problem is well defined whereas the solution must be determined. Therefore, the innovation thrust is to find acceptable solutions by generating new ideas and concepts. Type I innovation is most commonly thought of and virtually all the examples in Part III of this book belong to this category.

Type II innovation is the opposite of Type I. Here, a solution is known but the problem must be identified. In other words, one has a certain technological idea but the dilemma is to find an application for this technology. For example, when LASER (Light Amplification by Stimulated Emission of Radiation) was invented in 1960, it was called "a solution looking for a problem." Since then, lasers have been utilized in thousands of highly varied applications (Figure 4.29).

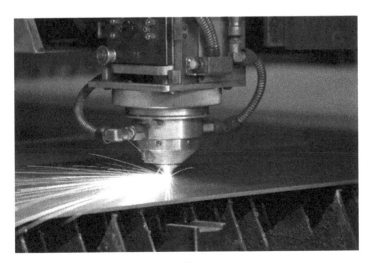

Figure 4.29 Metal cutting laser head[10]

---

10 Laserhead of an AMADA FO-4020NT industrial laser, installed at Metaveld BV.

Type III innovations represent situations where both the problem and the solution are well known. The challenge here is to determine what exactly needs to be done in order to bridge the gap between the two – in other words, how to implement a known solution to a known problem in the most effective manner. Resolving deficiencies in a manufacturing assembly line is a good example of Type III innovation (Figure 4.30).

Figure 4.30 Deficiencies in manufacturing assembly line

Type IV innovation is the opposite of Type III innovation. Here, both the problem and the solution are not known. So the aim here is to leisurely explore different ideas and solution concepts. For example, Google encourages employees to spend 20% of their time working on what they think will most benefit the company. This activity, at least in the early stages, matches Type IV innovation. Later, if and when the concept demonstrates potential impact, more people will join, until it becomes a "real project" (normally Type I innovation).

4.3.2 Technological Innovation Process

4.3.2.1 Pre-Innovation Process

Innovation should be a core element of every organization's survival strategy. By clearly defining strategic intents a firm gravitates towards the development of new added-value products and services. Pre-innovation process should include the following steps:

Step 1: Strategic thinking. Strategic thinking involves contemplating, identifying, and specifying the most promising innovation goals for the firm. That is, how innovation will add value to ones' organization's strategic intents as well as how this will lead to business opportunities with the greatest potential. Then, these insights should be translated into a set of specific firm's intents and expectations.

Step 2: Innovation portfolio policy. An innovation portfolio describes a set of ideal future products and services a firm aims to fulfill. One problem that must be addressed within an organization is the management policy of the firm's innovation portfolio (i.e. dealing with the entire collection of ideas and innovative projects that constitute the set of innovations-in-progress).

Such policy must balance the inherent innovation failure risks and the targeted rewards of success. That is, balancing the pursuit of innovative efforts with the realities of learning, risking, and failing in this endeavor. This is particularly difficult due to the fundamental personality differences between many operating executives and many practicing engineers. The first ones often seek stable, certain, and riskless environment, whereas the later ones often possess an adventurous spirit and are quite comfortable with uncertain circumstances. Establishing a viable innovation portfolio policy calls for bridging the chasm between risk-averse and creative-minded forces as well as instituting the right metrics according to which the firm could assess its efforts and correct its course as needed.

4.3.2.2 Original Innovation Models

The two original views regarding the nature of the innovation process evolved from the 1950s. They are depicted in Figure 4.31.

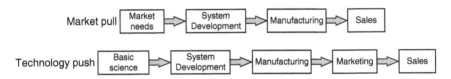

Figure 4.31 Early linear innovation models

Market pull. Early innovation processes were modeled as a linear set of operations pulled by market forces. In this model, innovations stem from perceived market needs that influence the direction and rate of technological development. In principle, research and engineering is undertaken only to support the innovation process.

Technology push. A similar, linear innovation process model, based on technology push rather than on market pull, evolved in parallel. These technological innovations are initiated on the cusp of fundamental research, leading to a creative concept (i.e. systems and/or service, operational business process, business organizations model, or marketing practice). Thereafter, these concepts are exploited by the firm.

In both models, the innovation process consists of sequential and unidirectional phases without control or feedback mechanisms. The market pull model is generally compatible with Type I innovation (i.e. the problem is defined but the solution must be determined), whereas the technology push model is generally compatible with Type II innovation (i.e. the solution is known but the problem must be identified).

4.3.2.3 Controlled Innovation Model

Several, more realistic innovation models, containing control mechanisms but still linear, have been proposed (e.g. Stage-Gate model).[11] These models divide the innovation process into phases with defined formal gates acting as decision points between each phase. After each phase, an end-of-phase review is conducted to ensure that the preceding phase was successfully completed. If the results meet the phase objectives, then work proceeds to the next phase. If not, then work continues within that phase or the project is terminated. Naturally, fewer and fewer innovative ideas flow through this innovation funnel due to its natural attrition process (Figure 4.32).

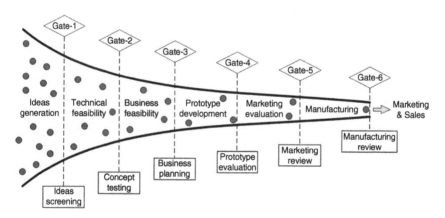

Figure 4.32 Controlled linear innovation model

4.3.2.4 Cyclic Innovation Model

A major drawback in most earlier theoretical innovation models is their linear nature. In addition, these innovation models do not integrate the full

---

11 Stage-Gate® is a registered trademark of Stage Gate, Inc., Business System: Product Innovation System.

.

innovation spectrum like scientific exploration, technological research, product development, market transitions, as well as the fundamental contribution of entrepreneurship in the innovation process. This chasm between earlier innovation models and real innovation practices renders these models somewhat inadequate.

The Cyclic Innovation Model (CIM) was developed by A.J. (Guus) Berkhout and his colleagues in the 1990s at the Delft University of Technology in the Netherlands.[12] It offers a nonlinear, feedback-rich framework that can help firms and policy makers to better understand the iterative nature of real innovation processes. CIM describes a circular rather than a chain process. It does not begin with technology push or market pull and it does not end with sales. Both are part of a perpetual creative process along a circular dynamic path with no fixed starting or ending points. In CIM, new technologies and changes in markets continuously influence each other in a cyclical manner. Together with the central role of entrepreneurship, it is considered to be the key characteristic of fourth-generation innovation model (Figure 4.33).

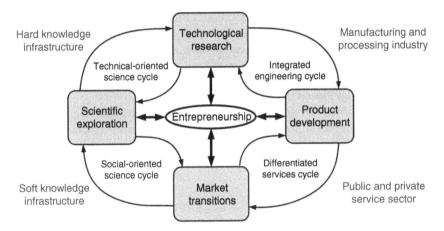

Figure 4.33 Cyclic Innovation Model

The upper part of Figure 4.33 shows two linked cycles in which technological research plays a central role. The technical-oriented science cycle involves interactions between scientific exploration and technological research, advancing the hard knowledge infrastructure. In addition, the integrated engineering cycle involves interactions between technological research and product development advancing the manufacturing and processing industry.

---

12 Portions of Berkhout et al. (2006) paper have been adapted and included in this chapter with author's permission. See https://www.researchgate.net/publication/228657169_Innovating_the_innovation_process. Accessed: May 2017.

Both these cycles are achieved by utilizing a wide range of disciplines from the hard sciences.

Similarly, the lower part of Figure 4.33 shows two linked cycles, but in this case, it is the world of human needs rather than the world of technology, which plays the central role. The social-oriented science cycle involves interactions between scientific exploration and market transitions. These interactions advance the soft knowledge infrastructure by creating new insights into emerging and receding socioeconomic trends. With these insights, new sociotechnical solutions can be developed faster and with less economic risk. In addition, the differentiated services cycle involves interactions between market transitions and product development, advancing the public and private service sector. Anticipating successful market transitions is very much a multi-disciplinary activity. Both these cycles are achieved by utilizing a wide range of disciplines from the soft sciences. In particular, soft sciences can explain and predict transitions in markets utilizing scientific way and to establish new solutions while knowing what the underlying socioeconomic forces are.

Overall, Figure 4.33 defines a system of cyclical change processes and their interactions as they take place in a successful innovation arena: hard and soft sciences as well as engineering and commercialization are brought together in a cohesive system of creative processes. Also, as can be seen in the figure, entrepreneurship plays a central role, without which innovation will not occur.

4.3.2.5 Technology Readiness Level (TRL)

Technology readiness levels (TRL) is a method for classifying the degree of maturity related to an innovative concept or system during its development and acquisition process.[13] The method is based on a scale from one to nine. One represents the least mature technology and nine represents the most mature technology.[14]

TRL was originally developed in the 1970s by Stan Sadin at the National Aeronautics and Space Administration (NASA). Over the years, the TRL scale has evolved and matured and is now widely used throughout the international engineering community. More specifically, it has been adapted by large companies, professional organizations as well as national and international bodies, for example: United States (US) Department of Defense (DoD), US Department of Energy (DoE), US Federal Aviation Administration (FAA), European Commission (EC), European Space Agency (ESA), The Oil & Gas

---

13 The reader should note that the TRL model is fundamentally different from the S-curve model. Whereas TRL deals with the degree of innovation maturity during systems' development and acquisition, S-curve models the evolution of technological systems from initial concept all the way to technology decline.

14 For more information about technology readiness levels (TRLs), see K. Bakke master's thesis: Technology readiness levels use and understanding (2017).

Industry (API 17 N) and more. In the adaptation process, the original terminology related to NASA space technology has been altered to reflect the specifics of relevant industries but the basic concept remained unchanged. As a result, using TRLs enables a fairly consistent discussions regarding technical maturity across different types of technologies.

The following is a set of TRLs using terminology which combines US DoD and other TRLs variants. Figure 4.34 depicts these TRLs and their corresponding innovation spectrums.

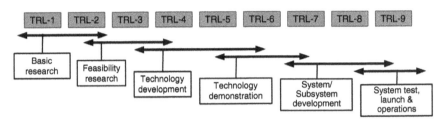

Figure 4.34 Technology readiness levels and innovation spectrums

TRL 1: Basic principles observed and reported. This TRL level is the lowest level of technology readiness. Scientific research begins to be translated into applied research and development. Examples might include paper studies of a technology's basic properties or principles that underlie this technology.

TRL 2: Technology concept and/or application formulated. In this TRL level, practical applications may be proposed or speculated but there may be no proof or detailed analysis to support the assumptions. Examples might include studies or other references that outline the application being considered which provide analysis to support the concept.

TRL 3: Analytical or experimental proof of concept. In this TRL level, active research and development has been initiated. This may include analytical studies and laboratory studies to physically validate the analytical predictions of the technology. Examples may include results of laboratory tests performed to measure parameters of interest and comparisons with analytical predictions for critical subsystems.

TRL 4: Component validation in a laboratory environment. In this TRL level, basic technological components and subsystems are integrated in order to establish that they will work together. This may be a simplified system version relative to the eventual, final system. Examples may include integrating several critical components and subsystems in the laboratory and evaluating their overall joint behavior.

TRL 5: Component validation in relevant environment. In this TRL level, the basic technological components are integrated with reasonably realistic supporting elements so they can be tested in a simulated environment.

As a result, the fidelity of breadboard technology increases significantly. Examples may include "high fidelity" laboratory integration of components and the results from testing laboratory breadboard system are integrated with other supporting elements in a simulated operational environment.

TRL 6: System/subsystem model or prototype demonstration in a relevant environment. In this TRL level a representative model or prototype system, which is well beyond that of TRL 5, is tested in relevant environment. This represents a major step up in a technology's demonstrated readiness. Examples may include testing a prototype system that is near the desired configuration in terms of performance, weight, volume, etc. in a high-fidelity laboratory environment or in a simulated operational environment.

TRL 7: System prototype demonstration in an operational environment. In this TRL level, a prototype near or at planned operational system is demonstrated. This prototype must represent a major step up from TRL 6 by requiring demonstration of an actual system prototype in an operational environment (e.g. in an aircraft or in space environment).

TRL 8: Actual system completed and qualified through test and demonstration. In this level, technology has been proven to work in its final form and under expected real-life conditions. In almost all cases, this TRL represents the end of true system development. Examples include developmental test and evaluation of the system in its intended super-system to determine if it meets its design specifications.

TRL 9: Actual system proven through successful mission operations. In this level, the actual application of the technology in its final form is proven under mission conditions, such as those encountered in operational test and evaluation.

The reader should be aware of some significant limitations related to the use of TRLs.

Nonlinearity in technology maturity. In contrast with the implicit linear character of the TRL scale, sometimes an increase in maturity also exposes new problems requiring additional research. Thus for example, a transition to TRL-8 may reveal new technological issues (e.g. manufacturability) that temporarily may throw the system back to, say, TRL 7 or TRL 6 levels.

Single technology maturity approach. A central characteristic of the TRL scale is its focus on a single technology. However, the higher TRL levels may deal with multiple technologies, which may exhibit different maturities levels. In these circumstances, using a single TRL scale is problematic because a single TRL value may not reflect the true status of the system at hand.

Focus on product development. The original TRL scale dealt with product oriented technologies. However, adaptations of the TRL scale into non-technological readiness levels in other domains may prove difficult. For example, creating a TRL scale for commercialization (e.g. readiness of an innovation to go to the market), manufacturability (e.g. readiness of an organization to manufacture a new product), or organization (e.g. readiness of an organization to sell and support a new product) is not straightforward.

Software aging. As mentioned, TRL scale measures the maturity of technologies in order to gauge their readiness for use in their specified context. The fundamental assumption, made in connection with TRL philosophy, is that a technology that has been evaluated as being at a given TRL level, absent any change, will remain at the same TRL level. This is not always the case. Software for example, continually ages as a result of maintenance activities[15]. Furthermore, software systems using non-developmental items (NDI) are subjected to repeated releases of these NDI components with limited information as to the specific nature of the NDI software changes. This phenomenon inevitably, requires modifications at the system level, which compounds the software decay problem.

4.3.3 Innovation Funding

4.3.3.1 Funding Sources
Innovation funding is a significant concern for any organization involved in creative efforts. Innovations are considered key to foster growing and maintain countries' and firms' competitive position in the global economy. Therefore, it is essential that small and large actors (i.e. firms as well as universities and research institutes) will collaborate in a meaningful way to leverage their individual advantages. Considerable resources are required in order to pay for highly skilled personnel, purchase research equipment as well as maintain infrastructure like laboratories, libraries, computer systems, office space and the like.

Fundamentally, research funding is available from two sources: public and private (Figure 4.35). Federal-level research & development (R&D) funding in the United States, for example, is based on authorizations by Congress to provide federal funding to universities, large industrial firms, and small and medium enterprises (SMEs). Similarly, the European Union funds many research projects at different TRL levels through Europe and associate countries in order to enhance the European competitive edge.

Along this line, state and regional bodies promote academic and commercial research activities in order to attract skilled personnel, maintain high employment levels and create wealth for their citizens. Privet R&D funding sources include corporations, venture capital, angel investors (i.e. individuals using

---

15 According to Eick et al. (2001) three mechanisms of maintenance induce software aging over time: (1) Reduction in software architectural integrity over time (2) Increased number of software files affected by a single software change, and (3) increased probability that modifications will introduce new software failures.

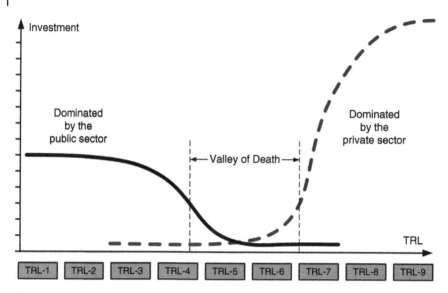

Figure 4.35 Innovation funding sources

their wealth to invest in promising high-tech businesses) and philanthropists (i.e. persons who seek to promote the welfare of others by providing generous financial donations). Most research universities are unique in that they receive public as well as private research funding.

As can be seen in Figure 4.35, most of the available funding for the initial innovation activities is provided by the public sector and then, once systems component are validated in their relevant environments, privet sector funding begins to be available. A funding gap, often referred to as the "Valley of Death," exists between basic research and proof of potential commercialization of new products. Many innovation projects were abandoned due to lack of funding during this particular period.

4.3.3.2 Global Innovation Levels

According to the National Science Foundation (NSF)[16], the total US R&D funding for 2013 was $456 billion (in current PPP dollars). This included $297 billion funded by business and $122 billion funded by the federal government through federal agencies and federally funded R&D centers as well as state and local governments. In addition, $37 billion was funded by other sources including support from universities and colleges, nonfederal government, and nonprofit organizations (Figure 4.36). As can be seen, late in the 1980s private industry's share of R&D investment increased dramatically whereas government's

16 Source: National Center for Science and Engineering Statistics, National Patterns of R&D Resources, Science and Engineering Indicators 2016.

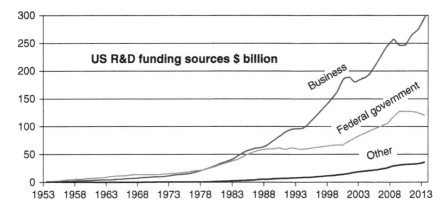

Figure 4.36 United States R&D funding sources (1953–2013)

contribution increased moderately. This shift in the public versus privet R&D funding seems to reflect diminishing congressional concerns regarding the competitive position of the United States in the global economy.

According to the same NSF source, the gross domestic expenditures on R&D in 1981–2013 (in billions of current PPP dollars) is depicted in Figure 4.37. The figure represents the United States, the EU, Russia, and four Asian countries (China, Japan, South Korea, and India).

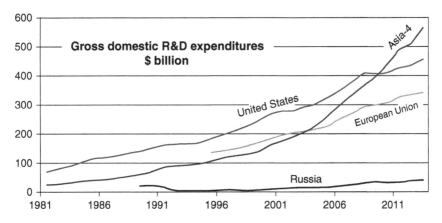

Figure 4.37 Comparison: Gross domestic R&D funding (1981–2013)

The reader should note that, starting in the early 2000s the rate of R&D funding in the Asia-4 countries increased dramatically, surpassing the United States in total yearly R&D expenditures sometime around 2010. This phenomenon reinforce the concern that, during the last two decades, the US Congress ignored some of its obligation to provide sufficient financial R&D support to ensure

US dominance in basic science and innovation. Over time, this affects the competitiveness of the United States in world markets, leading inexorably to a reduction in the quality of life for its citizens.

Some corporations invest heavily in R&D while others are not too aggressive in this regard. For example, the top 20 global R&D spenders in 2015 are depicted in Table 4.2.

Table 4.2 Top 20 global R&D spenders in 2015[17]

Rank	Corporation	Industry	2015 R&D Spending [$B]	Intensity [%]
1.	Volkswagen	Auto	15.3	5.7%
2.	Samsung	Computing and Electronics	14.1	7.2%
3.	Intel	Computing and Electronics	11.5	20.6%
4.	Microsoft	Software/Internet	11.4	13.1%
5.	Roche	Healthcare	10.8	20.8
6.	Google	Software/Internet	9.8	14.9%
7.	Amazon	Software/Internet	9.3	10.4%
8.	Toyota	Auto	9.2	3.7%
9.	Novartis	Healthcare	9.1	17.3%
10.	Johnson & Johnson	Healthcare	8.5	11.4%
11.	Pfizer	Healthcare	8.4	16.9%
12.	Daimler	Auto	7.6	4.4%
13.	General Motors	Auto	7.4	4.7%
14.	Merck	Healthcare	7.2	17.0%
15.	Ford	Auto	6.9	4.8%
16.	Sanofi	Healthcare	6.4	14.1%
17.	Cisco	Computing and Electronics	6.3	13.4%
18.	Apple	Computing and Electronics	6.0	3.3%
19.	GlaxoSmithKline	Healthcare	5.7	15.0%
20.	AstraZeneca	Healthcare	5.6	21.4%

However, innovation success is far from being assured. Different organizations publish reports regarding innovation success rates. For example, in 2010,

---

17 Source: Strategy& 2015 Global Innovation 1000 analysis, Bloomberg data, Capital IQ data, 2015. See: http://www.strategyand.pwc.com/media/file/2015-Global-Innovation-1000-Fact-Pack.pdf. Accessed: Sep., 2017.

Strategyn examined success rate reports from 12 different sources.[18] Accordingly, the average success rate for all 12 sources is 17%, and if the low and high outliers are removed, the average success rate is only 8.5%. So, a conservative estimate is that traditional innovation processes succeed in about 10% of the time.

4.3.4 Further Reading

• Bakke K., 2017.	• Markham and Mugge, 2014.
• Berkhout et al., 2006.	• NSB, 2016.
• Berkhout et al., 2010.	• OECD, 2005.
• Branscomb, 2002.	• Pohle and Chapman, 2006.
• Cooper, 1990.	• PWC, 2015.
• DOD, 2011.	• Shalley et al., 2016.
• Dodgson and Rothwell, 1996.	• Shavinina, 2003.
• EARTO, 2014.	• Silverstein et al., 2012.
• Eick et al., 2001.	• Skogstad, 2010.
• Jain et al., 2010.	• Smith, 2004.
• Mankins, 1995.	• Strategyn, 2010.

4.4 Measuring Creativity and Innovation

4.4.1 Defining Innovation Objectives

Different innovation activities are ongoing processes within many organizations. Firms routinely make changes to products, services, and other processes. The overall innovation goal should be visionary and impressive – ideally, something that has not been seen before. Equally important, the relative impact and importance of the firm's innovation efforts should be measured at reasonable intervals. However, in order to measure whether an organization is successful in its creativity and innovation efforts, the organization should identify its particular innovation objectives.

Broadly speaking, organizations may engage in one or several of the following classes of innovations: (1) product or services innovations, (2) business process innovations, (3) organizational innovations, or (4) marketing innovations. The Organization for Economic Cooperation and Development (OECD) defines a set of objectives that are relevant to these four classes of innovations (Table 4.3). Of course some of these factors are relevant to more than one class of innovation.

---

18 Source: Strategyn, Innovation Track Record Study, 2010. See: http://www.strategyn.at/sites/default/files/uploads/TrackRecord_07.pdf. Accessed: August 2017.

Table 4.3 Categories, objectives, and classes of innovation[19]

Categories and objectives	Product or services innovations	Business process innovations	Organizational innovations	Marketing innovations
Competition, demand, and markets				
Replace products being phased out.	X			
Increase range of goods and services.	X			
Develop environment-friendly products.	X			
Increase or maintain market share.	X			X
Enter new markets.	X			X
Increase visibility or exposure of products.				X
Reduce time to respond to customer needs.		X	X	
Production and delivery				
Improve quality of goods and services.	X	X	X	
Improve flexibility of production or service provision.		X	X	
Increase capacity of production or service provision.		X	X	
Reduce unit labor costs.		X	X	
Reduce consumption of materials and energy.	X	X	X	
Reduce product design costs.		X	X	
Reduce production lead times.		X	X	
Achieve industry technical standards.	X	X	X	
Reduce operating costs for service provision.		X	X	
Increase efficiency or speed of supplying and/or delivering goods or services.		X	X	
Improve IT capabilities.		X	X	

(Continued)

19 Adapted with minor modification from OECD, 2005. See: https://www.oecd.org/sti/inno/2367580.pdf. Accessed: August 2017.

Table 4.3 (Continued)

Categories and objectives	Product or services innovations	Business process innovations	Organizational innovations	Marketing innovations
Workplace organization				
Improve communication and interaction among different business activities.			X	
Increase sharing or transferring of knowledge with other organizations.			X	
Increase the ability to adapt to different client demands.			X	X
Develop stronger relationships with customers.			X	X
Improve working conditions.		X	X	
Other				
Reduce environmental impacts or improve health and safety.	X	X	X	
Meet regulatory requirements.	X	X	X	

4.4.2 Measuring the Innovation Process

Two categories of innovation-related parameters pertain to the measurement of innovation: (1) resources devoted to the innovation process, and (2) patent statistics and other complementary bibliometric information (e.g. statistical analysis of written publications, such as books, articles, and the like).

Innovation-related data may be collected at the enterprise, national, or international levels according to the guidelines defined in the Frascati Manual (OECD, 2015). Patent statistics are much easier to obtain and may be used to measure the product of research activities (e.g. the number of patents granted to a given firm or country and their classes). However, patent statistics are inferior indicators of an overall innovation level. This is because many innovations are not patented, and some innovations may be covered by multiple patents. Also, some patents have no technological or economic value, and others have very high value.

4.4.2.1 Breadth of Innovation

According to the Frascati Manual just mentioned, innovation activities are all the scientific, technological, organizational, financial, and commercial activities intended to lead to implementing innovation. Therefore, this set of activities is substantially wider than mere R&D activities. In other words, beyond R&D, innovation activities may include actions like testing, production, marketing, sales, training and more. Quantitative measurements of expenditures as well as return on investment (ROI) related to innovation activities provide important data as to the level of innovation activity at the national, industry, or enterprise levels.

The breadth of innovation activities is described in the following sections: (1) research and experimental development, (2) product and process innovations, and (3) marketing and organizational innovations.

Research and experimental development. Internal R&D is defined as research and development work that is done within an enterprise. This includes all R&D activities performed by an enterprise that are, by definition, innovation activities. Similarly, software development related to making scientific or technological advances and/or resolving scientific or technological uncertainties on a systematic basis is also classified as R&D. Also, building and testing of prototypes in order to make further innovative improvements is classified as R&D. In addition, one should include internal R&D, which comprises R&D and services acquisitions from other organizations.

Product and process innovations. In addition to specific R&D activities, enterprises may acquire technology and knowhow in the form of patents, non-patented inventions, and licenses, disclosures of knowhow, trademarks, designs, and patterns. Similarly, it may include computer services and other scientific and technical services for product and process of innovation activities. Innovation activities also involve the acquisition of capital goods for the purpose of conducting innovation work. This may consist of land and buildings, machinery, instruments, and equipment, as well as computer hardware and software.

Enterprises' development of innovations may also include activities related to the introduction of new product and process innovations, planning and designing procedures, technical specifications and other user and functional characteristics. In addition, innovation activities may include engineering efforts, production setup and adjustment, quality control follow-up and the like, required to produce or use new or improved products or processes.

Finally, preliminary market research, market tests, training, and launching advertising for new or significantly improved goods or services is defined as innovation activity when it is required for supporting a product or process innovation.

Marketing and organizational innovations. Preparation for organizational innovations is considered innovative activities. It includes the development and planning of new organizational methods, structures, and the work needed to actually implement it. Similarly, preparation for marketing innovations is considered an innovative activity. Here, too, it comprises activities related to the development and implementation of new marketing methods that have not been practiced previously by the said organization.

4.4.2.2 Global Innovation

The Global Innovation Index (GII)[20] for the year 2016 ranks world economies according to their innovation capabilities, using the Frascati Manual. Each of 128 economies, representing 92.8% of the world's population and 97.9% of the world's GDP, is ranked utilizing the following seven categories: (1) institutions, (2) human capital & research, (3) infrastructure, (4) market sophistication, (5) business sophistication, (6) knowledge & technology outputs, and (7) creative outputs. Each category is then further divided into subcategories and finally, the global innovation category of each country is computed based on a total of 82 indicators (Table 4.4).

Table 4.4 Global innovation indicators[21]

1		**Institutions**
1.1	Political environment	1.1.1 Political stability & safety
		1.1.2 Government effectiveness
1.2	Regulatory environment	1.2.1 Regulatory quality
		1.2.2 Rule of law
		1.2.3 Cost of redundancy dismissal, salary weeks
1.3	Business environment	1.3.1 Ease of starting a business
		1.3.2 Ease of resolving insolvency
		1.3.3 Ease of paying taxes
2		**Human capital & research**
2.1	Education	2.1.1 Expenditure on education, % GDP
		2.1.2 Government expenditure/pupil, secondary, % GDP/cap
		2.1.3 School life expectancy, years
		2.1.4 PISA scales in reading, math & science
		2.1.5 Pupil–teacher ratio, secondary
2.2	Tertiary education	2.2.1 Tertiary enrolment, % gross
		2.2.2 Graduates in science & engineering, %
		2.2.3 Tertiary inbound mobility, %

(Continued)

---

20 Adopted from Doutta et al., 2016.
21 Abbreviations in Appendix C: List of Acronyms.

Table 4.4 (Continued)

2.3	Research & development (R&D)	2.3.1 Researchers, FTE/mn pop.
		2.3.2 Gross expenditure on R&D, % GDP
		2.3.3 Global R&D companies, avg. expend. top 3
		2.3.4 QS university ranking, average top score
3		**Infrastructure**
3.1	Information & communication technologies (ICTs)	3.1.1 ICT access
		3.1.2 ICT use
		3.1.3 Government's online service
		3.1.4 E-participation
3.2	General infrastructure	3.2.1 Electricity output, kWh/cap
		3.2.2 Logistics performance
		3.2.3 Gross capital formation, % GDP
3.3	Ecological sustainability	3.3.1 GDP/unit of energy use, 2005 PPP$/kg oil eq
		3.3.2 Environmental performance
		3.3.3 ISO 14001 environmental certificates/bn PPP$ GDP
4		**Market sophistication**
4.1	Credit	4.1.1 Ease of getting credit
		4.1.2 Domestic credit to private sector, % GDP
		4.1.3 Microfinance gross loans, % GDP
4.2	Investment	4.2.1 Ease of protecting minority investors
		4.2.2 Market capitalization, % GDP
		4.2.3 Total value of stocks traded, % GDP
		4.2.4 Venture capital deals/bn PPP$ GDP
4.3	Trade, competition, & market scale	4.3.1 Applied tariff rate, weighted mean
		4.3.2 Intensity of local competition
		4.3.3 Domestic market scale, bn PPP$
5		**Business sophistication**
5.1	Knowledge workers	5.1.1 Knowledge-intensive employment, %
		5.1.2 Firms offering formal training, % firms
		5.1.3 GERD performed by business, % of GDP
		5.1.4 GERD financed by business, %
		5.1.5 Females employed w/advanced degrees, % total
5.2	Innovation linkages	5.2.1 University/industry research collaboration
		5.2.2 State of cluster development
		5.2.3 GERD financed by abroad, %
		5.2.4 JV–strategic alliance deals/bn PPP$ GDP
		5.2.5 Patent families 2+ offices/bn PPP$ GDP

Table 4.4 (Continued)

5.3	Knowledge absorption	5.3.1 Intellectual property payments, % total trade
		5.3.2 High-tech imports less re-imports, % total trade
		5.3.3 ICT services imports, % total trade
		5.3.4 FDI net inflows, % GDP
		5.3.5 Research talent, % in business enterprise
6		**Knowledge & technology outputs**
6.1	Knowledge creation	6.1.1 Patents by origin/bn PPP$ GDP
		6.1.2 PCT patent applications/bn PPP$ GDP
		6.1.3 Utility models by origin/bn PPP$ GDP
		6.1.4 Scientific & technical articles/bn PPP$ GDP
		6.1.5 Citable documents H index
6.2	Knowledge impact	6.2.1 Growth rate of PPP$ GDP/worker, %
		6.2.2 New businesses/th pop.
		6.2.3 Computer software spending, % GDP
		6.2.4 ISO 9001 quality certificates/bn PPP$ GDP
		6.2.5 High & medium-high-tech manufactures, %
6.3	Knowledge diffusion	6.3.1 Intellectual property receipts, % total trade
		6.3.2 High-tech exports less re-exports, % total trade
		6.3.3 ICT services exports, % total trade
		6.3.4 FDI net outflows, % GDP
7		**Creative outputs**
7.1	Intangible assets	7.1.1 Trademarks by origin/bn PPP$ GDP
		7.1.2 Industrial designs by origin/bn PPP$ GDP
		7.1.3 ICTs & business model creation
		7.1.4 ICTs & organizational model creation
7.2	Creative goods & services	7.2.1 Cultural & creative services exports, % of total trade
		7.2.2 National feature films/mn pop.
		7.2.3 Global ent. & media market/th pop.
		7.2.4 Printing & publishing manufactures, %
		7.2.5 Creative goods exports, % total trade
7.3	Online creativity	7.3.1 Generic top-level domains (TLDs)/th pop.
		7.3.2 Country-code TLDs/th pop.
		7.3.3 Wikipedia edits/mn pop.
		7.3.4 Video uploads on YouTube/pop

Table 4.5 depicts the Global Innovation Index (GII) ranking and innovation score of the top 20 countries for 2016.

Table 4.5 Global Innovation Index 2016 rankings (Top 20 countries)

Rank	Country/Economy	Score (0–100)
1	Switzerland	66.28
2	Sweden	63.57
3	United Kingdom	61.93
4	United States of America	61.40
5	Finland	59.90
6	Singapore	59.16
7	Ireland	59.03
8	Denmark	58.45
9	Netherlands	58.29
10	Germany	57.94
11	Korea, Rep.	57.15
12	Luxembourg	57.11
13	Iceland	55.99
14	Hong Kong (China)	55.69
15	Canada	54.71
16	Japan	54.52
17	New Zealand	54.23
18	France	54.04
19	Australia	53.07
20	Austria	52.65

4.4.3 Innovation Capability Maturity Model

Innovation management generally combines high costs with a high rate of failure. The resulting risk can be reduced by assessing the organization's overall innovation capability maturity. One way to evaluate and then to improve the capability of a firm to undertake innovative projects is to utilize the Innovation Capability Maturity Models (ICMM). This approach builds on two successful capability maturity models, the Capability Maturity Model (CMM) related to software development processes and the Capability Maturity Model Integration (CMMI) related to systems' development processes.

CMM has been evolving since the end of the 1980s at the Software Engineering Institute (SEI) located at Carnegie Mellon University. CMMI, a successor to CMM, has been evolving since early 2000, also at Carnegie Mellon in collaboration with members of industry and government representatives. Both models gained substantial followers throughout industry and academia and are explained in a comprehensive manner (e.g. Chrissis et al., 2011).

A number of researchers have proposed different variants of innovation maturity models. All these models share a similar structure, attempting to evaluate how well an organization deals with creative ideas from inception to fruition. The level of innovation capability maturity assigned to an organization provides a broad indication as to the organization's ability to succeed in its innovation efforts. However at this time, ICMM is in its infancy.[22] Different researchers propose different ICMM variants, and ICMM research is not supported by government agencies or academia. Given this precaution, an ICMM is described here based on an amalgamation of several ICMMs proposed by Heinz Erich Essmann (2009), Darrell Mann (2012), Robynne Berg (2013), and other researchers.

4.4.3.1 ICMM Description

ICMM[23] utilizes similar concepts practiced by CMM for the software domain and CMMI for the systems domain, namely:

- Create a universal innovation model of best practice.
- Evaluate innovation level and certify organizations.
- Create innovation training materials so that the best practice becomes universally observable and widely understood concept.
- Create an innovation infrastructure of professionals and peer-group network.

A proposed ICMM is composed of five levels (Figure 4.38).

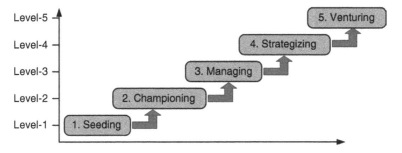

Figure 4.38 Proposed Innovation Capability Maturity Model (ICMM) levels

Level 1: Seeding. During this level, managers and staff talk about creativity and innovation but are doing very little and often, in fact, tend to extinguish innovative efforts. Management does not appoint executive officer to oversee innovation efforts or appoint inappropriate persons with limited vision and minimal financial leverage to fund innovative projects. The net result is that new ideas are rarely acted on and the few that are undertaken are conducted on an ad-hoc basis, practically hidden from management.

Level 2: Championing. During this level, innovation activities have taken place but success is limited. The main reasons for this are the lack of acceptance across the organization that innovation is a key business process rather than a high-risk enemy. In other words, although, there is some openness to innovation and

---

22 See also: B. Knoke, A Short Paper on Innovation Capability Maturity within Collaborations. http://ceur-ws.org/Vol-1006/paper2.pdf. Accessed: Nov. 2017.
23 Adapted with changes from Essmann (2009), Mann (2012), and Berg (2013).

possibly some innovation measurements are conducted, there is a lack of strategic direction in terms of innovation. In addition, while there is now an executive sponsorship, he or she is not seen as actively pursuing innovation. Also, innovation work is typically assigned to a selected departments and is done in isolation from the rest of the organization. In short, the organization is not yet fully engaged in innovative culture and such processes are not replicable.

Level 3: Managing. During this level, the organization establishes innovation practices and strategies with some strategic success. This may include successful launching of new products or services and the creation of new markets. Some managers exhibit competence in innovation so the culture of the organization may be regarded as innovative. However, traditional projects are separated from innovative projects and innovation activities remain the domain of particular individuals or departments. Also, top-level management is not fully engaged in the firm's innovation enterprise.

Level 4: Strategizing. During this level, executive management adopts a unified innovation policy and intimately follows and support innovative efforts throughout the organization. A robust innovative business model and mechanisms to capture projects' lessons is established. Also, the organization utilizes creativity and innovation methods and tools in order to increase the likelihood of projects' success as well as reduce the duration of innovation cycles significantly. In addition, individuals, at all hierarchical levels, consider innovation as a core company strength and a path for quick ascent within the organization. As a result, the organization succeeds in numerous innovation projects across products, services, and processes.

Level 5: Venturing. During this level, innovation is considered a core capability of the organization, which becomes embedded in the corporate culture. Innovations extend beyond the organization's core business and the firm acquires a reputation for rapid and effective innovation cycles. Furthermore, innovative programs backed by robust policies and appropriate business models drive strategy across all departments. Employees are eager to be a part of this company due to its notable openness to creative ideas and innovation as well as its generous benefits package and company-wide enthusiasm.

4.4.3.2 McKinsey's 7-S Framework and ICMM

The McKinsey's 7-S framework[24] is a vehicle to analyze how well an organization is positioned to achieve its intended objectives. It was developed in the early 1980s by Tom Peters and Robert Waterman of the McKinsey & Company consulting firm. The basic premise of the model is that there are seven internal aspects of an organization that must be harmonized in order to ensure that an organization is successful. These include: (1) strategy, (2) structure, (3) systems, (4) shared values, (5) skills, (6) style, and (7) staff. The 7-S model can be used in a wide variety of situations where an alignment perspective is useful, for example, Table 4.6 depicts harmonization between McKinsey 7-S framework and ICMM.

---

24 The McKinsey 7-S Framework: Ensuring That All Parts of Your Organization Work in Harmony. https://www.mindtools.com/pages/article/newSTR_91.htm. Accessed: November 2017.

Table 4.6 McKinsey 7-S framework and ICMM[25]

McKinsey 7-S element	ICMM Level 1 (Seeding)	ICMM Level 2 (Championing)	ICMM Level 3 (Managing)	ICMM Level 4 (Strategizing)	ICMM Level 5 (Venturing)
Strategy	Organization reacts to external events.	Organization adopts hosts of fashionable fads.	Innovation becomes a central pillar of strategy.	Innovation becomes a predictable business process.	Innovations extend beyond the organization's core business.
Structures	Organizational structure is rigidly maintained.	Limited collaborations decrease likelihood of innovation success.	Traditional projects are separated from innovation projects.	Flexible structure allows assignment of individuals to specific projects.	Teams assemble and disassemble seamlessly to meet emerging needs.
Systems	Few innovative projects are successful.	Projects are measured to determine innovation improvements.	Broad communication channels and training infrastructures reduce innovation cycles.	Broad capturing of projects' lessons reduce innovation cycles.	Lean system allow rapid innovation cycles applicable to wide range of business endeavors.
Style	Management persistently resists any change.	Management maintains status quo but "tolerates innovators."	Managers are expected to lead traditional projects while supporting innovation.	Management assigns full-time innovation director and adopt unified innovation policy.	Management supports and funds innovative efforts, accepting failures as basis for improvement.
Staff	Carry out what has been instructed.	Oppose innovation efforts on multiple grounds.	Take small and smart innovation steps, pause, and learn by doing.	Promote individuals based on their innovation success.	Imagine how the future will be and work toward this vision.
Skills	Do the current job.	Accept stagnation and routine work environment.	Seek to accomplish more with the available resources.	Utilize creativity and innovation methods and tools.	Take innovation risks and learn from failures.
Shared values	Accept what is happening and hope for a brighter future.	R&D is ineffectual but "necessary indulgence."	Innovation is important and will contribute to the organization.	Respect innovators as they ensure the organization's future.	Innovations will bring continued change and improvements.

25 Adapted from: Mann (2012).

4.4.4 Further Reading

- Achi et al., 2016.
- Berg, 2013.
- Cooper et al., 2002.
- Corsi and Neau, 2015.
- Doutta et al., 2016.

- Essmann, 2009.
- Mann, 2012.
- OECD, 2005
- OECD, 2010.
- OECD, 2015.

4.5 Obstacles to Innovation

A number of factors can inhibit or literally block innovation activities within an organization. These may be divided into the following five groups: (1) human habits, (2) cost factors, (3) knowledge factors, (4) market factors, and (5) institutional factors (Figure 4.39).

Figure 4.39 Common obstacles to innovation[26]

4.5.1 Human Habits Factors

Human habits seem to be the most common obstacle to innovation. This obstacle is difficult to overcome because it manifests deep-rooted human psychic and apprehensions. Figure 4.40, for example, depicts a nominal psychological reaction to change.[27]

---

26 Background image: William Blake, John Bunyan, Christian Reading in His Book, Plate 2, in *The Pilgrim's Progress: From This World To That Which Is to Come*, 1678.
27 Inspired by Bridges and Bridges (2017).

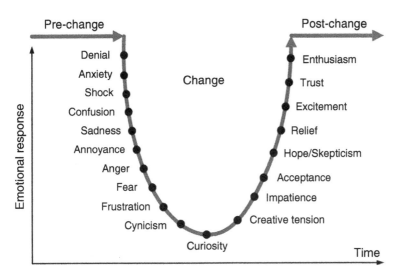

Figure 4.40 Emotional response to change

Resistance to change manifests itself as a tendency to reject new ideas and innovations. First, people have emotional ties to the old ways of doing things. These are their comfort zones. Changing daily routines make people unsecure. So they are bound to exhibit resistance whenever change requires them to do things differently. Second, fear of the unknown incentivizes people to resist change. In fact, overcoming this resistance can only be achieved by convincing people that the risks associated with stagnation are greater than those of moving forward in a new direction. Finally, people's resistance to change may stem from their feeling that they will be worse off at the end of the change or they are suspicious that the change favors other individuals or other groups.

Sometimes, people resist changes for other reasons. For example, poor communication within the organization leads members of the staff to misunderstand the need for change, or they don't believe that the company can competently manage the change. In addition, they could be exhausted from a barrage of management's changes and consider this one as just another temporary fad.

At times, resistance stems from simple tendency to conform to the ongoing process with minimal considerations. In other times, a "not invented here" attitude prevails where people automatically reject anything having external origin flavor. Finally, resistance arises when the benefits and rewards for making a change are not seen as adequate for the difficulties involved.

A whole slew of resistive forces hamper an inventor or a creative person vis-à-vis his colleagues, managers, partners, and potential financial backers. This may include the credibility of the concept as well as the inventor's, underestimation of the concept's significance, along with skeptical underestimation of its potential. In such an environment, many people tend to demonstrate narrow vision, limited imagination, and, often, arrogance.

4.5.2 Costs Factors

Costs factors seem to be the second most common obstacle to innovation. This may come about due to either lack of internal financial resources and/or difficulties in obtaining external funding (e.g. obtaining bank loans, obtaining public funding or venture capital). SMEs as well as new entrepreneurs with no track record face nearly insurmountable problems in securing bank loans. Public sources have dried up in many Western countries over the last decade or two, and venture capital may only be available when the perceived innovation risks seem low. In addition, as was mentioned before, the interval between basic research and demonstration of potential commercialization of new products or process (known as the Valley of Death) presents a particular funding innovation crisis, as many innovation projects are abandoned at this stage due to lack of funding.

Sometimes, secondary cost factors affect companies that misuse core financial tools within the innovation domain. This is due to the very special nature of innovative projects relative to traditional projects. Specifically, utilizing discounted cash flow (DCF) and net present value (NPV) metrics in a traditional project is very beneficial. However, using these tools to evaluate innovation investments should be avoided or carried out judiciously. In fact, innovation investments should be contemplated for the long haul. Overemphasis on short-term earnings per share is detrimental to innovation.

4.5.3 Institutional Factors

Many companies pay lip service to innovation but, in fact, resort to conventional thinking and processes that cannot meet innovative management needs. Their executives tend to focus on day-to-day issues like resolving routine problems, meeting schedule and budget constraints, following quarterly earnings, and the like. By and large, business culture in its quest for order, control, standard operating procedures, and efficiency is diametrically opposed to innovative culture. Unfortunately, creative insights rarely arrive on schedule or comply with business-as-usual philosophy.

As a result, quite often, companies minimize the importance of organizational culture, especially, with regards to innovation. In addition to lack of management attention to innovation, these organizations often do not realize that innovation failures are a part of the game, so they stigmatize unsuccessful risk takers. The more astute organizations encourage their employees to spend some of their time advancing their own ideas. The vast majority of these ideas will not result in successful products or services, but a few will, and exploiting them will make the company successful.

Organizations that, rightfully, implement rigorous processes in traditional projects often tend to apply the same rigid rules when dealing with innovative projects. However, the latter need a healthy dose of flexibility, patience, and constant nourishment in order to thrive. Or, to put it more specifically, organizations should allow innovative projects to: (1) diverge and explore different directions from the original mandate and (2) use experiments to evaluate critical assumptions and refine their technical and business strategy.

Other institute-level obstacles to innovation could be lack of infrastructure needed to undertake innovative projects. These may be experimentations facilities (e.g. environmental test lab, supercomputers for advance simulations, robotics laboratory and the like). Still other obstacles to innovation could be specific legislation and regulation, particularly in the medical field as well as in areas related to protection of animals and the environment. Along this line, adherence to standards can also hinder innovation in certain domains, especially where human safety is concerned. Finally, certain monopolies, patents, intellectual property rights (IPRs), and taxation structures, for example, could impose substantial institute-level obstacles on innovation.

4.5.4 Knowledge Factors

Other obstacles to innovation may stem from lack of knowledge regarding essential technology needed to implement certain creative ideas as well as neglect to study the market for relevant products. Similarly, difficulties in finding partners for developing products or processes as well as sharing marketing efforts could put insurmountable barriers to innovative efforts.

Sometimes, lack of qualified personnel and technical expertise, either within or outside the enterprise, in the labor market or in academia stops an innovation project in its tracks. This is especially apparent in SMEs located in remote areas, as well as in larger organization where management policies tend to restrict innovation to a central R&D departments.

4.5.5 Markets Factors

When dealing with Type I innovation (i.e. the problem is known and the solution must be determined), there must be a demand for the new product or a process. Uncertain demand for particular innovative products or services constitutes a formidable barrier for innovation. And, of course, developing product or services for which the potential market is dominated by other established enterprises may face substantial market obstacles.

4.5.6 Innovation Obstacles and Classes of Innovations

The Organization for Economic Cooperation and Development (OECD) defines a set of innovation obstacles and their relevance to each of four classes of innovations (Table 4.7).

Table 4.7 Innovation obstacles versus four classes of innovations[28]

Innovation obstacles	Product or services innovations	Business process innovations	Organizational innovations	Marketing innovations
Human habits:				
Resistance to change	X	X	X	X
Cost factors:				
Excessive perceived risks	X	X	X	X
Cost too high	X	X	X	X
Lack of funds within the enterprise	X	X	X	X
Lack of finance from sources outside the enterprise:				
• Venture capital	X	X	X	X
• Public sources of funding	X	X	X	X
Institutional factors:				
Lack of infrastructure	X	X		X
Weakness of intellectual property rights	X			X
Legislation, regulations, standards, taxation	X	X		X
Knowledge factors:				
Insufficient Innovation potential (R&D, design, etc.)				
Lack of qualified personnel:				
• Within the enterprise	X	X		X
• In the labor market	X	X		X
Lack of information on technology	X	X		
Lack of information on markets	X			X
Deficiencies in the availability of external services	X	X	X	X

28 Adapted with modifications from OECD (2005). See: https://www.oecd.org/sti/inno/2367580.pdf. Accessed: Feb., 2017.

Table 4.7 (Continued)

Innovation obstacles	Product or services innovations	Business process innovations	Organizational innovations	Marketing innovations
Difficulty in finding cooperation partners for:				
• Product or process development	X	X		
• Marketing partnerships				X
Organizational rigidities within the enterprise:				
• Attitude of personnel toward change	X	X	X	X
• Attitude of managers toward change	X	X	X	X
• Managerial structure of enterprise	X	X	X	X
Inability to devote staff to innovation activity due to production requirements	X	X		
Market factors:				
Uncertain demand for innovative goods or services	X			X
Potential market dominated by established enterprises	X			X

4.5.7 Further Reading

• Bridges and Bridges, 2017.	• Kasser, 2015.
• Christensen, 2010.	• OECD, 2005.
• Corsi et al. (Editors), 2006.	• Shteyn and Shtein, 2013.

4.6 Promoting Organization's Innovative Culture

4.6.1 Introduction

Numerous articles and books explain extensively how to promote organizations' innovative culture. For example Jain et al. (2010) is a massive 440-page book describing in detail how to manage research, development, and innovation within organizations. Clearly, organizations can promote innovative culture and succeed in self-renewal if they identify their specific reasons for innovations and focus on relevant promising innovative areas.

The purpose of this chapter is to provide cursory description of the vast literature promoting innovative culture in organizations. Innovative culture is based on the following eight interrelated ingredients: (1) leadership, (2) organization, (3) people, (4) assets, (5) culture, (6) values, (7) process, and (8) tools. Each of these elements affects the other elements and, in turn, is affected by them all (Figure 4.41).

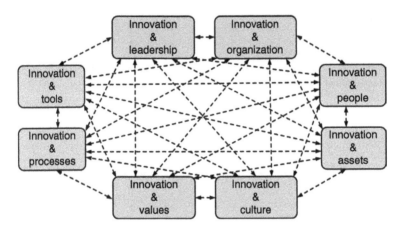

Figure 4.41 Innovative culture foundation

4.6.2 Innovation and Leadership

Researchers involved in organization leadership found that leaders of institutions that truly promote innovation emphasize the importance of collaboration amongst all members of the stuff and, by and large, tolerate higher levels of innovation failures. In such organizations, exchange of information among team members provides a continuous learning experience and also brings people to feel like they are a part of the innovation team.

4.6.2.1 Innovation Leadership Styles

Leaders involved in innovation should strike a delicate balance between meeting individuals' personal research goals and meeting the aims of the organization as a whole. Thus, employees should have the sense that they are contributing to the organization as well as for themselves and for society at large. Various researchers in management have identified several characteristic management styles and their effects within an R&D organization (Table 4.8).

In summary, collaborating management style seems to be the most effective way to lead innovative projects.

Table 4.8 Management styles and effects within R&D organization

Style	Description	Effects within R&D organization
Abdicating	The manager neglects to allocate specific assignments and responsibilities to subordinates, thus neglecting to deal with on ongoing issues.	Neither the manager nor the employees have much influence on any particular decision. This management style is often the worst option, as no one seems to lead the R&D organization in any reasonable manner.
Domineering	The manager makes decisions and tells his subordinates what to do.	Subordinates have no influence on the decision process. In the rare occasions that the manager is truly a brilliant engineer, he may lead his team to considerable achievements. However, not utilizing the team's collective abilities is a recipe for failure and discouragement
Delegating	The manager provides information to the subordinates about the problem and suggests possible solutions. The responsibility for the decision is given to the subordinates.	The subordinates are given full responsibility for the decisions, and the manager does not provide much of a guidance expected of him. This management style often causes conflicts between the aims of the innovation team and the stated aims of the organization.
Negotiating	The manager asks subordinates for information and suggestions on what to do and then makes decisions on the basis of these suggestions.	Subordinates have influence on the decision process. However, not consulting with the team as part of making decisions may cause team's discouragement as well as increase the likelihood of failure.
Collaborating	The manager asks subordinates for information, as well as suggested solutions. Thereafter, the manager negotiates with the team, and together they reach mutually satisfying decisions.	Both the manager and the subordinates have a great deal of influence on making decisions. In most R&D organizations, this management style is the most optimal to achieve significant innovative results.

4.6.2.2 Innovation Leader Skills

Scholars argue about the specific characteristics of innovation and leadership for many years. Basically, an innovation leader is not expected to act as an engine of ideas. His/hers business should be to recognize promising creative ideas, foster an environment of innovation, and share these ideas with managers, employees, suppliers, and business partners. Ultimately, the goal is to bring creative ideas into fruition. Finally, an innovation leader must have sufficient courage to fight for his convictions and act in accordance with the best interests of the organization as he sees it. The following is a set of skills and personality attributes that can benefit a leader in promoting innovative culture within an organization.

Cultivate strategic business perspective. Innovation leaders should cultivate understanding of societal and industry trends as well as business, marketplace, and their relevant customer base. They should be able to articulate how these external dynamics may affect their organization and its business future. In addition, these innovation leaders should be able to describe their vision of the future and succeed by showing where they want to go and how to get there. A strong customer focus is important, but often visionary persons see beyond what customers are expecting (e.g. the iconic Steve Jobs). They should also be involved, as well as encourage their stuff to participate in an ongoing, organized learning about strategic issues related to the company and its competitors.

Seize opportunities and manage risks. Innovation leaders should create a climate of reciprocal trust. This includes making themselves highly accessible, initiating warm, collaborative relationships with their management and the innovators who work for them and, no less important, backing colleagues under all circumstances. In addition, innovation leaders should be proactive in seizing opportunities when they present themselves. This includes making quick decisions and working independently for extended periods of time with minimal support. While being bold when it comes to considering new ideas, innovation leaders should be aware of the risks involved in choosing any innovative course of action. Also, they should initiate plans to identify risk and minimize their effects.

Possess communication skills. Innovation leaders should have solid communications skills. That is, they should be sufficiently persuasive vis-à-vis their management as well as colleagues so they will accept and follow the concepts presented by their subordinates with enthusiasm and conviction. Innovation leaders should inspire and motivate through their own personal action. If this means sitting late at night with the team in order to deal with a problem that emerged or getting involved in a particular thorny issue at the bottom of the heap, then so be it. Finally, innovation leaders are expected to provide honest and occasionally harsh feedback to subordinates so they can always count on straight answers from their leaders.

Possess relevant professional expertise. Innovation leaders should have reasonable expertise in the subject matter of their innovative projects. Therefore, the most appropriate people to assume the roles of innovation leaders within technical enterprises should be scientists, engineers, designers, and the like. In addition, innovation leaders are expected to demonstrate curiosity and desire to know more. Advancing their professional knowledge could give them an important competitive edge in order to lead effectively, and also stimulate colleagues in their quests to excel.

Able to lead courageously. Innovations require breaking down old patterns and creating new ones. This means innovation leaders and their team members must believe in themselves and trust each other. Accomplishing this trust helps innovation leaders to become more confident in embracing the role of a change agent in support of constructive disruption within the organization. Innovation leaders may often engage in high-stakes meetings, and under these circumstances, they should not refrain from stating their views even if it conflicts with others' opinions. Along this line, leaders must often make difficult decisions in order to steer the team in a more appropriate direction. Sometimes, it means a leader must kill an innovative project when it is clearly going astray. On occasions, team members must be demoted or let go altogether. In short, innovation leaders should reject any temptation to please upper management for the sake of garnering good points, but rather, exhibit unwavering loyalty to doing what's right for the organization, the customers, and society at large.

4.6.3 Innovation and Organization

Researchers who have investigated large number of companies offer the following measures to encourage a climate of participative management within R&D organizations:

Treating employees. People at all hierarchical levels should be treated fairly and with respect. This is because they are a valuable resource so they can contribute ideas and knowledge to the innovation process. By and large, people want to participate in the decision process, and when they are encouraged to do so, they accept change and are more committed to the organization. In addition, when people have an input in the decisions process, the agreed resolutions tend to be supported by all.

Long-term commitment. Organizations should make a long-term commitment to the development of people because that makes them more valuable to the organization. People who have developed over the years can be trusted to make important decisions about the scientific, engineering, and management of their work activities, resulting in high satisfaction and organizational effectiveness.

Dual management structure. However, in reality, not all managers can embrace collaborative style. Therefore, under certain circumstances, it is possible to utilize a dual management hierarchy within the organization. Under this policy, organizations could develop two branches of hierarchical authorities, typically with technical positions parallel to management positions. The technical hierarchical structure comprises a professional echelon that has the same degree of control and authority as the corresponding positions in the management hierarchical structure.

It should be noted, however, that dual management hierarchy within the organization can succeed only if the top officers of the company ensure that the status, authority, career opportunities, and reward systems of the two hierarchical branches are maintained on an equal footing. Often, this parity is not maintained, to the detriment the technical hierarchical branch.

4.6.4 Innovation and People

Researchers in the area of psychology and human development point out that the difference between creative process and innovative process is that the first one deals with conceiving creative ideas whereas the latter is the process of turning these ideas into successful products, services, or ventures. Naturally, the core personalities of creative people are often different than that of innovative people. In general, innovative people tend to exhibit the following attributes:

Formal education or training. Education or professional training is essential for noticing and understanding potential creative ideas. Well-trained experts are more likely to distinguish between relevant and irrelevant information.

An opportunistic mindset. Opportunistic-alert individuals are more likely to identify gaps and weak areas in the availability of product or services within the marketplace. Such persons often seek novelty, variety, and pioneering endeavors in all aspects of life.

High emotional quotient (EQ). Despite popular accounts of crazed, individualistic geniuses who invented, singlehandedly, some gadgets, inventions are most often achieved by a group of people. Therefore, innovation people need to have high EQ in order to communicate with other people and convince managers, staff, backers, and other bodies of the viability of their ideas as well as the strategy to achieve the stated goals.

Prudent personality. Prudent people tend to plan their every move in order to ensure a successful outcome in their endeavors. By and large, such individuals are organized, cautious, and reasonably risk-averse.

Persistent personality. Persistent persons do not give up too easily in the face of hardships. By and large, they are driven, resilient, and energetic in perusing their goals. This tenacity enables them to exploit potential opportunities effectively.

Virtually all companies have a formal organization structure in place. This structure is primarily concerned with the relationship between authorities and subordinates. The hierarchical organization begins at the top with the most senior officer of the company and then flows down to the subordinate managers and then subordinate employees below (Figure 4.42).

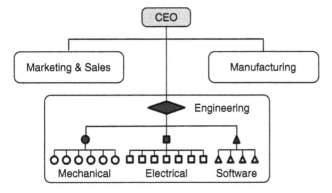

Figure 4.42 Example: Formal organizational chart

However, organizations also have informal structures that, in many ways, are as important as the formal ones. For example, some informal social networks may be revealed by their communications metadata (e.g. email source and destination, additional recipients of emails, etc.). Analyzing such informal social networks provides a rather blurry view. Nevertheless, researchers in the area of social networks identify four common role players whose performance is critical to the productivity of any R&D organization:

Central connectors. The central connectors interact with most individuals in the informal network. They are not necessarily the formal managers within a unit or a department, but they know who can provide needed information or expertise at any given point in time. For example, two individuals identified as (A) in Figure 4.43 assume the roles of central connectors in this informal engineering social network.

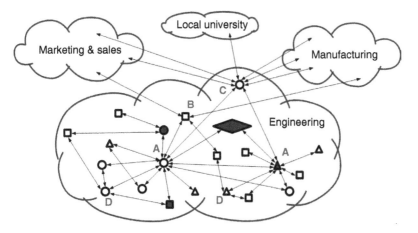

Figure 4.43 Example: Informal social networks

Boundary spanners. The boundary spanners connect the informal network with other parts of the company or with other networks outside the organization. These individuals undertake the job of identifying and nurturing relevant new external connections, which could provide or disseminate valuable information regarding the company. For example, an individual identified as (B) in Figure 4.43 assumes the role of boundary spanner in this informal engineering social network.

Information brokers. The information brokers provide information to different individuals within and outside the core informal network. In essence, these information brokers are instrumental in transforming several small informal networks that are quite isolated from one another into a single, large and dynamic informal network. Information brokers play a role similar to that of boundary spanners only they do it within a larger social network. For example, an individual identified as (C) in Figure 4.43 assumes the role of information broker in the informal company-wide social network.

Peripheral specialists. The peripheral specialists often operate on the periphery of the informal social network. However, their contribution to the success of the firm is crucial because they possess specific kinds of information or technical knowledge. Also, they are sufficiently smart and willing to share it with other members of the group whenever needed. For example, two individuals identified as (D) in Figure 4.43 assume the roles of peripheral specialists in this informal engineering social network.

Managers and leaders of R&D organizations can improve the innovative power of the business by recognizing and then appreciating the value of these informal social networks. Thereafter, managing the informal networks can be achieved by way of social network analysis using graphical tools that map out the relationships in an organization.

For instance, management could identify the central connectors in the informal network and then determine whether they attempt to maximize the communication channels bandwidth. Or they can assess whether the boundary spanners interact with the optimal set of external organizations and individuals in order to obtain important information or optimize the standing of the firm. Similarly, leaders in the organization could review the effectiveness of the information brokers and, if needed, add more individuals to support the existing ones. Along this line, manager could name publicly the peripheral specialists, giving them more free time to expand their knowledge as well as assign new persons to provide other needed technical expertise to users of the informal network.

4.6.5 Innovation and Assets

A range of local and external innovation assets are available to the organization in order to promote its innovative culture. This mainly includes: (1) human resources, (2) universities and research institutions, (3) industrial base, (4) physical infrastructure, (5) financial capital for innovation, and (6) legal and regulatory environment:

Human resources. Human resources are probably the most important component of the innovation potential in a company. This is simply because people, scientist and engineers, are at the core of any successful development and implementation of innovative ideas. This should incentivize corporations to constantly strive to attract and retain innovative persons and to invest in its workforce skills and technical knowledge. A lifelong-process of education and training should lie at the heart of improving professionals' skills as well as labor quality.

Universities and research institutions. Universities and research institutions should be key pillars in the innovation efforts of mature R&D organizations. This is because they represent important sources of human ingenuity and brain capital. As such, universities and research institutions are often creators, receptors, and, very often, interpreters of new knowledge. National and regional, as well as industrial, investments open up a slew of opportunities for partnerships between academia and industry that can significantly contribute to innovative efforts of attuned corporations.

Industrial base. An important innovation asset is a strong industrial base. This means that an innovative R&D corporation should be aware of its internal as well as the external industrial base. This should include awareness of the corporations' main products and services, the ongoing business models, market advantages, and more. Obviously, a fair percentage of the innovative efforts should harmonize with these core parameters. Nevertheless, management should encourage some disruptive innovation efforts as well.[29]

Physical infrastructures. Physical infrastructures are required for an economy to function and survive. This includes transportation and communication networks, public utilities, electricity, water, sewer, gas, building, and laboratories, for example. Physical infrastructure is a key assets needed to achieve beneficial innovation within R&D organizations.

Financial capital for innovation. Innovative activities require access to capital in order to transform ideas into products and services. However, most companies, universities and individual researchers are unable to finance the entire innovation development cycle using internal resources. Therefore, financial capital for innovation is considered a vital innovation asset. Such assets are usually more available in more technologically developed areas

---

29 The term "disruptive" is used in a rather loose manner to describe means by which a company is able to successfully challenge an established product or process by offering a new product or process with significantly improved degree of ideality. A more accurate description of the term "disruptive" is available in Christensen et al. (2015).

and large cities because of the significant presence of both financial institutions and risk investors.

Legal and regulatory environment. Government authorities provide (or deny) necessary innovation assets in the form of legal and regulatory framework. This framework facilitates the development of R&D and technology by forming development policies at both local and national levels. Innovation assets like government's tax reduction schemes or special financial assistance for innovation, incentivizes R&D corporations to undertake ambitious development projects. In contrast, negative innovation assets like inordinate bureaucratic barriers or lack of copyright protection burden tend to deter R&D corporations from undertaking innovative projects.

4.6.6 Innovation and Culture

Culture consists of shared beliefs, attitudes, self-perceptions, norms, role perceptions, and values that are transmitted from one generation to another in a given society. Culture facilitates stable behavior, because people do what is customary. Within organizations, culture consists of objective elements (e.g. office buildings, office furnishings, research laboratories, equipment, etc.) and subjective elements (rules, laws, values, norms). In addition, organizational culture encompasses a multitude of unstated assumptions concerning the way people interact with one another and the way things are accomplished within the organization.

From an R&D standpoint, some organizational cultures are more effective than others. A culture that emphasizes innovative behavior, participative climate, hard work, tolerance for disagreement, rewards commensurate with contribution as well as high-quality management-employee relations is more likely to be effective than other cultures. Innovative culture is one that leaders cultivate in order to nurture creative thinking and behavior. Such workplace promotes an attitude that innovation is the responsibility of all members of the organization rather than the domain of a few gifted individuals or the executive elite. Establishing and sustaining innovative culture in R&D organizations is considered a prerequisite for creating competitive advantage in the marketplace. Exemplary culture of R&D organizations may include the following characteristics:

Defining innovation strategies. Within innovation culture, the organization ensures that the employees are encouraged to think beyond the development and production of current products. That is, people are inspired to embrace different arenas in which individuals can be involved with innovation like profit models, processes, and policies. In addition, within such culture, an organization is expected to balance its emphasis on operational excellence with innovation dynamism, whether evolutionary, revolutionary, or disruptive.

Embracing creative Ideas. Within innovation culture, the organization ensures that members of an organization's internal or external community can freely provide insights and ideas that lead to new innovations. The culture is such that management is receptive to ideas from experts as well as from novices alike. Encouraging ideas throughout the company is one of the best ways to leverage existing talent within or outside the organization. In such innovative culture, employees at all positions are eager to offer their talents and skills to the company, because they know that their ideas will be valued and that good concepts will be adopted if they work.

Encouraging innovative communication. Within innovation culture, the organization ensures that the door of each member of the organization (Yes, including the CEO) is open for positive communication regarding innovation issues. Varied communications opportunities are available to each employee, from snail mails to emails to phone conversations to face-to-face meetings on an individual basis or on the basis of a collaborating team.

Encouraging innovative collaboration. Within innovation culture, the organization encourages intensive collaboration within internal organizations or outside bodies like other companies, universities, government agencies, and the like. This brings new perspectives and ideas to the innovation process. Such culture relies first on collaboration among isolated business and functional units, which can leverage the full range of expertise across the organization by pulling capabilities from across the company. Similarly, within such innovation culture, external collaborations across geographies, cultures, and time zones can harness thousands of smart persons, providing a significant competitive advantage.

Empowering innovative champions. Within innovation culture, the organization ensures that all employees are allowed to experiment with new and creative ideas. This means that the innovation culture within the organization prevails despite the ongoing, relentless pressure on managers to meet optimal performance in the business's core activities.

Implementing flat management structure. Within innovation culture, the organization ensures that the management structure maintains relatively short approval processes as well as coherent lines of communications that encourage innovation. Of course, management can achieve similar results by empowering employees to act independently. Unfortunately, both alternatives (i.e. "flat management" and "empowerment") are possible only in very special circumstances.

Holacracy, which was devised by Brian Robertson in 2007, is a third way. Its advocates claim that *holacracy* is a revolutionary management style that distributes authority and decision-making throughout an organization. As a result, holacracy supports fast and agile organization that is ready to make quick decisions that empower individual members of the organization. This approach is described in Brian's book *Holacracy: The New Management*

System for a Rapidly Changing World. It has been adopted by several organizations throughout the world, but the transition from a hierarchical structure into a flat distributed control structure is not trivial.

Providing innovation training and tools. Within innovation culture, the organization ensures that employees receive specific training and tools in order to propel their creative ideas into products or services. For example, people undergo a series of innovation workshops in which they learn what to do with their creative ideas and who, within the organization, should hear about it. In addition, employees practice how to create business presentations and, if possible, simplified demonstrations of their ideas, emphasizing its contribution to the company. Some companies provide "free" time for employees to experiment with new technologies, products, or processes; however, researchers are concerned about this "overengineering of the innovation process." Instead, they propose, providing some structure and support to help people navigate uncertainty and tap into the creative process in the most effective way.

Measuring success and rewarding innovators. Within innovation culture, the organization ensures that innovation performance metrics is created and publicized. Such metric should be based on appropriate criteria like the business value an employee has generated, how sustainable that value has been, what new ideas he or she brought forth, and how many of them actually were successful. Recognizing employees for innovative success is a hallmark of an excellent innovation culture. Such recognitions may be expressed in formal or informal ways through a symbolic means or outright, in monetary form. Acknowledgments may come in many forms: mentioning the achievement in the company news bulletin, putting up posters in the hallways, issuing letters of commendations, awarding medals, cups, etc. or outright cash.

Committing resources. Within innovation culture, the organization ensures that sufficient company resources are appropriated for the innovation process. These resources are allocated to employees who need them to define, implement, and test prototypes of their creative ideas. Since resources are always limited, the organization should have in place mechanisms to disburse the resources wisely and select the best utilization of those resources.

Accepting failures. Within innovation culture, the organization is prepared for innovation failures and accepts such risk as a necessary part of the innovation process. Innovative failures occur when the underlying concepts are incorrect, when the process is not managed properly, when the funding is inadequate, or when the expectations are unrealistic. However, innovation failures are not altogether lamentable because failures lead to learning, adaptation, and the creation of improved new ideas and attack strategies. In the final analysis, almost all innovations are the result of prior learning from failures.

4.6.7 Innovation and Values

Human values may be defined as the moral principles and beliefs or accepted standards of persons or a social group.[30] Values have major influences on a person's behavior and attitude and serve as broad guideline in one's life.[31] Culture of innovation requires further guidance in making individual and group decisions. Innovation values embody one's attitudes about innovation and also help shape the organization's day-to-day operations. Therefore, most organizations define their innovation values and operate accordingly. These values serve as a unifying force within the organization, so long as everyone is committed to fulfill them. Typical innovation values shared by many organizations are described here:

Quality. Each person in the organization is committed to produce sustainable and excellent work that pushes the company forward. They all embrace their responsibilities and strive to provide quality solutions and added value to the company, its customers, and society as a whole.

Leadership. Each leader in the organization is committed to guide his/her team along the company's goals, project positive example to others and invest in others so that they can follow the guidelines of leadership. In addition, leaders are dedicated to empowering those around them, openly share what they know, respect all people, and accord them reasonable freedom to conduct their own assignments as they see fit.

Individuality. Each person in the organization respects the knowledge, skills, ideas, and capabilities of each employee within the company. Each leader is committed to determining and utilizing the strong points of each employee while helping support their weaker ones. Each person can hone unique abilities, becoming experts at their individual fields so they can better support the organization, while also appreciating the value of teamwork.

Integrity. Each person in the organization is committed to acting in accordance with the highest standards of professional behavior. That is, all members of the organization are open, transparent, honest, and respectful in their dealings with other people, clients, consumers, vendors, and the public at large. In addition, each person in the organization operates in a spirit of cooperation and dignity, accepting other individuals' unique talents and work styles.

Accountability. Each person in the organization cares about the company and its constituents. Members of the organization take responsibility for their actions and act in the best interest of the people around them. In addition,

---

30 See: The Collins dictionary, https://www.collinsdictionary.com/dictionary/english/values. Accessed: March 2017.
31 See: The BusinessDictionary, http://www.businessdictionary.com/definition/values.html. Accessed: March 2017.

they are willing to take reasonable risks and trust their own ideas, knowing they will be judged objectively and fairly. They know they can rely on their managers and colleagues to support them in their day-to-day activities.

Creativity. The company is composed of individuals with diverse sets of perspectives and experience. As a source of new ideas, each person in the organization is committed to promote company's creativity, ushering better products, services, and processes.

Innovation. Innovation creates a sustainable future for the company even though it involves some risk. Therefore, each person in the organization is committed to promote innovative work through collaboration with other people. In addition, all leaders of the organization are committed to ensure the future well-being of the company by boldly cultivating innovative initiatives as well as ensuring a consistent and well-paced investment of resources.

Collaboration. Each person in the organization is committed to collaborate with others and work as teams. By and large, collaboration yields more accomplishments at a faster rate, producing better results than laboring individually. In addition, a collaborative effort provides learning opportunities, enriching the lives of those involved, as well as sharing skills and resources across organizational boundaries.

Continuous Improvement. Each person in the organization is committed to tracking involvement in innovation activities, evaluating results, and incorporating lessons learned into the next phase of a project and into new projects. In addition, members of the organization openly share findings and insights with management and colleagues in order to pass on what has already been achieved into the future.

4.6.8   Innovation and Processes

See Section 4.3.2, Technological Innovation Process.

4.6.9   Innovation and Tools

According to the European Industrial Research Management Association (EIRMA[32]), mechanized capturing, leveraging, and utilizing knowledge generated during innovation processes (i.e. from idea generation through prototype development, testing, and manufacturing) can provide substantial benefit to organizations engaged in innovative processes.

This undertaking can be enhanced by the use of a variety of commercial-off-the-shelf (COTS) software tools and/or tools developed within the

---

32 See: http://www.eirma.org/. Accessed: August 2017

organization. The purpose of such tools is to support easy access to R&D information throughout the organization. Typical data could include projects' management and technical documentation as well as contact information for persons involved in these projects. Also, these tools could provide information not customarily available in formal projects' documents. Furthermore, such tools could support R&D collaboration among people dispersed geographically or across organization boundaries, forming effective "virtual teams."

Kohn et al. (2003) have divided these software tools designed to support innovation management into three classes: (1) innovation acquisition tools, (2) innovation management tools, and (3) innovation corporate tools. Many of these software tools are available commercially, most often incorporating multiple and overlapping functionalities (Figure 4.44).

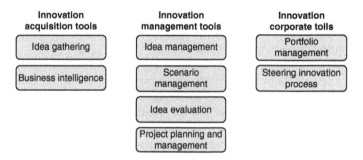

Figure 4.44 Typical innovation management software tools

4.6.9.1 Innovation Acquisition Tools

Innovative acquisition tools may be further divided into the following categories:

Idea gathering software tools. These software tools enable companies to solicit innovative ideas from employees and managers within the company as well as from customers, partners, collaborators, stakeholders, and the public at large by means of emails, web-based applications, and the like. In addition, these tools facilitate appropriate level of transparency of a company and its innovative efforts.

Business intelligence software tools. These software tools are designed to retrieve, analyze, transform, and report data for business intelligence. Business intelligence provides up-to-the-minute information for appropriate levels of the organization. This information can be analyzed and visualized in order to reinforce convenient and optimal business and technical decisions.

4.6.9.2 Innovation Management Tools

Innovative management tools may be further divided into the following categories:

Idea management software tools. These software tools store creative ideas in a central database and then support the management of this data throughout the lifetime of these creative ideas. That is, from generation to concept development and finally, to project implementation. Beyond capturing innovation ideas, such tools provide structured means to evaluate, prioritize, and select ideas for actual implementation. In a nutshell, these tools provide an organizationwide innovation mechanism to gather, store, display, and manage creative ideas and their innovative status.

Scenario management software tools. These software tools provide interface that supports analysts as they input, manipulate, analyze, and work with innovation scenarios. Typically, these tools will support analysts in converting users' stories into scenarios, defining specific attributes for these scenarios, conduct scenario-based analysis, perform online collaborative reviews of innovative scenarios, and the like.

Idea evaluation software tools. Most organizations gather a lot of innovative ideas, but only a few of them hint of potential business opportunities and even less, turn into profitable ventures. Idea evaluation software tools provide means to evaluate these ideas and analyze whether they are feasible as well as technically and economically well-grounded. For example, virtual reality (VR) tools use computer technology to generate realistic images and relevant design parameters, replicating and simulating the behavior of conceptual and imaginative systems (Figure 4.45).

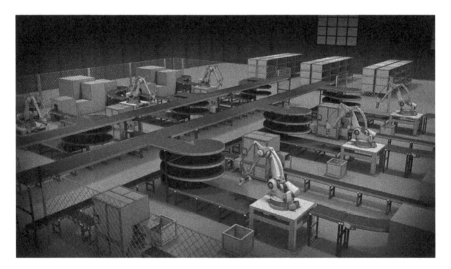

Figure 4.45 Evaluating production line concept using virtual reality tool

Project planning and management software tools. These software tools provide project status visibility. In particular, these tools accept, maintain and display relevant project status information like status of each task, action lists, spreadsheets, Gantt charts, and the like. These tools provide flexibility and allow customization so they can fit to the needs of different organizations. In addition, these tools support collaboration along geographically dispersed users by monitoring and managing all project updates. Finally, these tools provide a powerful change control facilities.

4.6.9.3 Innovation Corporate Tools

Innovative corporate tools may be further divided into the following categories:

Portfolio management software tools. These software tools support the process of managing projects' portfolios in order to identify and select individual projects that offer the greatest possible value. That is, to maximize the worth of projects' portfolio to the organization. These tools are able to track multiple active projects, identify the most promising projects as well as analyze the overall portfolio performance at the organization level.

Steering innovation process software tools. These software tools provide the means to achieve a sustainable innovation process throughout the entire company so that top officers of the company can observe and steer the innovation portfolio within the company. More specifically, the tools support: (1) the establishment and facilitation of innovation communities, each focusing on specific business area, (2) end-to-end management of the portfolio of innovations including innovative ideas, business opportunities, new technical and management concepts, ongoing and planned innovative projects, etc.

4.6.10 Conclusion: Ascent to Innovation: Practical Steps

Many researchers, scholars and entrepreneurs describe practical steps to promote organization's innovative culture. Figure 4.46 and the text below describe a set of practical steps needed by an organization to ascent into true innovative culture.

Step 1: Promote innovation culture. In this step, leaders should stress the importance of creativity and actively soliciting new ideas. This could be accomplished by installing an idea gathering software tools and encouraging individuals to use them. People at all levels of the organization should know that constant searching for new products or services increases the likelihood that the organization will maintain its competitive edge. Leaders should be truly available to hear creative ideas and visibly ensure that each idea gets an

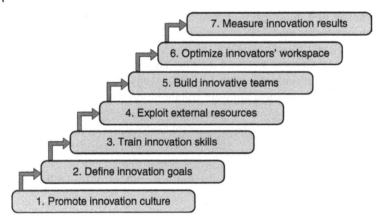

Figure 4.46 Ascent to innovation practical steps

expedient and fair hearing by the appropriate innovation organ of the company. In addition, leaders should offer a reasonable amount of free time to individuals wishing to explore their own creative ideas, and conduct regular group workshops or brainstorming sessions to discuss the status of different innovative projects and to raise new innovative ideas. Finally, it is a good idea for leaders to wander every now and then into the offices or laboratory spaces of the rank and file crowd, ask people about their current work and challenge their way of doing things. Did they think of alternative approaches? Is it possible to improve their effectiveness?

Step 2: Define innovation goals. In this step leaders as well as staff members should identify the most promising innovation goals of the firm. That is, sketch attractive directions for the company evolution and show how innovation will add value to the organization and how this will lead to business opportunities with the greatest potential.

Step 3: Train innovation skills. In this step, leaders or experts from outside the organization should train member of the staff in formal and informal innovation techniques. As much as possible, this process should be incorporated into the regular rhythm of the organization. Sometimes, training could be incorporate into projects' sprints, that is, a period during which specific work has to be completed prior to key milestones (e.g. review, software or system's release, project completion, etc.). Part III of this book provides a large set of creative methods; many of them are applicable to most companies and organizations as training material.

Step 4: Exploit external resources. Traditionally, innovative projects have been managed in isolation from external players. However, current thinking suggests that exploiting external knowledge outside organizational boundaries is critical to organizations' innovation success. In this step leaders are encouraged to engage in R&D cooperation with customers, partners, other companies, and universities as well as specialized small and medium enterprises (SMEs). Such cooperation may be beneficial for several reasons, including: (1) complementing internal resources, (2) enhancing technical capabilities quickly, (3) increasing the development speed, (4) reducing internal specialization needs, and (5) minimizing labor costs as well as capital assets costs.

Step 5: Build innovative teams. In this step, leaders should attempt to build diversified innovative teams throughout the organization. First, leaders should bring into the group engineers with different backgrounds and capabilities. Several studies show that a homogeneous group facilitates team-bonding but a group with diverse backgrounds, qualifications, experience, and problem-solving skills promotes creativity and helps in overcoming difficulties in innovation projects.

Second, leaders should match individuals to specific assignments because certain individuals are better suited to performing specific jobs than others. Job matching should take into consideration the nature of the job as well as the abilities and wishes of the person under consideration. Jobs may be characterized by the specific tasks to be fulfilled as well as the organization's culture, supervisory structure, career path, compensation plan, and the like. The majority of engineers have a built-in inner motivation that is manifested in their tendency to make consistent decisions while taking reasonable risks.

Third, leaders should publicly recognize and motivate individuals and teams on proposing creative ideas or successfully accomplishing innovative undertakings. Furthermore, suggestions must be taken seriously and acted upon within reasonable time or else the communal spring of ideas will dry up very quickly. Rewarding either by way of giving monetary incentives or through intangible public recognition is also a valuable motivating force.

Fourthly, leaders should be tolerant to creative ideas that, eventually, prove unsuccessful. A leader should stress that risk-taking is accepted as a norm within the organization. Individuals should be encouraged to express their ideas with no fear of retribution.

Step 6: Optimized innovators' workspace. The workspaces that people occupy have genuine consequences for their psychological well-being and creative performance. Fortunately, several psychology and neuroscience studies provide simple and effective steps that leaders can take in order to optimized innovators' workspace. First, a leader should encourage team members to make decisions regarding their own workspace arrangements (e.g. small offices, cubicles, work benches, etc.). This ability to a make one's mark on his/hers immediate environment provides an empowering effect that often improves individuals' productivity. Second, a leader should encourage team members to personalize their workspace by adding stuff like pictures, small trinkets, and a few green plants. Research corroborates that the presence of office plants helps lower stress levels and, if the office is illuminated, then plants can produce oxygen, thus reducing office pollution levels.

Step 7: Measure innovation results. Traditional firms measure their economic success in term of standard financial parameters like revenues, profits, return on investment, market share, and the like. Ultimately, innovation outcome should prove a financial success in the marketplace. However, measuring innovation processes is challenging because innovation seek to explore unknown territories, which leads to a higher rate of failures. Innovation measurement is a process that is best managed with a long-term perspective. That is, each innovative effort should not be measured individually, so it is important to measure success and failure using portfolio thinking – that is, across many projects. For example, a set of objective metrics could include:

1) Innovation deliverables (i.e. new product, new service, etc.)
2) Activities that enhance the organization's image (e.g. publications, conference presentations, interviews, etc.)
3) Creation of intellectual property (e.g. patents, trade secrets, etc.)

Savvy organizations could also add to the above list assortment of intangible innovation measurements. This could include learning obtained during the innovation process, patience, and wisdom gained through the commercialization process, as well as firm's reputation and attractiveness to external talents.

4.6.11 Further Reading

- Christensen et al., 2015.
- Cross and Prusak, 2002.
- Cross and Parker, 2004.
- De Bonte and Fletcher, 2014.
- Dyer et al., 2011.
- Deschamps, 2008.
- Farris, 1982.
- Forbes, 2012.
- Gostick and Christopher, 2008.
- Govindarajan and Bagchi, 2008.
- Jain et al., 2010.
- Kohn et al., 2003.
- Lawler, 1973.
- Lawler, 1991.
- Lawler, 2006.
- Lee, 2012.
- Midgley, 2009.
- Owen, 2016.
- Pelz and Andrews, 1976.
- Prather, 2009.
- Robertson, 2015.
- Shellshear, 2016.
- Smith, 2016.
- Start up Donut, 2017.
- Sternberg and Davidson, 1996.
- Summa, 2004.
- Tellis, 2013.
- The HR Observer, 2015.
- Vroom and Yetton, 1973.

4.7 Pushing Creative Ideas by Individual Engineers

4.7.1 Large Organizations Seldom Innovate

4.7.1.1 Organizations Life Cycle Model

Starting in the mid-1970s and vigorously continuing today, Ichak Adizes and his colleagues have developed a pioneering organizations' life cycle model. Accordingly, organizations follow similar organisms-like life cycles. That is, they generally undergo predictable life-stages, growing, aging, and then dying. A simplified adaptation of Adizes' model is depicted in Figure 4.47 and in the following text.

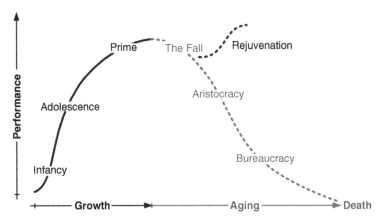

Figure 4.47 Organizations life cycle model[33]

Infancy stage. This stage starts with a dream of a few people and continues in setting up the financial and physical infrastructure as well as human resources of the infant organization. Naturally, the organization is action-oriented with the aims of getting to work and producing results.

Adolescence stage. In this stage, the organization manufactures successful products or provides needed services. Sales are booming and the company's cash flow is positive and stable. The company is expanding and customers as well as investors are satisfied.

Prime stage. In this stage, the organization has reached its prime vitality and is optimally position along its life cycle. In addition, it has achieved an excellent balance between control and flexibility. However, the continued success often tends to drive management and staff into arrogance and overconfident behavior.

---

33 Simplified adaptation from Adizes, 2017.

The fall stage. In this stage, an organization starts to lose its vitality. At the beginning, the symptoms will not show up because the firm may be cash rich but the problems will slowly surface as time goes on and the company ages. If a deliberate rejuvenation effort does not start (now or in previous stages), then, slowly but surely, the infrastructures that holds the organization together will start to collapse, sales decline, costs explode, and profits slump.

Rejuvenation. Rejuvenation is a practical means of reversing the organization's aging process. It often involves a study of the specific aging causes and trying to create mechanism to oppose them. So, like natural organisms, the intent is to ensure life extension to the organization by repairing or replacing damaged or aged components of the organization.[34]

Aristocracy stage. In this stage, the expectations for continued growth are quietly abandoned. Management loses interest in obtaining new markets or acquiring state of the art technologies. Often, managers are suspicious of any change proposal and punish those who do not follow the day-to-day routine. As managers are unable to reverse the company's downward spiral, they are likely to engage in recrimination and office squabbling.

Bureaucracy stage. In this stage, the organization is largely incapable of producing and selling products or providing any needed services. However, it is kept alive by artificial life support provided by interested political or business parties.

Death stage. Organization's death occurs when no one is interested to keep the organization as a living, producing, and functioning entity. The company may declare bankruptcy and its creditors and shareholders may acquire or sell its remaining assets and infrastructure.

4.7.1.2 Anecdotes of Rejuvenation and Extinction

Sometimes, an organization is able to revitalize itself and, on occasion, a company is able to achieve rejuvenation on a repeated basis. A good example of an organization that constantly rejuvenates itself is the International Business Machines (IBM) Corporation. IBM is a US-based multinational technology

---

34 Attempting to rejuvenate a company is always tricky and tenuous at best. One of the most topsy-turvy rejuvenation stories relates to Apple Inc. Steve Jobs, then chairman, chief executive officer (CEO), and co-founder of Apple wanted to rejuvenate the company by recruiting a talented and experienced manager to lead Apple. In 1983, Jobs offered the position of Apple's CEO to John Sculley but this appointment proved to be a failure because Sculley's and Jobs's visions for Apple were very different. Two years later, in 1985, Sculley decided to rejuvenate Apple. Following a charged power struggle, Steve Jobs was forced out of Apple. The departure of Jobs from Apple proved disastrous. A series of major product failures and missed deadlines destroyed Apple's reputation, bringing the company to the brink of bankruptcy. Then, in 1997, Apple invited Jobs to become CEO of his former company; he successfully rejuvenated Apple to grow into one of the most successful American multinational technology companies. In 2003, Jobs was diagnosed with a pancreatic tumor and died on October 5, 2011, from relapse of his cancer disease.

company, which was established in 1911. Despite some major debacles[35] and more than 100 years after its inception, the company demonstrates tremendous rejuvenation powers. For example, as of 2017, IBM has a large and diverse portfolio of products and services, including: Cloud computing infrastructure, cognition computing technologies, analytics insight tools, securely managing remote devices, IT systems and services operations, integrated development and deployment tools, systems and data security tools, and more (Figure 4.48). IBM is also heavily involved in scientific and engineering research. In fact, IBM Research constitutes the largest industrial research organization in the world, with 12 laboratories in 10 countries: (i.e. United States, Australia, Brazil, China, Ireland, Israel, India, Japan, Kenya, and Switzerland).

IBM System/360 Model 50   IBM Blue Gene/Q supercomputer

Figure 4.48 IBM flag products (1965 and 2012)

Unfortunately, the vast majority of organizations pay lip service to the importance of innovation but, in fact, do little to promote it. For example, in a McKinsey poll,[36] 84% of the managers surveyed agreed that innovation is critical for their business growth but only 6% of them were satisfied with the innovation performed within their organizations. Very few managers knew what exactly the problem was and how specifically, to promote innovation. Thus, many companies, like biological organisms, rise economically and then

---

35 A personal anecdote: In 1971, the author was a junior engineer at the IBM Endicott laboratory when a call came to join the Boca Raton, Florida, "Skunk works" facility, in order to develop the new IBM Personal Computer system. The grand vision of the project called for an open hardware architecture that would devastate the many competitors, each of whom was committed to its own hardware platforms. On a pre-project engineering meeting, the author, interested in software development, asked about software and was told that the intent was to purchase the software from a small company located in California (i.e. Microsoft). The author proceeded to point out that the open hardware architecture would enable other companies to manufacture the machines more economically than IBM and the software would be produced by another company. What was there for IBM? This question was politely ignored and the author did not join the IBM-PC project.
36 See: http://www.mckinsey.com/business-functions/strategy-and-corporate-finance/how-we-help-clients/growth-and-innovation.Accessed: March 2017.

slowly decay and eventually die off. Many of them, small to medium businesses rise and fall quietly, events that hardly surface in the media.

Sometimes however, the meteoric ascent and decline of a major corporation plays out a dramatic Greek-like tragedy. This, for example, was the case for the Digital Equipment Corporation (DEC) that was founded in 1957 by a legendary engineer, Ken Olsen and his co-founder Harlan Anderson. DEC became a dominant US-based multinational company in the computer industry until its death-spiral in the late 1990s. Ken Olsen's strategy was to develop minicomputers competing on price and performance with mainframe computers, such as the IBM/360 and IBM/370 systems. DEC's computer peripherals as well as their PDP and VAX minicomputer series (Figure 4.49), were selling in large numbers. In its heyday, DEC had a valuation of more than $12 billion and employed over 120,000 people worldwide, making the company second only to IBM. Ken Olsen managed DEC with minimal intrusion, giving each division of the company near total freedom to innovate and expand as they saw fit. DEC's unique corporate culture contributed both to its phenomenal successes and later, to a company-wide ossification, which led to its ultimate downfall.

PDP11/40
computer

VAX 780 computer

Figure 4.49 DEC flag products (mid-1970s)

As the 1980s approached, DEC was running business along its philosophical roots in the 1950s. Its products were elegantly designed, requiring extensive support and service and, most importantly, were expensive. DEC completely missed the advent of the personal computer, which was a true disruptive innovation, because Olsen could not fathom using a computer for home applications. DEC's business model supported microprocessor systems at $50,000 and up, whereas an IBM-PC system was offered at a list price of about $2,000. In the aftermath, DEC's shares plunged, Olson was forced to leave his position, and the spirit of DEC died with his departure. DEC was sold to Compaq in 1998 and subsequently merged with Hewlett-Packard in 2002, practically disappearing from the market.

4.7.1.3 Noninnovative Organizations

According to Dougherty et al. (2000), "Noninnovative organizations restrict [innovative behavior] by framing knowledge as separate, bounded subsets of operations, and defining their links in terms of the optimization of ongoing operations. [They] limited new knowledge to that which improves existing operations; [or] confirms or ratifies current operations," Careful examination of many large companies would reveal that they match the description of noninnovative organizations, exhibiting the following attributes.

Innovation vision. To manage innovation in a systematic way, each manager and staff member must understand and agree with a corporate-wide definition of innovation. However, many large companies do not define a shared, widely known and agreed upon innovation vision. Without this, it's nearly impossible to know the innovative direction the company is to precede, how much actual innovation is occurring and whether innovation actually adds to the bottom line of the company.

Innovation metrics. By and large, companies measure just about everything that affects their financial status. However, many large companies do not have innovation metrics in place and thus do not measure their innovation process. Such companies cannot follow their financial innovation investment and staffing, often resulting in terminating potentially promising innovative projects, and vice versa.

Innovation-friendly processes. Over the past decades, virtually every company has reorganized its operating model for efficiency and speed. This is also true regarding optimization of the supply and distribution chains. Few companies however, have invested in innovation and, despite the hype; many companies did not establish a truly permeating innovation-friendly processes. This phenomenon is often reflected in the company's innovation model, which is not tuned for innovation; thus, little achievements could be expected. For example, many companies' budgeting processes are inherently conservative, making it impossible for employees and first-line managers to get funding for small-scale experiments. Similarly, many companies do not assess new ideas in an effective and timely fashion and, more often than not, do not reward innovation performers fairly.

Ongoing learning. By far, most companies do not invest systematically in promoting and improving the innovation skills of their employees and managers. It seems that, despite evidence to the contrary, many senior managers do not see the value of such life-long learning. This attitude may be emanating from managers' belief that creativity is an innate attribute of a few gifted individuals.

Risk taking. Under existing market competition, many companies place inordinate emphasis on removing risk from new projects. Within this atmosphere, many companies exhibit risk-averse culture. So people at all levels of the organization tend to stay away from high-risk projects and unsafe creative ideas. As a result, management is reluctant to invest in new innovative projects or tends to terminate ongoing innovative projects before they have a chance to produce real results.

Innovative leaders. Few companies have in place enthusiast, trained, and capable leaders who are responsible for the design, construction and maintenance of the company's innovation engine. Absence of such individuals inhibits the organization from pushing its innovation process and blocks people at all levels from ongoing innovation training and access to the right tools and resources. In addition, no one ensures that innovation projects will be monitored and adequately funded. Finally, no one at the upper executive echelon ensures that hiring and promotion criteria should include ingredients to strengthen the company's creative talent pool.

4.7.2 Characteristics of Innovative Engineers

4.7.2.1 Career and Vocational Categories

Some people gravitate naturally to the various subfields of science and engineering, other do not. A pioneering researcher, John Holland, proposed in the 1950s a theory of personality that focuses on career and vocational choice. This theory, known as the "Holland Occupational Themes[37]" group people on a broad basis related to their personality and suitability for different types of occupations. The text below describes the following six personality categories defined by Holland: (1) realistic, (2) investigative, (3) artistic, (4) social, (5) enterprising, and (6) conventional. Most engineers however, are affiliated with the realistic or investigative categories.

Realistic personality. Individuals affiliated with this category are generally independent, persistent, practical, and down-to-earth. They usually enjoy hands-on and manual activities such as working with tools or machines as well as building or repairing things as oppose to dealing with ideas and people. By and large, they prefer learning by way of doing, working with their hands producing tangible results, within practical, task-oriented settings. In addition they often value practical things one can see, touch, and use. Most engineers exhibit realistic or investigative personality.

Investigative personality. Individuals affiliated with this category are generally intellectual, introspective, inquisitive, curious, methodical, rational, analytical, and logical. They often understand and are drawn to abstract mathematical, or scientific challenges and like to study and solve math and science problems. Invariably, these individuals enjoy scientific / technical research and place a high value on science and learning. They prefer independent work environment

---

37 See more: https://en.wikipedia.org/wiki/Holland_Codes. Accessed: January 2018.

focusing on solving complex problems in original ways. Most engineers exhibit realistic or investigative personality.

Artistic personality. Individuals affiliated with this category are generally original, intuitive, sensitive, expressive, unstructured, spontaneous, and nonconforming. By and large, they value aesthetics, seeking opportunities for self-expression through artistic creation and enjoy creative activities such as composing or playing music, writing, drawing or painting, and acting in or directing stage productions. Typically, they tolerate ambiguity, have an aversion to convention and conformity and prefer work environment, which fosters flexibility and encourages originality as well as imagination.

Social personality. Individuals affiliated with this category enjoy helping others and are generally focused on human relationships and interpersonal dynamics. They tend to be cooperative, patient, caring, helpful, empathetic, tactful, and friendly and concerned with the welfare of others. By and large, these individuals prefer a team-based work-environment, which encourages significant interactions with others so problem are solved mainly through discussions and interpersonal skills.

Enterprising personality. Individuals affiliated with this category are generally energetic, assertive, ambitious, dominant, adventurous, persuasive, sociable, and self-confident. Invariably, such individuals seek to attain money, power, and status by engaging in activities such as leadership, management, sales, marketing and the like. However, enterprising individuals tend to avoid activities that require careful observation and scientific, analytical thinking.

Conventional personality. Individuals affiliated with this category are often efficient, thorough, careful, organized, orderly, and conscientious. They are comfortable working within a well-defined work environment, following well-defined instructions. In fact, they tend to exhibit distinct aversion to undertaking ambiguous or unstructured assignments. Often, these individuals like to work with numbers, records, or machines in a set, systematic, and orderly way

4.7.2.2 Engineers versus Other Occupations

In today's competitive marketplace, engineers are expected to participate in all phases of the innovation process. This means that engineers should master entrepreneurial competences as well as managerial and leadership skills. That includes the abilities to identify, motivate, and guide colleagues in innovation-related assignments.

A 2013 study by Williamson et al., based on data collected between 2004 and 2010 from approximately 5,000 engineers and 76,000 nonengineers, assessed: (1) how personality traits of engineers differ from other occupations and (2) the correlation between these personality traits and engineers' career satisfaction. The Williamson study referred to the following set of relevant personality traits:

Assertiveness. A person's disposition to speak up on matters of importance, expressing ideas and opinions confidently, defending personal beliefs, seizing the initiative, and exerting influence in a forthright, but not aggressive manner.

Conscientiousness. A person's disposition to be dependable, reliable, and trustworthy, inclined to adhere to societal norms, rules, and values.

Customer service orientation. A person's disposition to provide highly responsive, personalized, quality service to colleagues and people at large, putting the person first; and trying to make him or her satisfied, even if it means going above and beyond the normal job description or policy.

Emotional stability. A person's disposition to being objective, maintaining an overall level of adjustment and emotional resilience in the face of job stress and pressure.

Extraversion. A person's disposition to be sociable, outgoing, gregarious, expressive, warmhearted, and talkative.

Image management. A person's disposition to monitor, observe, regulate, and control his/hers self-presentation and image he/she project during interactions with other people.

Intrinsic motivation. A person's disposition to be motivated by intrinsic work factors such as challenge, meaning, autonomy, variety, and significance. These traits are in contrast with extrinsic factors such as pay, earnings, benefits, status, recognition, etc.

Openness. A person's disposition to see the big picture as well as receptivity/openness to change, innovate, experience and learn new things.

Optimism. A person's disposition to maintaining an upbeat, hopeful outlook concerning situations, people, prospects, and the future, even in the face of difficulty and adversity. That is, a tendency to minimize problems and persist in the face of setbacks.

Teamwork disposition (agreeableness). A person's disposition to communicate and cultivate good human relationships, as well as a propensity for working cooperatively within a team.

Tough-mindedness. A person's disposition to appraise information and make work decisions based on logic, facts, and data versus feelings, values and intuition.

Visionary style. A person's disposition to focus on long-term planning, strategy, and envisioning future possibilities and contingencies.

Work drive. A person's disposition to work for long hours and an irregular schedule; investing high levels of time and energy into job and career as well as being motivated to extend oneself, if necessary, to finish projects, meet deadlines, be productive, and achieve job success.

Table 4.9 depicts a rather unflattering comparison between engineers and nonengineers based on the mean scores on the Williamson personality traits.

Table 4.9 Personality traits: Engineers versus nonengineers

Personality traits	Engineers scored significantly higher	No significant differences	Engineers scored significantly lower
Assertiveness			X
Conscientiousness			X
Customer-service orientation			X
Emotional stability			X
Extraversion			X
Image management			X
Intrinsic motivation	X		
Openness		X	
Optimism			X
Teamwork disposition		X	
Tough-mindedness	X		
Visionary style			X
Work drive			X

4.7.2.3 Innovative Engineers versus Traditional Engineers

Another study (Ferguson et al. 2014) examined how innovative engineers perceive themselves in comparison with traditional or non-innovative engineers. Traditional engineers use scientific knowledge (i.e. mathematics physics, chemistry, etc.) as well as engineering knowhow (i.e. Mechanical, Electrical, etc.) in order to design, test, produce, maintain and eventually dispose man-made artifacts. The Ferguson study, conducted in 2011–2012, interviewed 53 senior engineers, each averaging about 30 years' innovation and engineering experience. Fundamentally, researchers focused on the following research question: "What are the characteristics or knowledge, skills, and attributes that enable or inhibit engineers from translating their creative ideas into innovations that benefit society?" In addition, the Ferguson study identified five personal characteristics that differentiate innovative engineers from traditional or noninnovative engineers:

Challenger. Innovative engineers question or dispute the current way of doing things.

Collaborator. Innovative engineers work with other persons or group in order to achieve certain goals or do things.

Persistent. Innovative engineers continue to pursue things even though it is difficult or other people want them to stop. By and large, they continue beyond the usual, expected, or normal time.

Risk takers. Innovative engineers accept the possibility that something unpleasant may happen (such as project failure, loss of face, etc.).

Visionary. Innovative engineers have clear ideas about what should happen or be done in the future.

The study also analyzed how innovative engineers describe the difference between themselves and the traditional or noninnovative engineers (Table 4.10). Not surprisingly, this view reflects the tension that often exists within a firm between innovating new products and processes as compared with traditional project engineering problem solving.

Table 4.10 Comparison: An innovative versus a noninnovative engineer

An innovative engineer	A noninnovative engineer
Is a collaborator	Focuses narrowly on his area of expertise
Is a risk taker	Minimizes risk for himself and the firm
Has a long-term view	Focusing on solving immediate problems
Is persistent	Easily abandon efforts when encounter difficulties
Challenges rules	Sticks to existing rules

4.7.2.4 Innovative Engineers in Traditional or Noninnovative Organizations

It is the observation of this author, as well as other researchers, that the majority of companies and organizations are traditional, noninnovative entities. Some noninnovative engineers employed by these organizations embrace traditional practices. Most of them excel in their profession and contribute immensely to their organizations and to society at large. Quite often, however, innovative engineers find themselves employed in traditional or noninnovative organizations. They come up with numerous creative ideas, believing their companies could adopt them and prosper, but instead, their ideas are rejected quickly and on the flimsiest of grounds (Figure 4.50).

It would be interesting to ascertain just how much companies and society as a whole could have gained if these closet-innovators were listened to. After all, most of these individuals exhibit many innovation-desirable personality attributes, such as intelligence and knowledge, ability to see what others miss, as well as intrinsic motivation and an inclination to work hard. On the other hand, failure of these closet-innovators may also stem from certain personality vulnerabilities like: insecurity, lack of stubbornness, lack of tenacity and nonconformity as well as limited ability to influence other people

Figure 4.50 Quickly rejecting new ideas

4.7.3 Innovation Advice to Creative Engineers

Often creative engineers employed in traditional or noninnovative organizations struggle to convince their organizations to adopt new creative ideas. As seen in previous sections, this undertaking tends to be frustrating and, more often than not, creative individuals give up because they encounter too many obstacles. The purpose of this section is to provide some innovation advice to creative engineers who are engaged in traditional or noninnovative organizations. In particular, the author is convinced that moving ideas through an organization could be more successful if such individuals expand their skills and engage in the following efforts: (1) expanding their professional and intellectual horizons, (2) reducing risks inherent in new ideas, (3) dealing with colleagues, (4) dealing with management, and (5) adopting entrepreneurial attitude.

4.7.3.1 Expanding Horizons

Any innovative person should recognize that his/her current knowledge is, by definition, limited and should be expanded throughout one's life.

First, an innovative person should expand his vision as regard to the company or the organization he/she is working for. This could include learning about the company, its products, its structure, its key leaders and managers who run the company as well as their individual responsibilities. In addition, one should be familiar with the company's public policies and declared mission statements, particularly those related to innovation, new products and future aspirations. Finally, one should gain knowledge about the declared objectives of the company versus its actual performance. The objective of this learning is to identify supporters and allies as well as available resources (and obstacles), to be utilized (or avoided) in the quest to implement his/hers creative ideas.

Second, engineering-related knowledge is constantly changing and expanding. Staying ahead of the curve and seeking out new technologies and opportunities is vital for a creative person. Furthermore, despite the trend toward professional specialization, the need for cross-fertilization among engineering disciplines is important. For example, innovations in new materials practiced in the chemical industry could substantially impact the design of aircraft propulsion systems within the aerospace domain.

Another important expansion of an innovative person's horizon, relates to a better understanding of customers' needs in terms of new products, services and other expectations. Despite the tendency exhibited by some large companies to limit the interaction between rank and file engineers and consumers, creative engineers should talk with marketing people and more so, with customers. Preferably in an informal way, one should overcome potential communication barriers in order to gain customers' trust and learn about their needs and expectations. Based on such exchange, a creative person could discover hidden needs that could benefit both the company and customers.

Undoubtedly, creative persons could identify several additional domains in which horizon expansion could contribute immensely to the likelihood of becoming successful innovators in any company or organization.

4.7.3.2 Reducing Risks

Creative engineers often get an early "no" from their direct supervisors, because managers often have little desire to risk core activities for unproven innovation efforts. As a result, creative engineers often return to their regular jobs, and give up on their creative initiatives. One reason for such scenario is that managers are so involved in putting out fires and chasing short-term targets that most can't devote much energy to creative ideas. Another reason is that innovation is inherently a risky business. First, either the idea or the solution or both are new to the marketplace. Second, innovators often push the technological envelope, utilizing an immature technology or processes.

New products give rise to many potential risks related to their performance, durability, maintainability, affordability, safety, and the like. Similarly new processes raise potential problem issues like how the process is implemented? What could be the undesirable effects of the new process? and the like. At a higher level, organizations must always deal with deeper and more fundamental risks. For example: Are we correctly solving the wrong problem? Are we solving the right problem in a wrong way? Is the planned solution going to be available at the correct time frame and address the real needs of our customers?

By and large, managers and colleagues will not accept mere documents (e.g. Word, PowerPoint, Excel, etc.) as proof that a new product or a process will successfully make it to the market. The creative engineer should gather evidence to prove the proposed innovative effort is viable. Hopefully, this information will increase management's confidence in the proposed idea and make a potential business investment seem less risky. In addition, most people take some time to warm up to new ideas so they tend to reject them initially in a rather automatic manner. The flip side of these tendencies is that stakeholders who feel they've been personally involved in an innovative project tend to support it in the long run.

One way to mitigate the real or perceived risks is to create a prototype or a simulator, emulating the creative idea and its application. A prototype is an evolving instrument reflecting a dynamic perception of a new idea. At an early stage a creative engineer should build a rudimentary prototype that has limited capabilities like showing screen shots of systems' operations. At a more advance stage the prototype should be able to demonstrate basic functioning of the proposed creative idea.

Another way to reduce risks is to prepare a viable business plan. A business plan is a document describing how a creative idea will be transformed into a viable technical and financial product or service. An honest and realistic business plan could be an important asset for an innovative engineer in term of verifying the viability of his idea from a business standpoint. Furthermore, a convincing and affirmative business plan will go a long way to lessen the risk inherent in any innovative project. Typically, a business plan should cover the following topics: (1) executive summary, (2) business description, (3) market strategies, (4) competitive analysis, (5) design and development plan, (6) operations and management plan, and (7) financial consideration.

Of course, innovation needs time to develop. So, creative engineers should strive to reach a position within the organization in which they get some free time to experiment with new ideas. Some other innovative engineers quietly highjack research time from other projects in order to strengthen their innovative concepts.

4.7.3.3 Dealing with Colleagues

Innovative engineers who encounter difficulties in implementing their creative ideas can collaborate with their colleagues and friends, strengthening their leverage within a noninnovating organization. First, working under some type of collaborative effort reduces the innovative load, normally carried out by the creative engineer. This is done by way of spreading tasks among several people. In such shared setting, several people contribute their time and resources to the completion of a particular undertaking. Although members of a team may have conflicting ideas regarding the direction of the innovative process, team collaboration often produces improved overall solutions to a problem.

Another benefit to innovative collaboration is that when a creative idea is advanced by a group of engineers rather than by an individual, it will receive substantially more attention as managers will find it more difficult to reject it outright. When a creative engineer establishes an innovation group to support his creative idea, he should obtain appropriate consensus around goals, limitations, and evaluation criteria for the group. In addition, and as much as possible, group participants should have the appropriate combination of knowledge, skills, and authority to collaborate productively. Another advantage of creative collaboration among colleagues is the prospect of building lasting friendships with some of one's co-workers. Friendship offers companionship as well as a chance to discuss professional and personal issues in relative confidentiality. Another valuable aspect of friendship is that one can offer and receive a helpful hand or a valuable advice when needed.

4.7.3.4 Dealing with Management

This section describes (1) typical interactions between creative engineers and their immediate supervisor, (2) pushing ideas through a company's bureaucracy layers, and (3) dealing with the opposition.

Convincing the immediate supervisor. Most often, a creative engineer starts his/her quest for company support by presenting the creative idea and proposed innovative project to their immediate supervisor. Ideally, this suggestion is aligned with the long-term expansion strategy of the organization so managers are more likely to support it. However, more often than not, such alignment does not exist since creative ideas spring up in a rather random manner. In addition, in most noninnovating companies, managers preach the importance of innovation, but, in fact, do not have a clue regarding where the organization seeks to go and find no time and inclination to truly support innovative efforts. As a result, the most likely response will be to reject the stated request, under one guise or another.

At this stage, the creative engineer is advised to try to put himself in his supervisor's shoes and identify the true motivations for such a rejection. This analysis will place him in a better position to understand the goals of the organization vis-à-vis the proposed innovative project. This now is the make-or-break moment, which separates the boys from the men. If the engineer believes in his creative idea then he is advised to continue his battle up the hierarchical chain, relentlessly. However, he/she should know that by pursuing this advice, they escalate the situation so it is wise to discuss this decision in a positive manner with the immediate supervisor and avoid bad-mouthing him/hers under any circumstances.

Pushing ideas through the bureaucracy. Pushing ideas through the bureaucracy is not a simple undertaking. One is now swimming with the big fish. Any unwarranted blunder and one may finds himself out of the door. Most creative engineers are not familiar with the piranha-infested world of the strong and mighty. So, a creative engineer is strongly advised to start by investigating about the individual managers, their management style, and the culture of the organization in order to avoid mistakes and achieve maximum impact. Likewise, he should study both the formal and informal rules of the company and carefully abide by them. Otherwise, he might expose himself to difficult situations with someone seeking his sculp. Another aspect a creative engineer should get familiar with is the specific funding decisions criteria used within the company. Understanding what senior managers look for and tailoring his request accordingly will increase the likelihood of project approval.

Dealing with the opposition. A creative engineer should expect to encounter numerous objections to his ideas as well as his innovation strategy from several quarters. He/she is advised to listen, learn, and deal with this criticism calmly, rationally, and positively. In fact, one should consider such feedback, a gift freely offered. Thus, nudging him to improve one's creative idea, innovation strategy or the manner he/she presents it to the world. Furthermore, an innovative engineer is advised to seek collaborative relationships with individuals who oppose the proposed innovative efforts. Such collective efforts will show that all objections and concerns are seriously being addressed and resolved. Hopefully, such strategy, on the part of the creative engineer, will bring an opposing individual into the supporting camp.

4.7.3.5 Adopting Entrepreneurial Attitude

A creative person is essentially a dreamer in the sense that he/she dreams up creative ideas out of thin air. An innovative person is, first and foremost, a leader of people and a visionary. So, for the purpose of this section, we describe an entrepreneur as a person who is able to combine effectively creativity and innovative competences. That is, he/she is able to conceive new ideas as well as organize and manage risky projects and enterprises.

Therefore, adopting entrepreneurial attitude starts with the ability to listen to other people without putting forward ones' own point of view. That is, asking people to explain exactly what they mean and trying to see the good in what someone else says. Often time, asking questions forces one to consider different aspects of an issue, extending ones' horizon. Furthermore, entrepreneurs should train themselves to expect and welcome criticism of their creative ideas. Healthy entrepreneurial attitude should be: "If my ideas are not criticized and labeled as impossible then my listeners must have lost me somewhere along the way."

In addition, adopting entrepreneurial attitude includes improving ones communication skills. That is the ability to work with colleagues, top company brass, outside customers and suppliers as well as making formal presentations in front of large audiences. In general, an engineer-entrepreneur should adopt more business-like thinking than exhibited by many engineers. This means getting involved in strategy issues, planning and negotiating thus gaining organization-wide point of view.

Finally, adopting entrepreneurial attitude should include the development of leadership skills. This topic is outside the scope of the book but, ideally, it should include: (1) having a clear sense of where one wants to go and how one plans to do so, (2) making difficult and sometimes unpopular decisions because one has confidence in himself and in his abilities, (3) having courage to act in risky situations where desired results are not assured, (4) ability to build productive and committed team that draws the best from each of its members, (5) capacity to motivate others by articulating an achievable vision and ideals, (6) exhibiting personal integrity so others will be able to listen and follow, (7) ability to communicate effectively so diverse individuals could understand and follow a message, and (8) generosity in helping others to achieve their full potential.

4.7.4 Further Reading

- Adizes, 1990.
- Adizes, 2017.
- Baden-Fuller and Stopford, 1994.
- Carroll, 1993.
- Dougherty et al., 2000.
- Dyer et al., 2011.
- Ferguson et al., 2014.
- Grant, 2017.
- Holland, 1997.
- Isaacson et al., 2015.
- Keeley et al., 2013.
- Kingdon, 2012.
- Reed, 2016.
- Schein, 2004.
- Siegle et al., 2002.
- Williamson et al., 2013.

4.8 Human Diversity and Gendered Innovation

Human diversity as well as gender issues have a profound effect on creativity and innovation within organizations. The purpose of this chapter is to acquaint readers with this subject.[38]

4.8.1 Human Diversity

Humans are different from one another.[39] This is easily deduced, for example, from Figure 4.51.

Within an organization, individuals differ from one another along a wide range of visible and invisible dimensions. For example, individuals differ along ascribed and achieved characteristics. Ascribed characteristics include demographic attributes such as gender, age, ethnicity, as well as cognitive models, attitudes,

---

38 Parts of this chapter, related to gender issues, were adapted from: Danilda and Thorslund book - Innovation & Gender, 2011. Reproduced with permission of the authors.
39 This section was inspired by Østergaard C.R. presentation: "Innovation and employee diversity: Does diversity really matter for innovation?"

Figure 4.51 Human diversity – Asiatiska folk

values, and norms. In contrast, achieved characteristics include educational background, functional background, work experience, and knowledge base.

By and large, diversity of individuals within an organization has three dimensions: (1) types, i.e. the number of different types or groups, (2) balance, i.e. the shares of the different groups, and (3) disparity, i.e. the distance between the different groups. The measure of individuals' diversity is the distribution of differences among these individuals within the organization with respect to a common attribute.

There are several methods to measure the diversity of individuals within an organization. For example, the Shannon index has been a popular diversity index, where, p_i is the proportion of individuals associated with the ith characteristic in the relevant domain.

$$Diversity = \sum_{j=1}^{m}\left[\sum_{i=1}^{n}p_i\left(ln\frac{1}{p_i}\right)\right]$$

Let us examine two synthetic examples based on the following domains and characteristics:

1) Gender diversity domain composed of: (1) males and (2) females.
2) Ethnic group domain composed of: (1) Africans, (2) Caucasians, (3) Oceanian, (4) East Asian, and (5) Native American.

3) Academic discipline composed of: (1) engineering, (2) exact sciences, (3) life sciences, (4) social sciences, (5) management, (6) humanities, and (7) arts.
4) Academic level domain composed of: (1) bachelor's degrees, (2) master's degrees, and (3) PhD degrees.

Table 4.11 and Table 4.12 depict synthetic examples of two organizations composed of 100 persons, of which, 70 are holding academic degrees. The first organization, with Shannon diversity index of 1.38, is quite homogeneous whereas the second organization, with Shannon diversity index of 4.67, is a lot more diversified.

Now, the above synthetic examples may exaggerate the situation but there are clearly differences in diversification levels among the different organizations. Interestingly, there is a substantial body of knowledge showing that firms composed of diverse individuals in terms of knowledge, experience, and skills, benefit from complementarities that foster development and innovation.

This phenomena stems from several reasons: First, education, experience, and demographic characteristics affect the interpretation of problems and their solutions strategy. In fact, diversity of management composition proved a predictor for firm's performance. Differences in the knowledge base within a firm create opportunities for learning and insights. Similarly, employee diversity creates a broader search space, making firms more open towards new creative ideas. At the same time, too much diversity can hinders innovation because diverse individuals and groups could fracture healthy competition, creating conflict and distrust, which may increase transaction costs leading to reduced cooperation and internal communication.

In conclusion: employee diversity generally improves a firms' innovative performance. The strongest effect emanates from diversity in education. Next, employee diversity in terms of gender, age, and education has a secondary effect on creativity and innovation. Ethnicity seems to have no effect on creativity and innovation whereas diverse age distribution tends to have a negative effect on creativity and innovation.

4.8.2 Shift in Gender Paradigm

Gender diversity as well as gender equality contributes to innovativeness within an organization. The term *gender* may be defined as: "A concept that refers to the social differences, as opposed to the biological ones, between women and men that have been learned, are changeable over time and have wide variations both within and between cultures."[40] The same source defines

---

40 See: 100 words for equality A glossary of terms on Equality between Women and Men, European Commission.

Table 4.11 Synthetic example: Homogeneous organization

Gender diversity	Males	Females						Total	Index
	99	1						100	0.06
Ethnic group	Africans	Caucasians	Oceanian	East Asian	Native American			Total	Index
	4	80	4	10	2			100	0.74
Academic discipline	Engineering	Exact Sciences	Life Sciences	Social Sciences	Management	Humanities	Arts	Total	Index
	65	5	0	0	0	0	0	70	0.26
Academic level	Bachelor's	Master's	PhD					Total	Index
	62	6	2					70	0.32
Shannon diversity index									**1.38**

Table 4.12 Synthetic example: Diversified organization

Gender diversity	Males	Females						Total	Index
	55	45						100	0.69
Ethnic group	Africans	Caucasians	Oceanian	East Asian	Native American			Total	Index
	15	54	10	15	6			100	1.30
Academic discipline	Engineering	Exact Sciences	Life Sciences	Social Sciences	Management	Humanities	Arts	Total	Index
	25	15	4	6	10	6	4	70	1.72
Academic level	Bachelor's	Master's	PhD					Total	Index
	40	20	10					70	0.96
Shannon diversity index									**4.67**

the term *gender equality* as: "The concept meaning that all human beings are free to develop their personal abilities and make choices without the limitations set by strict gender roles; that the different behavior, aspirations, and needs of women and men are considered, valued, and favored equally." The quantitative aspect implies an equal distribution of women and men in all areas of society such as education, work, recreation, and positions of power. The qualitative aspect implies that the knowledge, experiences, and values of women and men are given equal weight and used to enrich and direct all areas of society.[41]

However, these lofty ideas are not yet translated into facts on the ground. For example, in Sweden, an advanced country, as far as gender equality is concerned, the occupational disparity between men and women is apparent. Figure 4.52 depicts a gender distribution per occupation (Source: Women and Men in Sweden Facts and Figures 2016). This figure provides a subset of occupations where either men or women constitute over 80% of the given occupation's work force. Clearly, men's occupations are significantly more lucrative then women's. According to this figure and theories of the gender system, women and men are segregated horizontally as well as vertically in the labor market. The horizontal level implies that women (as a group) and men (as a group) are active within different sectors of working life. In addition, in most

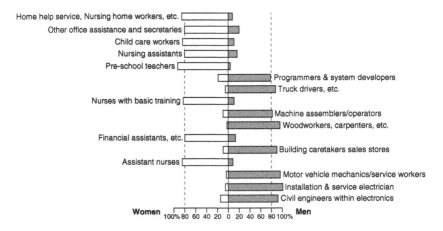

Figure 4.52 Gender distribution per occupation (Sweden, 2016, Subset of data)[42]

---

41 See also: Women and men in Sweden: Facts and figures 2012. https://masculinisation.files. wordpress.com/2015/05/women-and-men-in-sweden-facts-and-figures-2012.pdf. Accessed November 2017.

42 The reader should note that, although names of some occupation categories are similar, each employment category is mutually exclusive.

countries, only few occupations exhibit balanced participation by women and men. Vertical segregation is evident in the fact that women and men hold different management and leadership positions in working life.

However, over the last decade or two, considerable number of researchers across the world found that the economic case for gender equality is, in fact, much more pronounced than the prevailing social case. While the social case stresses the need for equal treatment of all individuals within the organization, the economic case stresses the wider economic benefits that span individuals, enterprises, regions and nations as well as addressing inequalities in the wider labor market. Such economic benefits include, among others, greater level of innovation, and enhanced marketing opportunities, enhanced employee recruitment, and retention as well as improved corporate image and reputation.

The shift in gender paradigm highlights a new gender perspective, which involves looking at the impact of gender on women's as well as men's opportunities, social roles, and interactions. Increased female participation in the workforce contributes to quality of life and social well-being as well as economic development through increasing national and global gross domestic product (GDP). Thus, applying gender perspective becomes important in contributing to innovation and sustainable growth. More specifically, this initiative opens new vista for tackling some of humanity's "grand challenges," like global warming, scarcer energy, food and water supplies, aging population, public health, pandemics, security, and the like.

4.8.3 Gender Disparity and Innovation Implications

4.8.3.1 General

Several recent studies have focused on how gender is integrated within innovation policy. These studies concluded that innovation and technology are generally male-dominated fields, resulting in an interpretation of men as technically or scientifically skilled and women as unskilled in these areas. These studies also show that, in theory, current definitions seem gender neutral, but in practice, the way they are operationalized and measured, they are strongly male-gender biased. That is, interpretation of the technology as a male field gives innovation a masculine connotation.

Other studies examined how the proportion of women and men within a team affects creativity and innovation performance of individuals within the team. Researchers found out that teams operate most effectively when the gender composition is balanced, i.e. about 50–50 percent. As gender balance shifts either way, individuals in the minority faction tend to exhibit less psychological safety, less self-confidence, less tendency for experimentation, and less efficiency at work. In addition, men in a minority tend to bond with one another, whereas women in the minority tend to go outside the group for interactions and networking.

4.8.3.2 Gender Disparity

Innovation research shows clear and unequivocal cognitive and emotional differences between males and females. Probably the most pronounced difference relates to risk taking. Women, by and large, are more risk-averse than men. More specifically, women tend to minimize loss and men tend to maximize gain. One explanation is that men tend to overestimate their potential abilities, which leads them to underestimate risks. At the same time, women either side-step or overcome risky situations in order to avoid the consequences that may emerge. Innovation is always associated with risk and uncertainty, as one cannot innovate without taking risky decisions. As a result, women as individuals tend to be less effective.

Another gender difference relates to social skills. Most women have better facilities for networking, collaboration, empathy, inclusion, and sharing power. Along this line, researchers found that women tend to introduce innovations that are socially oriented, such as improving environmental issues, developing local economic actors, and increasing employment. By and large, men tend to work longer hours than women. This often has a bearing on the overall innovative output of men versus women. Researchers speculate that this is due to different cultural pressures on each of the genders, where women appear to undertake the majority of domestic labor, including parenting responsibilities and keeping track of the family's activities (e.g. sports, clubs, play dates, and doctor's appointments).

Another substantial differentiator between the two genders relates to the contrast in professional biases between men and women. Such biases are clearly expressed in the number and composition of universities' graduates. For example, Table 4.13 and Figure 4.53 depict the number of bachelor's degrees conferred in 2013–2014 by the US postsecondary institutions, segmented by sex of students and academic disciplines.[43] As can be seen, although more women attain bachelor's degrees than men, they are more likely to gravitate toward education, health professions, and psychology, whereas much more men than women seem to gravitate toward computer and information as well as sciences and engineering. In fact, according to the US Census Bureau of September 2013, women were significantly underrepresented in the science, technology, engineering, and mathematics (STEM) fields. For example, in the United States, women made up less than one-quarter of those in the lucrative STEM jobs.

Finally, researchers point out that traditional gender roles related to women's motherhood and greater responsibility at home negatively affect women's work-related contribution and consequently, their advancement within the

---

43 Based on the National Center for Education Statistics (NCES), the primary federal entity for collecting and analyzing data related to education in the United States and other nations. See: https://nces.ed.gov/programs/digest/d15/tables/dt15_318.30.asp. Accessed: April 2017.

Table 4.13 Awarded bachelor's degrees by sex and discipline: US 2013–2014

Academic discipline	Abbreviation	Males	Females	Total
Agriculture and natural resources	AGR	17,249	17,867	35,116
Architecture and related services	ARC	5,173	3,971	9,144
Area, ethnic, cultural, gender, and group studies	ARE	2,466	5,809	8,275
Biological and biomedical sciences	BIO	43,427	61,206	104,633
Business, management, marketing, and personal and culinary	BUS	188,418	169,661	358,079
Communication and communications technologies	COMM	34,370	58,221	92,591
Computer and information sciences and support services	COMP	45,393	9,974	55,367
Education	EDU	20,353	78,501	98,854
Engineering and engineering technologies	ENGI	88,938	20,031	108,969
English language and literature/letters	ENGL	15,809	34,595	50,404
Family and consumer sciences/human sciences	FAM	3,014	21,708	24,722
Foreign languages, literatures, and linguistics	FOR	6,266	14,069	20,335
Health professions and related programs	HEA	30,931	167,839	198,770
Homeland security, law enforcement, firefighting, and related	HOM	33,383	29,026	62,409
Legal professions and studies	LIG	1,456	3,057	4,513
Liberal arts and sciences, general studies, and humanities	LIB	16,485	28,775	45,260
Mathematics and statistics	MAT	11,967	9,013	20,980
Multi/interdisciplinary studies	MUL	16,119	32,229	48,348
Parks, recreation, leisure, and fitness studies	PAR	24,713	21,329	46,042
Philosophy and religious studies	PHI	7,582	4,415	11,997
Physical sciences and science technologies	PHY	17,802	11,502	29,304
Psychology	PSY	27,304	89,994	117,298
Public administration and social service professions	PUB	5,917	27,566	33,483
Social sciences and history	SOC	88,233	84,863	173,096
Theology and religious vocations	THE	6,594	3,048	9,642
Transportation and materials moving	TRA	4,053	535	4,588
Visual and performing arts	VIS	38,081	59,165	97,246
Total		**801,496**	**1,067,969**	**1,869,465**

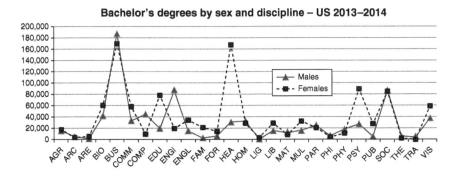

Figure 4.53 Awarded bachelor's degrees by sex and discipline: US 2013–2014

organization and earning power. This phenomenon, often termed *gender pay-gap*, is derived from several factors. First, the birth of a child causes many women to leave their work for certain duration or even withdraw from the labor force altogether. Second, anticipating such possibility, some employers are reluctant to make large investments in hiring and training women of child-bearing age. Third, motherhood and traditional family responsibilities often reduce women's productivity at work. For example, they may expend less efforts and initiatives, have constraints on work and travel schedules, and may exhibit reluctance to be promoted to more demanding jobs.

4.8.3.3 Innovation Implications

The relative absence of women in certain professions and industries, specifically in the mathematically intensive STEM fields, coupled with the motherhood wage penalty, plays a major role in the endemic gender pay-gap. On the other hand, several studies show that gender-related differences can be leveraged to advanced innovative ideas, especially in areas related to the quality of life. In particular, forming mixed and roughly balanced teams of men and women in technological contexts is a critical component in increasing productivity. This finding is explained by the observation that gender diversity provides different perspectives and insights and the combination of these different viewpoints offers a wider range of ideas. More specifically, within a climate of balanced proportions of men and women, individual team members exhibit optimal psychological safety, readiness to experiment as well as self-confidence. This, in turn, increases the likelihood that the team as a whole will be more innovative and productive.

4.8.4 Advancing Gendered Innovation

In their book *Innovation & Gender* (2011), Danilda and Thorslund discuss six core statements describing how gender perspective can help strengthen weak links in the innovation milieus and provide a track towards an innovation case for gender diversity. In this section, the author adapts these core statements into a set of six strategies designed to advance gendered innovation (Figure 4.54).

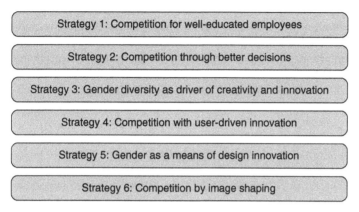

Strategy 1: Competition for well-educated employees

Strategy 2: Competition through better decisions

Strategy 3: Gender diversity as driver of creativity and innovation

Strategy 4: Competition with user-driven innovation

Strategy 5: Gender as a means of design innovation

Strategy 6: Competition by image shaping

Figure 4.54 Six strategies for advancing gendered innovation

The rationale for integrating a gender perspective into the innovation milieu can have different sources. Nevertheless, the common factor is change driven by the need for competencies, better effectiveness and new products or markets. The point here is that to be a leading organizations within their domain, companies must "think new" and be innovative. Applying a gender perspective is one way to achieve the two objectives of equality and growth at the same time.

By applying a gender perspective, companies will be able to develop existing (or create new) innovation processes as well as commercialization models for alternative value propositions and modes of consumption. Integrating a gender perspective into the innovation milieu will afford the actors the necessary gender-awareness to identify potential improvements for gender equality and sustainable growth. Driven by sustainability, the "third industrial revolution" signifies new research and innovation opportunities in green technology and sustainable design.

To sum it up, gender perspective can help highlight normative[44] thinking about women and men that is an obstacle to innovative thinking. In fact, less normative thinking about women and men will lead to new ideas, new business opportunities, the identification of new markets, and excellence in innovation milieus.

---

44 A feminist theory that asserts that most languages employ the male category as the norm and the corresponding female category as a derivation and thus less important. This, according to proponents of the theory, reflects a social gender bias.

4.8.4.1 Strategy 1: Competition for Well-Educated Employees

The ability of companies to attract and retain talented human capital is crucial to their success and competitiveness in the market. Companies are looking for technology workers with more experience and a broader skill set. Competition for these employees, combined with a drop in science graduates and the impending retirement of the baby-boom generation, has led to fierce recruitment competition amongst enterprises. Companies should adopt effective diversity-inclusion practices in order to benefit from reduced absenteeism and employee turnover.

Organizations should actively recruit and promote women into top executive board membership positions in order to balance the male–female composition so women's talents will be tapped to their full potential.

If the global world and especially, its Western part, is to achieve its goal of becoming a dynamic and competitive knowledge-based economy in a globalized world, then all concerned, governments, corporations, universities and other, must make better use of women's talents and skills. In addition, companies nowadays recognize their need for more leaders with varied skills and women have the skills to meet the new demands of technological work in terms of technical, business, and interpersonal skills. Therefore, companies should understand the gendered opportunities in their markets and among their customers. This should lead to promoting women into leadership positions from in-house as well as from outside resources, thus building a talent pipeline and achieving gender-balanced leadership.

Studies in Europe and the United States highlight the fact that senior managers often point to insufficient experience among potential female leaders and board members as an obstacle to women's advancement. Women, on the other hand, often point out the main obstacles to their advancement is gender stereotypes, lack of role models, and adverse attitudes within the organizations.

Be as it may, the talent shortage is the most easily recognized case of innovative opportunity for gender equality. Therefore, companies should recognize that the underrepresentation of women in engineering and technology studies threatens their future recruitment pool. Therefore, companies in the innovation milieu should offer financial and other benefits to university female students in the STEM fields in order to find and attract them as future high-tech employees.

4.8.4.2 Strategy 2: Competition through Better Decisions

For companies as well as innovation milieus, gender diversity makes for improved decision-making at all organizational levels and results in better decisions. This has been shown in a variety of settings, occupations, and organizations, and also applies to group task performance, creativity, and innovation. Gender diversity is especially important and beneficial to problem-solving tasks. A recent industry report estimates that by 2012, teams with balanced participation of women and men will double their chances of exceeding performance expectations when compared to all-male teams.

Research has also found a correlation between the presence of women in higher management and financial echelons of the organization, as measured by total return to shareholders and return on equity. For instance, a Finnish study reveals a positive and significant correlation between female leadership and firm profitability. Even if the researchers do not prove causality, their findings have several important implications. The study suggests that a firm may gain a competitive advantage over its peers by identifying and eliminating obstacles to women's advancement to top management. Besides being fair, gender awareness in the provision of career opportunities is also in companies' best interests.

For example, in the case of the Swedish innovation milieu Fiber Optic Valley,[45] the change agents in a gender network and top managers from the organizations involved were offered tailor-made training in innovative leadership. This training provided a theoretical understanding of how the lack of a gender perspective affects results and limits profitability. As a result of the program, the managers involved increased their capacity to lead a change process, applied their knowledge of gender equality and created an innovative environment.

4.8.4.3 Strategy 3: Gender Diversity as a Driver of Creativity and Innovation

Innovation is about creating something new and is enhanced by diversity in gender, experiences, perspectives, knowledge, and networks. Individuals – women and men – who are allowed to develop their full potential will be creative, engaged, and willing to take risks.

Companies should institute a gender diversity policy because there is a positive relationship between gender diversity in the firm's knowledge base and its innovative capabilities. Human capital resources have different dimensions such as education, training, and experience. Demographic dimensions such as gender, age and cultural background also affect the application and combination of existing knowledge and the communication and interaction between employees. Employee diversity is often considered positive since it could create a broader search base and make the firm more creative and more open toward new ideas. Ideally, gender diversity should increase a firm's knowledge base and increase the interaction between different types of competencies and knowledge.

Furthermore, innovation often depends on groups of individuals in the organization. It is in the context of a complex social system in an organization, where different types of knowledge come into play to generate new knowledge

---

45 See: http://fiberopticvalley.com/english/. Accessed: Apr. 2017.

or ideas. Therefore, the composition of individuals within the enterprise is an important factor for understanding innovation, since a diverse workforce also contributes to diversity in the knowledge base. Enterprises with a balanced workforce (50–60% of same gender) are almost twice as likely to innovate compared to those with the most segregated workforce (90–100% of same gender). Thus, a balanced gender distribution has a strong effect on the likelihood to innovate and, in general, the innovative performance of the enterprises.

4.8.4.4   Strategy 4: Competition with User-Driven Innovation

User-driven innovation creates successful new concepts, products, and services for companies and organizations. Companies should exploit external actors like customers, suppliers, and other stakeholders, including them in the innovative process. This tapping of knowledge regarding their problems and needs will, eventually, generate successful and profitable products and services.

The complicated nature of innovation renders it almost impossible for a single company to achieve the next breakthrough on its own. Nowadays, companies need to open their innovation processes to include their users, partners, or suppliers and ensure they snap up the next bright idea of relevance to the company. One way for ensuring that outside ideas and knowledge find its way into the companies is to include users in the innovation process. Valuable insights can be gathered by the company at the front end of the innovation process by tapping users' tacit knowledge and understanding of their needs and the challenges they face.

Companies and organizations can work with user-driven innovation in many ways, including gender as an important clue to revealing user needs. For example, according to Bloomberg,[46] women make up 85% of all consumer purchases (Figure 4.55), yet 90% of technological products and services are designed by men.

So as an important engine of the worldwide economy, companies should seek women's inputs in innovation, production, distribution, and sales cycle.

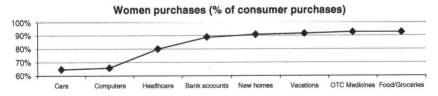

Figure 4.55 Women items purchases as percent of consumer purchases (Bloomberg)

---

46 Bloomberg Television, July 22 2016, Interpublic Group's Michael Roth discusses women and consumer purchasing.

Involving more women can bring new markets and new technological applications to the design process. It is a fair assumption that effective marketing to women, without a reliance on stereotypes and normative prejudices, would open up new lines of business.

For example, the Swedish innovation milieu Skane Food Innovation Network (SFIN)[47] has been influenced by ideas on open innovation and user-driven innovation. Its gender initiative is part of the foresight activities aimed at identifying future challenges for the food industry and areas where the milieu needs to focus. An important point of departure for the network is a gender perspective on food production innovation and how this perspective could help develop the industry.

4.8.4.5 Strategy 5: Gender as a Means of Design Innovation

Consumers are interested in the individualization of products as well as products that break with traditional gender stereotypes. Commercial interests have been criticized for designing for the dominant group in society, notoriously for the young, white, able-bodied, highly educated male. This approach excludes, of course, other groups who do not fit these criteria. It has been argued that in designing for everybody, designers are actually subconsciously following the male norm in society.

There is a positive and a negative aspect to the companies' focus on the ways women differs from men. On the one hand, a focus on what women want could serve to strengthen and give value to feminine-connoted skills and preferences. On the other hand, by developing a product based on "typical women's interests," designers run the risk of reinforcing and re-inscribing perceived gender differences rather than "transforming gender." Gender design might reproduce stereotyped patterns and a view that women and men are different. Moreover, designers' beliefs about women do not often conform to the skills, preferences, and experiences of most women.

The exception to this rule might be what has sometimes been called *feminist direct user-involvement techniques,* where potential males and females users are involved in the design from an early stage. Companies are encouraged to adopt this approach in order to match female end-users' desires. Again, some practical problems come up with these techniques, as designers need to pay more attention to the selection of potential test-users in order to ensure that they represent what the end-users' designers seek to reach.

For example, in the case of the "All Aboard" project run by the Swedish Marine Technology Forum (SMTF),[48] the long-term goal was to develop new products and services for the leisure boat market. A further goal was to increase

---

47 See: https://ec.europa.eu/growth/tools-databases/regional-innovation-monitor/organisation/ sydsverige/sk%C3%A5ne-food-innovation-network. Accessed: April 2017.
48 See: http://smtf.se/en/. Accessed: April 2017.

the number of nontraditional product developers working in the boat industry, especially women. The All Aboard case strived toward design-driven innovation and inclusive design approaches. Increased innovative capacity was sought in order to improve the industry's competitiveness. The main motivation for the cluster was to utilize the potential for increased sales, innovation, and product development by engaging demanding customers and taking into account their ideas. One of several hypothesis based on findings from the car industry (such as Volvo's "Your Concept Car – by women for modern people" project[49]) was that women (as a group) would be more interested in environmental, safety, and security aspects of leisure boat life as opposed to men (as a group).

4.8.4.6 Strategy 6: Competition by Image Shaping

Socially progressive companies are advised to adopt initiatives to hire, promote, and retain gender diversity at all levels of the organization. Such initiatives are not driven solely by the desire to increase revenue, but rather, to achieve other goals. In particular, companies with diverse workforces convey their commitment to tackle social equality and thus benefit from a better image in the marketplace. In the context of sustainable development, organizations are taking a different approach to doing business. Financial performance is no longer regarded as the exclusive driver.

Instead, economic, environmental, and social factors, including gender equality, should play an increasing role in management's decisions. Under greater scrutiny than ever before, organizations should be encouraged by their stakeholders to account for and take responsibility for their economic, environmental, and social (EES) performance. Sustainability reporting has emerged as an effective mechanism by which organizations communicate EES information transparently with employees, suppliers, customers, investors, and others. Although gender equality is being recognized globally as a priority, many organizations should make extra efforts to transform this recognition into practice and then communicate it.

4.8.5 Gendered Innovation Example

The following example describes the "All Aboard" project that was initiated by the Swedish Marine Technology Forum (SMTF), aimed at improving the design of leisure boats (Figure 4.56) through gender innovation and stimulating cooperation between companies in the maritime industry. This example is derived, with minor editorial work, from the Danilda and Thorslund book, *Innovation & Gender* (2011).

---

49 See: https://www.media.volvocars.com/global/en-gb/models/ycc/0. Accessed: April 2017.

Figure 4.56 Typical small leisure boat

4.8.5.1 The "All Aboard" Gendered Innovation Project

As with many other industries, the norm in the boat industry has been "men" and "male," but surveys carried out in Sweden indicated a majority of non–boat owners interested in buying or renting a boat are women. A survey in 2007 by the Stockholm International Fairs showed that there are many potential leisure boats buyers among women aged 25–45. The main driving force behind All Aboard was to utilize the potential for increased sales through engagement with demanding customers and taking their ideas into account. One of several hypothesis based on the findings from the car industry was that women (as a group) more than men (as a group) would be interested in environmental, safety, and security aspects of leisure boating. Research findings indicated that women are more predisposed than men to buying eco-labeled products. In fact most projects initiated by Swedish Marine Technology Forum have a sustainability perspective, and one of the main challenges for the maritime industry is minimizing its environmental impact.

The All Aboard project goal was to find innovative new products and services for the leisure boat market and, taking a long-term perspective, the innovation milieu aimed to break the gender segregation in the Swedish maritime industry. As of 2011, estimates are that this industry employs fewer than 10% women, making it one of the most horizontally segregated. More female designers, product developers and employees could contribute to broader perspectives and views. Design is at the core of the All Aboard project, and cooperation with industrial designers was perceived as a way of attracting women and of innovating and stimulating cooperation between companies in this cluster as well as academia. In addition, increased innovation capacity was expected to improve the competitiveness of the maritime industry.

4.8.5.2 Communication with End Users

The first stage for All Aboard was a survey carried out in cooperation with a Swedish network – Kvinnor pa Sjon (Women at Sea) – for women engaged in (leisure) boating. The network promotes women as role models and entrepreneurs and arranges courses and activities specifically targeting women. Over 300 female and some male boat owners completed a questionnaire in two sections. One section targeted sailors and motorboat drivers with topics on comfort and safety, whilst the other targeted sailors with topics linked specifically to sailing. The respondents came up with numerous ideas on how to improve leisure boats and requirements for new products and services for leisure boating.

Boat owners were interested in such things as better solutions for storage, damp, safety on board, solar energy, accessibility and boarding, and leaving the boat. Based on this survey, a number of key development areas were identified, and some of the problems encountered by the respondents were visualized in situation diagrams.

In the next step, the survey and diagrams were presented at the Scandinavian Boat Show, one of the region's most important trade fairs for the leisure boat industry. Companies from the Swedish maritime cluster demonstrated products to solve some of the problems encountered by boat owners and presented possible future solutions. During the fair, visitors – women, men and children – were able to contribute still more new ideas and suggestions based on the diagrams and demonstrations. This contact provided further input for All Aboard, reinforcing dialogues between companies and potential customers and between different companies within the cluster.

Not surprisingly, the situations, problems, and development areas identified in the All Aboard project were familiar to many boat owners, whether female or male. According to one of the respondents in the survey, "most of the things on boats seem to be geared to the average man" with "men" and "male" the norm for leisure boats, as stated above. Nevertheless, as experienced by many consumer product industries, the "average man" seldom exists, and such man is actually extremely difficult to find. Subgroups of boat owners have different expectations and demands. In addition, aspects such as age, lifestyle orientation, socioeconomic status, and cultural contexts influence customer preferences alongside gender.

After the initial steps of All Aboard, students from the University College of Arts, Craft and Design in Stockholm were asked to interpret and come up with ideas for solutions based on the material from the project. During the Stockholm International Boat Show, a design workshop was arranged with students from University College. This workshop generated dialogues on ways of involving end users, utilizing the creativity of (potential) boat owners and taking into consideration the problems they experienced.

4.8.5.3 Toward Gender-Aware Design

The All Aboard project has been dubbed "the most exciting boat project in Sweden" by actors in the cluster, and so far, the initiative has generated input for new products and services, designs, and strategic plans. It has contributed to a positive image for the innovation milieu and potential customers appreciated the opportunity to express their views on possible product development areas. The project engaged with boat owners, women's networks, boat companies, national agencies, and the media.

At the next stage, a virtual boat was expected to be presented based on material from surveys, designers and companies in the cluster. Discussions are underway about whether or not to build a full-scale model of a leisure boat. Even though a model has not been launched, the mere idea of a leisure concept boat has resulted in much press coverage. This can partly be explained by the cooperation with former YCC[50] team members and the opportunity to benefit from their strong "brand."

Swedish boatbuilding has old traditions, and there are several strongly branded companies on the global market. Internationalization is generating new opportunities as well as fierce competition in the leisure boat segment. Therefore, attracting more female customers is seen as way to gain market share. Swedish Marine Technology Forum is convinced that it will be advantageous on the national and international markets for individual companies, the cluster and the boat industry to consider the preferences of both women and men. Designing boats for and responding to requirements of customer subgroups as well as involving women and men – designers and users – in product development will provide a competitive advantage.

All Aboard has raised awareness among cluster companies of the inherent market opportunities when women influence leisure boat purchases. Few companies had considered the ingrained norms of the boat industry prior to this project. The potential of inclusive, gender-aware design has not yet been fully recognized by the boat industry. There are still some obstacles to overcome before everyone can be brought on board and companies in the industry can connect with sizable market segments.

---

50 Volvo's project: "Your Concept Car, a project with women in the driver's seat," See: https://www.media.volvocars.com/global/en-gb/media/pressreleases/4934. Accessed: September 2017.

4.8.6 Further Reading

- Alsos et al., 2016.
- Blau and Kahn, 2016.
- Catalyst, 2017.
- Danilda and Thorslund, 2011.
- Díaz-García et al., 2013.
- Landivar, 2013.
- Ostergaard, 2017.
- Page, 2008.
- SCB, 2016.
- Schiebinger et al., 2011–2015.

4.9   Cognitive Biases and Decision-Making

Ovadia Harari (1943–2012), a revered aerospace engineer at the Israel Aerospace Industries (IAI) and a director of the Lavi fighter aircraft program at IAI (Figure 4.57), used to equate system engineers to eagles soaring at 10,000 feet above the earth, seeing for many miles in all direction and, simultaneously, able to spot and capture a small rabbit scurrying in the grass below. "System Engineers should be able to grasp the entire system and also be able to focus on its most minute component. In essence, System Engineers should listen to other engineers and then make decisions. Then they should lead the entire engineering team towards the stated goal," he used to say.

Figure 4.57 Ovadia Harari and the Lavi fighter aircraft[51]

The name of the game, though, is to make correct strategic decisions, which, in turn, depends on the cognitive abilities of the given system engineer. Cognitive abilities relate to psychological processes involved, among other, in acquisition and understanding of knowledge, formation of beliefs and attitudes, and problem solving, as well as decision-making.

Unfortunately, the human brain is wired in such a way that it often exhibits a systematic deviation from rational judgment. As a result, inferences about

---

51 The Lavi project was carried out by the Israel Aerospace Industries (IAI). The project started in 1979 and the first Lavi flight took place on December 31, 1986. The project was terminated by the Israeli government on August 30, 1987.

the world around us tend to be distorted in a somewhat illogical fashion. This phenomenon, known as cognitive bias, operates at a subconscious level, often leading engineers to make unintended mistakes or even disasters in their professional work (as well as in their personal life). So the purpose of this chapter is to acquaint engineers and other persons with the nature and effects of cognitive biases as well as to provide readers with some measure to mitigate its outcome.

4.9.1 Cognitive Biases

This section describes the concept of cognitive biases and provides a selected list of cognitive biases. In addition, it chronicles a bitter disaster that might be partially attributed to cognitive bias failures. Lastly, this section provides some advice on cognitive debiasing.

4.9.1.1 Concept of Cognitive Biases

In 1972, the late Amos Tversky and Daniel Kahneman[52] identified the phenomena of cognitive biases. They conducted a series of experiments and empirical studies that proved that traditional economic assumptions of rationality theory are, often, not valid. Framing bias, for example, arises when a problem is presented with two equal solutions that are expressed in different ways. People tend to choose one solution, depending on the way it is stated. For instance, participants in an experiment were asked to choose between two treatments for 600 people affected by a deadly disease. Treatment A is expected to save 200 lives whereas treatment B is predicted to result in 400 deaths. The "positive" treatment (A) was chosen by 72% of the participants whereas only 22% of them chose the "negative" treatment (B).

Generally speaking, each person acquires during his lifetime a unique set of cognitive biases. This leads each individual to create his/her own subjective reality from their perception of themselves as well as their environment. As a result, an individual's subjective reality dictates that person's behavior in the world. Thus, cognitive biases are the root cause of certain level of irrationality (e.g. perceptual distortion, inaccurate judgment, and illogical interpretation) of many people. Systems engineers, responsible for large and complex systems, should do their utmost to recognize and mitigate their own cognitive biases. They should discover their tendencies to think along repeatable patterns, which may lead to systematic deviations from rationality and good judgment.

4.9.1.2 Selected Cognitive Biases

Since the pioneering work of Tversky and Kahneman, a continually evolving list of cognitive biases has been identified. Interested readers may refer to Pohl (2012) for a comprehensive analysis of cognitive illusions, their experimental

---

52 An Israeli-American psychologist awarded the 2002 Nobel Prize in Economic Sciences for his groundbreaking work with Amos Tversky.

context as well as conclusions about the wider implications of each cognitive bias effect. A more accessible, though somewhat less reliable, list of cognitive biases is available in Wikipedia (List of cognitive biases[53]). Accordingly, nearly 200 unique cognitive biases are identified that are generally divide into three groups:

Decision-making, belief, and behavioral biases

This group of cognitive biases affects individuals' decisions-making processes as well as belief systems and behavior within widespread domains like engineering, business, and economic. For example:

Anchoring. The tendency to rely too heavily, or "anchor," on one trait or piece of information when making decisions.

Availability heuristic. The tendency to overestimate the likelihood of events with greater "availability" in memory, which can be influenced by how recent the memories are or how unusual or emotionally charged they may be.

Bandwagon effect. The tendency to do (or believe) things because many other people do (or believe) the same.

Bias blind spot. The tendency to see oneself as less biased than other people, or to be able to identify more cognitive biases in others than in oneself.

Confirmation bias. The tendency to search for or interpret or focus on or remember information in a way that confirms one's original preconceptions.

Courtesy bias. The tendency to give an opinion that is more socially expected than one's true opinion, so as to avoid offending anyone.

Curse of knowledge. The tendency of better-informed people not to comprehend problems from the perspective of lesser-informed people.

Endowment effect. The tendency for people to demand much more in order to relinquish an object than they would be willing to pay to acquire it.

Expectation bias. The tendency for people to believe in the validity of data that agree with their expectations and to disbelieve and ignore data that conflicts with those expectations.

Focusing effect. The tendency to place too much importance on one aspect of an event.

Framing effect. The tendency to draw different conclusions from the same information, depending on how that information is presented.

Mere exposure effect. The tendency to express undue liking for things merely because of familiarity with them.

Neglect of probability. The tendency to completely disregard probability when making a decision under uncertainty.

---

53 See: List of cognitive biases. https://en.wikipedia.org/wiki/List_of_cognitive_biases. Accessed: May 2017.

Normalcy bias. The tendency to refuse to plan for, or react to, a disaster that has never happened before.

Not invented here. The tendency to reject or avoid the use of products, research, standards, or knowledge developed outside a group.

Ostrich effect. The tendency to ignore an obvious negative situation.

Pessimism bias. The tendency to overestimate the likelihood that negative things will happen.

Planning fallacy. The tendency to underestimate the time needed to complete a task or a project.

Pro-innovation bias. The tendency to exhibit excessive optimism regarding the prospect of an invention or its usefulness, while often failing to perceive its limitations and weaknesses.

Zero-sum bias. A tendency to perceive situations solely on the basis that one person's gains must be at the expense of another.

Attributional biases

This group of cognitive biases refers to the systematic errors made by individuals who evaluate themselves or try to find reasons for other persons' behavior. For example:

Authority bias. The tendency to attribute greater accuracy to the opinion of an authority figure and be more influenced by that opinion, unrelated to its content.

Egocentric bias (attributional variant). The tendency for people to claim more responsibility for themselves for the success of a joint action than an outside observer would credit them with.

Illusory superiority. The tendency for individuals to overestimate their desirable qualities, and underestimate their undesirable qualities.

Illusion of manageability. The tendency for individuals to attribute success to their past actions when, in fact, it was failure. In addition, this bias leads individuals to mistakenly think that, should a problem arise, they would be able to fix it.

In-group bias. The tendency for individuals to give preferential treatment to others they perceive to be members of their own group.

Naive realism. The tendency for individuals to believe that they see reality as it really is: objectively and without bias; that the facts are plain for all to see; that rational people will agree with them; and that those who don't are either uninformed, irrational, or biased.

Memory errors and biases

This group of cognitive biases refers to errors which systematically distort the recall of individuals' memory. For example:

Consistency bias. The tendency to incorrectly remember one's past attitudes and behavior as resembling present attitudes and behavior.

Egocentric bias (memory error variant). The tendency to recall the past in a self-serving manner.

False memory. The tendency to form a misattribution where imagination is mistaken for an actual memory.

Generation effect. The tendency for individuals to remember statements or actions that they have said or done in contrast with similar statements or actions uttered or performed by others.

4.9.1.3 Dramatic Example of Cognitive Bias

There are numerous investigations of incidents determining that human error was central in highly negative events or disasters, but few studies explicitly link cognitive biases to such occurrences. The loss of the space shuttle *Columbia* is an example of a major disaster plausibly exasperated by multiple cognitive biases.[54]

On February 1, 2003, the space shuttle *Columbia* disintegrated during reentry into Earth's atmosphere and all seven crewmembers of flight STS-107 perished (Figure 4.58).

(a)   (b)

Figure 4.58 (a) Space shuttle *Columbia* and (b) Flight STS-107 crewmembers (left to right): Brown, Husband, Clark, Chawla, Anderson, McCool, and Ramon (NASA pictures).

---

54 James Glamz and Edward Wong, "Engineer's '97 Report Warned of Damage to Tiles by Foam," *The New York Times*, February 4, 2003. See: http://www.nytimes.com/2003/02/04/national/ engineers-97-report-warned-of-damage-to-tiles-by-foam.html. Accessed: August 2017.

During the early phase of *Columbia*'s launch, engineers observed a large piece of thermal insulation foam dropping from the external tank and hitting the leading edge of the left wing. Small foam debris dislodging from the external tank was a common sight during shuttle launches; however, here, a relatively large piece hit a critical wing area made of reinforced carbon–carbon composite material at a fairly high speed.

The incident was analyzed immediately after launch by NASA engineers using a certified simulator called Crater and, according to the Columbia Accident Investigation Board (CAIB F6.3-11, 2003) findings: "Crater initially predicted tile damage deeper than the actual tile depth. ... But engineers used their judgment to conclude that damage would not penetrate the dense layer of tile." In other words, a certified tool (granted, with its own significant limitations) indicates a probable catastrophic failure, but this finding was overruled on the basis of an engineering hunch.[55]

A physical test approximating the dynamic conditions of the actual event was conducted shortly after the accident. A similar piece of foam weighing approximately 0.5 kg was catapulted at a speed of approximately 230 m/sec onto a similar leading edge of a shuttle wing, creating a hole of about 20 cm. This demonstrated that the "foam debris is, in fact, the most probable cause creating the breach that led to the accident of the *Columbia* and the loss of the crew and vehicle."

It is debatable whether the crewmembers could have been saved even if NASA management was fully aware of the real situation on board flight STS-107. But amazingly, no effort whatsoever was made to rescue the astronauts. One can plausibly hypothesize that many NASA organization-level cognitive biases contributed to this disaster[56] but the author speculates that the framing effect, the tendency to react to how information is presented, beyond its factual content, was at the heart of the problem.

The use of the word *foam* gave engineers the comfortable feeling of the soft and fluffy foam used, for example, in household furniture. Such substance can hardly damage a tile made of carbon–carbon composite material. In fact however, the "foam" covering the shuttle's external tank was made of relatively dens material. Weighing approximately 0.5 kg and moving at relative speed of 230 m/sec, such debris definitely posed a serious risk to any object it might encounter.[57]

---

55 This is not meant to denigrate engineering "gut feelings," which are important tools in any engineer's arsenal. However, such feelings cannot be trusted when contrary data is presented.
56 For example: anchoring, bandwagon effect, confirmation bias, courtesy bias, expectation bias, neglect of probability, normalcy bias, ostrich effect, authority bias, in-group bias.
57 This debris packed a lot of kinetic energy totaling: $\frac{1}{2}mv^2 = 13{,}225$ Joules. One Joule is the energy released when your smart phone (200 g) is falling vertically to the ground from a height of half a meter.

4.9.1.4 Cognitive Debiasing

A large body of evidence has established that cognitive biases unconsciously affect a wide range of human behavior. By and large, individuals are not aware of the existence of the phenomenon and are unable to detect and therefore mitigate it. Nevertheless, people who undertake cognitive debiasing training and playing debiasing games that teach mitigating strategies showed improved judgment and decision-making emanating from significant reductions in their cognitive biases.

So, any systems engineer should always adopt a healthy dose of suspicion regarding his own motives, opinions, and planned actions since his mind is biased in one way or the other. Another advice to systems engineers is to present their ideas to colleagues, friends, and, yes, rivals as well, and then listens intently to their views. As their set of cognitive biases is invariably different, their opinions could be priceless.

Lastly, systems engineers should take example from a distinguished political genius, Abraham Lincoln the sixteenth president of the United States. Lincoln succeeded in cobbling a cabinet including three gifted rivals of national reputation (Figure 4.59) and "together, created the most unusual cabinet in history, marshaling their talents to the task of preserving the Union and winning the war" (Goodwin, 2006).

Figure 4.59 Lincoln's US Cabinet (1862)[58]

---

58 Adapted from: The first reading of the emancipation proclamation before the cabinet on July 22, 1862. Engraving by Alexander Hay Ritchie; painting by Francis Bicknell Carpenter, 1864.

4.9.2 Cognitive Biases and Strategic Decisions

Cognitive biases are important factors in making strategic decision. Understanding how biases influence strategic decision processes will help engineers and managers to improve the process and make it more rational. Das and Teng (1999) proposed an intriguing integrative perspective for the relations between cognitive biases and strategic decision processes. The authors start with the notion that different types of cognitive biases affect decision processes in different ways. By examining these relationships, one can clarify the domain and the role of key cognitive biases in strategic decision-making as well as better differentiate amongst various strategic decision processes.

4.9.2.1 Categories of Cognitive Biases

In order to simplify the process of associating cognitive biases with strategic decision processes, the authors used existing research to combine cognitive biases into four categories: (1) prior hypotheses and focusing on limited targets, (2) exposure to limited alternatives, (3) insensitivity to outcome probabilities, and (4) illusion of manageability.

Prior hypotheses and focusing on limited targets. Research shows that decision makers tend to bring their previously formed beliefs or hypotheses into the decision-making process. As a result, they tend to overlook information and evidence that may prove the opposite.

Exposure to limited alternatives. Information is usually incomplete under most circumstances; therefore, decision makers tend to focus on a relatively small number of options or instead, use intuition to supplement rational analysis. As a result, decision makers often specify only a subset of the decision spectrum, thus generating a limited number of alternative courses of action.

Insensitivity to outcome probabilities. Research has also shown that decision makers do not trust and often do not understand and use estimates of outcome probabilities. Instead, engineers and managers tend to use a single or a few key values rather than to compute statistically based probabilities. Another reason decision makers do not use estimates of probability is that they see each problem as a unique event rather than a single case out of multiple events.

Illusion of manageability.[59] Another type of cognitive bias relates to engineers and managers who inappropriately perceive a success of a certain strategic decision, when, in fact, it was a failure. In addition, they often mistakenly

---

59 This cognitive bias has significant real-life implications. It is curious that this particular bias has not been sufficiently recognized in strategy research and practices.

assume that, should problems arise, they would be able to fix them. That is, they are convinced the outcomes of their decisions can be contained, corrected, or reversed, given sufficient efforts.

4.9.2.2 Strategic Decision Processes

Based on existing research, Das and Teng defined a set of strategic decisions processes, which include: (1) rational mode, (2) avoidance mode, (3) logical incremental mode, (4) political mode, and (5) garbage can mode.

Rational mode. This strategic decision process is an ideal and theoretical benchmark against which all the others decisions processes are considered. It is based on the assumption that human behavior is fully rational and not affected by cognitive biases. Under rational mode, decision makers are assumed to act with known objectives, diligently analyze both the external environment and internal operations. Therefore, decision-making is a comprehensive, fully rational process in which engineers and managers gather all relevant information, develop alternative decisions, and then objectively select the optimal one.

Avoidance mode. This strategic decision process relates to the fact that strategic decision-making processes often incur organizations' tendency to resist changes. This phenomenon relates to the organization's tendency to avoid uncertainty, maintaining the ongoing status quo.

Logical incremental mode. In this process mode, strategic decision-making occurs incrementally or in a step-by-step fashion. Since the overall environment is often either unknown or unstable and engineers' as well as managers' cognitive capabilities are limited, this is a desired strategy to choose this mode in order to achieve optimal strategic objectives Also, implementing strategic decisions incrementally allows organizations to move slowly so that they can remain flexible in terms of gradually assimilate the impacts of new decisions.

Political mode. Under this strategic decision process, decision makers are often unable to attain even a broad consensus on organizational objectives. More specifically, decision makers must confront different groups within the organization, each of whom fights for a decision favorable to itself. The outcome is therefore decided by those who can form the most powerful coalition. Thus, engineers and managers must deal with each party, which perceives the problem in light of its own sphere of interests.

Garbage can mode. This strategic decision process is the most uncertain and fluid mode of strategic decision-making. It has no inherent consistencies and there is no particular rationale for making a strategic choice. Nevertheless, although engineers and managers have little control over the process, their cognitive biases are still prevalent in the decision-making process.

4.9.2.3 Cognitive Biases Versus Strategic Decisions Processes

Das and Teng (1999) further analyzed relevant scientific literature and created a model depicting the type of strategic decision processes an engineer or a manager is likely to adopt. Here, each strategic decision is subject to each of the cognitive bias categories. In addition, the authors provide specific propositions (P*i*) for each of nine prevalent combinations of cognitive biases and strategic decisions processes (Figure 4.60 and Table 4.14).

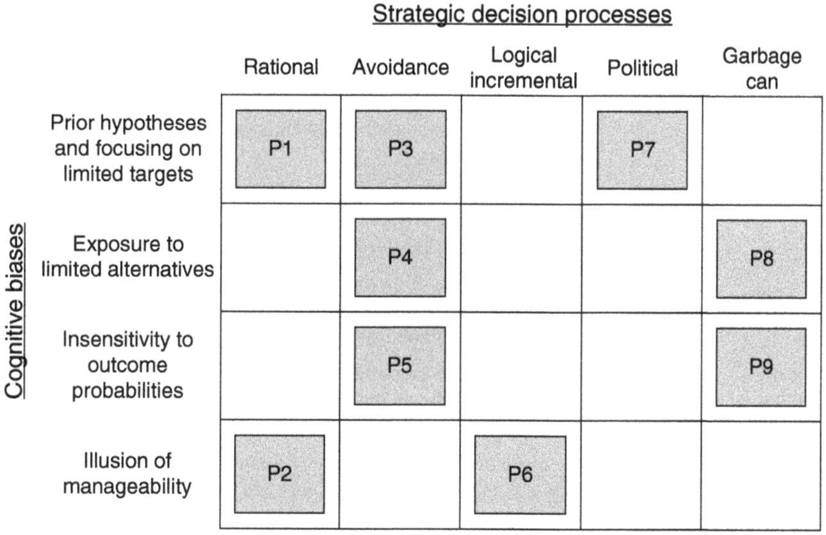

Figure 4.60 Cognitive biases versus strategic decision processes

To sum it up, Das and Teng show that engineers and managers involved in different decision processes are affected by different combinations of four basic types of cognitive biases. Distinct strategic decision processes can be better differentiated and understood by considering relevant cognitive biases. The suggested set of nine propositions can reveal the key relationships between the four types of cognitive bias and the five modes of strategic decision processes.

Thus, the integrative framework described above could be utilized to enhance strategic decisions, especially the ones that are highly uncertain and need to be made in a timely fashion. More specifically, engineers and managers could become more aware of the assumptions, heuristics and biases employed in their decision-making processes. This self-awareness of cognitive biases inherent in their decision-making should mitigate or at least, reduce the systematic errors arising from decision makers' cognitive biases.

Table 4.14 Proposition related to strategic decisions processes

Strategic decision processes	#	Proposition
Rational mode	P1	The more rational and systematic the strategic decision process, the more likely the engineers and managers will bring prior hypotheses to decisions.
	P2	The more rational and systematic the strategic decision process, the more likely the engineers and managers will have an illusion of manageability.
Avoidance mode	P3	The more emphasis on maintaining the status quo in a strategic decision-making process, the more likely the engineers and managers will bring prior hypotheses to their decision processes.
	P4	The more emphasis on maintaining the status quo in a strategic decision-making process, the more likely the engineers and managers will be exposed to limited alternatives.
	P5	The more emphasis on maintaining the status quo in a strategic decision-making process, the more likely it is that engineers and managers will be insensitive to outcome probabilities.
Logical incremental mode	P6	The more logical incremental the strategic decision process, the more likely the engineers and managers will have an illusion of manageability.
Political mode	P7	The more political the strategic decision process is, the more likely the engineers and managers will bring prior hypotheses to decisions.
Garbage can mode	P8	The more disorderly and anarchical the strategic decision process is, the more likely the engineers and managers will consider limited alternatives.
	P9	The more disorderly and anarchical the strategic decision process is, the more likely the engineers and managers will be insensitive to outcome probabilities.

4.9.3 Further Reading

- Brown, 2005.
- Das and Teng, 1999.
- Gilovich et al. (Editors), 2002.
- Goodwin, 2006.
- Ishizaka and Nemery, 2013.
- Jahan et al., 2016.

- Kahneman et al. (Editors), 1982.
- Kahneman and Tversky (Editors), 2000.
- Kahneman, 2013.
- Lewis, 2016.
- Pohl (Editor), 2012.
- Wikipedia, Accessed: April 2017.

4.10 Bibliography

Achi, A., Salinesi, C., Viscusi, G. (2016). Information Systems for Innovation: A Comparative Analysis of Maturity Models' Characteristics. Advanced Information Systems Engineering Workshops. CAiSE 2016.

Adizes I. (1990). *Corporate Lifecycles: How and Why Corporations Grow and Die and What to Do About It*. The Adizes Institute.

Adizes, I. (2017). *Managing Corporate Lifecycles: Complete Edition*. Adizes Institute Publications.

Alsos, G.A., Hytti, U., and Ljunggren, E. (editors). *Research Handbook on Gender and Innovation*. Cheltenham: Edward Elgar Publishing.

Altshuller, G. (1996). *And Suddenly the Inventor Appeared: TRIZ, the Theory of Inventive Problem Solving*, 2nd ed. Technical Innovation Center, Inc.

Baden-Fuller, C., and Stopford, J. (1994). *Rejuvenating the Mature Business: The Competitive Challenge*, rev. ed. Boston: Harvard Business Review Press.

Bakke, K. (2017). Technology readiness levels use and understanding, Master thesis, University College South-East Norway.

Berg R. (2013). The Innovation Maturity Model: The strategic and capability building steps for creating an innovative organization. Berg Consulting Group. See: http://bergconsulting.com.au/_literature_144915/Innovation_Maturity_Model. Accessed: February 2017.

Berkhout, A.J., Hartmann, D., and Trott, P. (2010). Connecting technological capabilities with market needs using a cyclic innovation model, *R&D Management* 40 (5): 474–490.

Berkhout, A.J., Hartmann, D., van der Duin, P., and Ortt, R. (2006). Innovating the innovation process. *International Journal of Technology Management* 34 (3–4): 390–404.

Blau, F.D., and Kahn, L.M. (2016). The Gender Wage Gap: Extent, Trends, and Explanations, National Bureau of Economic Research (NBER), Working Paper No. 21913, January. See: http://www.nber.org/papers/w21913. Accessed: April 2017.

Branscomb, L.M. (2002). Between invention and innovation: An analysis of funding for early-stage technology development, U.S. Dept. of Commerce, Technology Administration, National Institute of Standards and Technology.

Bridges, W., and Bridges S. (2017). Managing Transitions, Da Capo Lifelong Books; 25th Anniversary edition.

Brown, R. (2005). *Rational Choice and Judgment: Decision Analysis for the Decider*, Hoboken, NJ: Wiley-Interscience.

Cañas, A.J., and Novak, J.D. (2013). What is a Concept Map?, 3.11.2013. See: http://www.the-aps.org/APS-Storage/APS-Education/Pedagogy-Resources/Concept-Map.pdf. Accessed: May 2017.

Carroll P. (1993). *Big Blues: The Unmaking of IBM*. New York: Crown.

Cascini, G., Rotini, F., and Russo, D. (2009). Functional modeling for TRIZ-based evolutionary analyses, *Proceedings of the International Conference on Engineering Design, ICED'09*. Stanford.

Catalyst. Women in Science, Technology, Engineering, and Mathematics (STEM) See: http://www.catalyst.org/knowledge/women-science-technology-engineering-and-mathematics-stem#footnote49_c02wukq. Accessed: April 2017.

Chong, A. (2010). Driving Asia – As automotive electronics transforms a region. Infineon Technologies Asia Pacific Pte Ltd. May.

Chrissis, M.B., Konrad, M., and Shrum, S. (2011). *CMMI for Development: Guidelines for Process Integration and Product Improvement*, 3rd ed. Addison-Wesley Professional.

Christensen, C.M. (2010). *Innovation Killers: How Financial Tools Destroy Your Capacity to Do New Things*. Boston: Harvard Business Review Press.

Christensen, C.M., Raynor, M.E., and McDonald R. (2015). What is disruptive innovation? *Harvard Business Review*, December.

Cooper, R.G., Edgett, S.J., and Kleinschmidt, E.J. (2002). *New Product Development Best Practices Study: What Distinguishes the Top Performers*. Houston, APQC.

Cooper, R.G. (1990). Stage-Gate systems: a new tool for managing new products – conceptual and operational model. *Business Horizons* (May, June): 44–53.

Corsi, P., Christofol, H., Richir, S., and Samier H. (eds.). (2006). *Innovation Engineering: The Power of Intangible Networks*. Hoboken, NJ: Wiley-ISTE.

Corsi, P., and Neau, E. (2015). *Innovation Capability Maturity Model*. Hoboken, NJ: Wiley-ISTE.

Cross, R., and Prusak, L. The people who make organizations go – or stop. Boston: *Harvard Business Review* (June).

Cross, R.L., and Parker, A. (2004). *Gets Done in Organizations*. Boston: Harvard Business Review Press.

Danilda, I., and Thorslund, J.G. (eds.). (2011). Innovation & Gender, Vinnova, Tillväxtverket, & Innovation Norway. See: http://www2.vinnova.se/en/Publications-and-events/Publications/Products/Innovation--Gender/. Accessed: October 2017.

Das, T.K., and Teng, B.-S. (1999). Cognitive biases and strategic decision processes: an integrative perspective. *Journal of Management Studies* 36 (6): 757–778.

De Bonte, A., and Fletcher, D. (2014). *Scenario-Focused Engineering: A Toolbox for Innovation and Customer-centricity*. Microsoft Press.

Deschamps J.P. (2008). *Innovation Leaders: How Senior Executives Stimulate, Steer and Sustain Innovation*. San Francisco: Jossey-Bass.

Díaz-García, C., González-Moreno, A., and Sáez-Martínez, F.J. (2013). Gender diversity within R&D teams: Its impact on radicalness of innovation. *Innovation: Management, Policy & Practice* 15 (2): 149.

DOD. (2011). Technology Readiness Assessment (TRA) Guidance, April. See: "Technology Readiness Assessment (TRA) Guidance." Accessed: January 2017.

Dodgson, M., and Rothwell, R. (1996). *The Handbook of Industrial Innovation*. Cheltenham: Edward Elgar Publishing.

Dougherty, D., Borrelli, L., Munir, K. and O'Sullivan, A. (2000). Systems of organizational sensemaking for sustained product innovation, *Journal of Engineering and Technology Management* 17 (3): 321–355.

Doutta, S., Lanvin, B. and Wunsch-Vincent, S. (2016). The Global Innovation Index 2016: Winning with Global Innovation, Cornell University, INSEAD, WIPO, 2016. See: https://www.globalinnovationindex.org/gii-2016-report. Accessed: January 2017.

Dyer, J., Gregersen, H., and Christensen, C.M. (2011). *The Innovator's DNA: Mastering the Five Skills of Disruptive Innovators*. Boston: Harvard Business Review Press.

EARTO (2014). The TRL Scale as a Research & Innovation Policy Tool, EARTO Recommendations, April 30. http://www.earto.eu/fileadmin/content/03_Publications/The_TRL_Scale_as_a_R_I_Policy_Tool_-_EARTO_Recommendations_-_Final.pdf. Last accessed January 2017.

Ehrlenspiel, K., Kiewert, A., and Lindemann, U. (2007). *Cost Efficient Design*. Springer.

Eick, S., Graves, T., Karr, A., Marron, J., and Mockus, A. (2001). Does code decay? Assessing the evidence from change management data. *IEEE Transactions on Software Engineering* 27 (1) January. http://users.ece.utexas.edu/~perry/education/SE-Intro/tse-codedecay.pdf.

Essmann, H.E. (2009). Toward Innovation Capability Maturity. PhD Dissertation, Dep. of Industrial Engineering, Stellenbosch University, South Africa, December.

Farris, G.F. (1982). The technical supervisor. Beyond the Peter Principle. In Tushman, M.L., and Moore, W.L. (eds.). *Readings in The Management of Innovation*, 337–348. Pitman Publishing.

Ferguson, D.M., Purzer S., Ohland M.W. and Jablokow K., The Traditional Engineer vs. The Innovative Engineer, 121st ASEE Annual Conference and Exposition, Indianapolis IN., Paper ID #8751, June 15–18, 2014.

Fey, V., and Rivin, E. (2005). *Innovation on Demand: New Product Development Using TRIZ*. Cambridge University Press.

Gilovich, T., Griffin, D., and Kahneman, D. (eds.). (2002). *Heuristics and Biases: The Psychology of Intuitive Judgment*. Cambridge University Press.

Goodwin, D.K. (2006). *Team of Rivals: The Political Genius of Abraham Lincoln*. New York: Simon & Schuster.

Gostick, A., and Christopher, S. (2008). *The Levity Effect: Why it Pays to Lighten Up*. Hoboken, NJ: John Wiley & Sons.

Govindarajan, V., and Bagchi, S. (2008). The emotionally bonded organization: why emotional infrastructure matters and how leaders can build it, Tuck School of Dartmouth College, Working Paper.

Grant, A. (2017). *Originals: How Non-Conformists Move the World.* New York: Penguin Books; Reprint edition.

Hellestrand, G. (2005). ESL development gets a leg up. *Chip Design Magazine* (January 1).

Holland, J.L. (1997). Making vocational choices: a theory of vocational personalities and work environments. *Psychological Assessment Resources.*

Isaacson, W. (2015). *Steve Jobs,* reissue edition. New York: Simon & Schuster.

Ishizaka, A., and Nemery, P. (2013). *Multi-criteria Decision Analysis: Methods and Software.* Hoboken, NJ: John Wiley & Sons.

Jahan, A., Edwards, K.L., and Bahraminasab, M. (2016). *Multi-criteria Decision Analysis for Supporting the Selection of Engineering Materials in Product Design,* 2nd ed. Butterworth-Heinemann.

Jain, R.K., Triandis, H.C., and Weick, C.W. (2010). *Managing Research, Development and Innovation: Managing the Unmanageable,* 3rd ed. Hoboken, NJ: John Wiley & Sons.

Kahneman, D., and Tversky, A. (eds.). (2000). *Choices, Values, and Frames.* Cambridge University Press.

Kahneman, D., Slovic, P., and Tversky, A. (eds.) (1982). *Judgment Under Uncertainty: Heuristics and Biases.* Cambridge University Press.

Kahneman, D. (2013). *Thinking, Fast and Slow.* New York: Farrar, Straus and Giroux.

Karasik, Y.B. (2011). On the causes of non-uniform pace of progression of technologies and subsystems of a system. *The Anti TRIZ-Journal* 10 (1), February.

Karasik, Y.B. (2008). Some doubts about the law of completeness, *Anti TRIZ-Journal* 7 (7), August.

Kasser, J.E. (2015). *Holistic Thinking: Creating Innovative Solutions to Complex Problems,* 2nd ed. CreateSpace Independent Publishing Platform.

Keeley, L., Walters, H., Pikkel, R., and Quinn, B. (2013). *Ten Types of Innovation: The Discipline of Building Breakthroughs.* Hoboken, NJ: John Wiley & Sons.

Kingdon, M. (2012). *The Science of Serendipity: How to Unlock the Promise of Innovation.* Hoboken, NJ: John Wiley & Sons.

Kohn, S., Levermann, A., Howe, J., and Hüsig, S. (2003). Software im innovationsprozess. *Insti Studienreihe.* 1 (1): 85.

Landivar, L.C. (2013). Disparities in STEM Employment by Sex, Race, and Hispanic Origin, ACS-24, American Community Survey, September. See: https://www.census.gov/prod/2013pubs/acs-24.pdf. Accessed: April 2017.

Lawler, E.E., and Worley, C. (2006). *Built to Change: How to Achieve Sustained Organizational Effectiveness.* San Francisco: Jossey-Bass.

Lawler, E.E., *High-Involvement Management: Participative Strategies for Improving Organizational Performance.* San Francisco: Jossey-Bass, 1991.

Lawler E.E. (1973). *Motivation in Work Organizations.* San Francisco: Jossey-Bass.

Lee B. (2012). *The Hidden Wealth of Customers: Realizing the Untapped Value of Your Most Important Asset.* Boston: Harvard Business Review Press.

Lewis, M. (2016). *The Undoing Project: A Friendship That Changed Our Minds.* New York: W.W. Norton & Company.

Magee, J.F. (1964). Decision trees for decision making. *Harvard Business Review* (July).

Mankins, J.C. (1995). Technology Readiness Levels, A White Paper, Advanced Concepts Office, Office of Space Access and Technology, NASA, April 6.

Mann, D. (2012). *Innovation Capability Maturity Model (ICMM) – An Introduction, Systematic Innovation.* IFR Press. See: http://store.systematic-innovation.com/innovation-capability-maturity-model-an-introduction/. Accessed: December 2017.

Markham, S.K., and Mugge, P.C. (2014). *Traversing the Valley of Death: A practical guide for corporate innovation.* Stephen K Markham.

McDonald, D., Bammer, G., and Deane, P. (2011). *Research Integration Using Dialogue Methods.* ANU E Press.

Midgley, D. (2009). *The Innovation Manual: Integrated Strategies and Practical Tools for Bringing Value Innovation to the Marke.* Hoboken, NJ: John Wiley & Sons.

NSB-2016-1. (2016). Science and Engineering Indicators 2016, National Science Foundation, National Science Board, National Center for Science and Engineering Statistics (NCSES), Arlington, VA, January 2016. See: https://nsf.gov/statistics/2016/nsb20161/#/. Accessed: Jan., 2017.

OECD-2005. (2005). *The Measurement of Scientific and Technological Activities Oslo Manual: Guidelines for Collecting and Interpreting Innovation Data,* 3rd ed. OECD Publishing.

OECD-2010. (2010). *Measuring Innovation: A New Perspective, Organization For Economic Co-Operation & Development.* OECD Publishing.

OECD-2015, (2015). *The Measurement of Scientific, Technological and Innovation Activities, Frascati Manual 2015: Guidelines for Collecting and Reporting Data on Research and Experimental Development.* OECD Publishing.

Ostergaard, C.R. (n.d.). Innovation and Employee Diversity Does Diversity Really Matter for Innovation?, Department of Business and Management, Aalborg University, Denmark. See: http://www.uis.no/getfile.php/Forskning/Senter%20for%20Innovasjonsforskning/UIS%20innovation%20days_CROstergaard.pdf. Accessed: April 2017.

Owen, D. (2016) Overcoming the biggest barriers to innovation. *The Huffington Post* (June 9, 2016).

Page, S.E., The Difference: How the Power of Diversity Creates Better Groups, Firms, Schools, and Societies, Princeton University Press, 2008.

Pelz, D.C., and Andrews, F.M. (1976). *Scientists in Organizations: Productive Climates for Research and Development,* revised ed. Institute for Social Research.

Petrov, V. (2002). The laws of system evolution. *The TRIZ Journal* (March 22). https://triz-journal.com/laws-system-evolution.

Pohl, R.F. (ed.) (2012). *Cognitive Illusions: A Handbook on Fallacies and Biases in Thinking, Judgment and Memory,* reprint. Psychology Press.

Pohle, G., and Chapman, M. (2006). IBM's global CEO report 2006: business model innovation matters. *Strategy & Leadership* 34 (5): 34–40.

Prather, C. (2009). *The Manager's Guide to Fostering Innovation and Creativity in Teams.* New York: McGraw-Hill Education.

PWC. (n.d.). 2015 Global Innovation 1000 Innovation's New World Order, See: http://www.strategyand.pwc.com/media/file/2015-Global-Innovation-1000-Fact-Pack.pdf. Accessed: January 2017.

Reed, T. (2016). 7 Habits of the Most Innovative Engineers, Bliley Technologies, Aug 24, 2016. See: http://blog.bliley.com/7-habits-of-the-most-innovative-engineers. Accessed: March, 2017.

Robertson, B.J. (2015). *Holacracy: The New Management System for a Rapidly Changing World.* New York: Henry Holt and Co.

Salamatov, Y. (1999). *TRIZ: The Right Solution At The Right Time.* The Netherlands: Insytec.

San Y.T. (2014) TRIZ – Systematic Innovation in Business and Management. *FirstFruits Sdn Bhd.*

Savransky, S.D. (2000). *Engineering of Creativity: Introduction to TRIZ Methodology of Inventive Problem Solving.* Boca Raton, FL: CRC Press.

SCB. Women and men in Sweden Facts and figures 2016, Statistics Sweden, Population Statistics Unit SE-701 89 Örebro, Sweden, 2016. See: http://www.scb.se/Statistik/_Publikationer/LE0201_2015B16_BR_X10BR1601ENG.pdf. Accessed: April 2017.

Schein, E.H. (2004). *DEC Is Dead, Long Live DEC: The Lasting Legacy of Digital Equipment Corporation.* San Francisco: Berrett-Koehler Publishers.

Schiebinger, L., Klinge, I., Sánchez de Madariaga, I., Paik, H.Y. (2015). Schraudner, M. and Stefanick, M. (Eds.), Gendered Innovations in Science, Health & Medicine, Engineering and Environment, 2011–2015. See: http://ec.europa.eu/research/gendered-innovations/. Accessed: April 2017.

Shalley C., Hitt M.A.,and Zhou Jing (eds.). (2016). *The Oxford Handbook of Creativity, Innovation, and Entrepreneurship,* reprint ed. Oxford: Oxford University Press.

Shavinina, L.V. (Ed.) (2003). *The International Handbook on Innovation.* Pergamon.

Shellshear, E. (2016). *Innovation Tools: The most successful techniques to innovate cheaply and effectively.* 7 Publishing.

Shteyn, E., and Shtein, M. (2013). *Scalable Innovation: A Guide for Inventors, Entrepreneurs, and IP Professionals.* Boca Raton, FL: CRC Press.

Siegle, D., et al. (2002). Scales for Rating the Behavioral Characteristics of Superior Students – Renzulli Scale. *Creative Learning Pr*; Revised edition.

Silverstein, D., Samuel, P. and DeCarlo, N. (2012). *The Innovator's Toolkit: 50+ Techniques for Predictable and Sustainable Organic Growth,* 2nd ed. Hoboken, NJ: John Wiley & Sons.

Skogstad, P. (2010). A Unified Innovation Process Model: A Process Model and Empathy Tool for Engineering Designers and Managers, LAP LAMBERT Academic Publishing.

Smith, J. (2004). An Alternative to Technology Readiness Levels for Non Developmental Item (NDI) Software, Technical Report CMU/SEI-2004-TR-013, ESC-TR-2004-013, April. See: http://repository.cmu.edu/cgi/viewcontent.cgi?article=1517&context=sei. Accessed: Jan., 2017.

Smith L. How to Motivate Creative Thinking in the Workplace, BeAuditSecure. Accessed: August, 2016.

Softpanorama (Groupthink) (n.d.). See: http://www.blendedbody.com/GroupThink/Groupthink-PatternOfThoughtCharacterizedbySelf-Deception.htm. Accessed: May, 2017.

Start up Donut: Ten ways to encourage creative thinking. (n.d.). See: https://www.marketingdonut.co.uk/marketing-strategy/ten-ways-to-encourage-creative-thinking. Accessed: March 2017.

Sternberg, R.J., and Davidson, J.E. (1996). *The Nature of Insight*. Cambridge, MA: The MIT Press.

Stewart, D.V. (1981). The Design structure system: A method for managing the design of complex systems. *IEEE Trans. Eng. Management* 28: 71–74.

Strategyn, Innovation Track Record Study. (2010). See: http://www.strategyn.at/sites/default/files/uploads/TrackRecord_07.pdf. Accessed: August 2017.

Summa, A. (2004). Software tools to support innovation process focus on idea management, Working Paper 29, Innovation Management Institute, Helsinki University of Technology.

Tellis, G.J. (2013). *Unrelenting Innovation: How to Create a Culture for Market Dominance* San Francisco: Jossey-Bass.

The HR Observer, 10 Ways to Promote Innovation at the Workplace, April 30, 2015. See: http://www.thehrobserver.com/10-ways-to-promote-innovation-at-the-workplace/ Accessed: March 2017.

Vroom V., and Yetton, P.W. (1973). *Leadership and Decision Making*. University of Pittsburgh Press.

Wikipedia, List of cognitive biases. See: https://en.wikipedia.org/wiki/List_of_cognitive_biases. Accessed: April 2017.

Williamson, J.M., Lounsbury, J.W., and Han, L.D. Key personality traits of engineers for innovation and technology development, *Journal of Engineering and Technology Management* 30 (2): 157–168 (April–June).

Young Entrepreneur Council (2012). 6 ideas to promote innovation in your workplace this year. *Forbes* (December 31).

Part V

Creative and Innovative Case Study

"Success is most often achieved by those who don't know that failure is inevitable."

Coco Chanel (1883–1971)

5.1   Introduction to Part V

The purpose of this part of the book is to tell the story of an exemplary creative and innovative case study. This case study is based on actual work undertaken by the author and a team of dedicated engineers and managers that started in 2003 and continued to 2016. Part V: Creative and Innovative Case Study.[1] is generally organized along chronological order as depicted in Figure 5.1.

Chapter 5.1: Introduction to Part V. This chapter describes the contents and structure of Part V.

Chapter 5.2: A problem seeking a solution. This chapter describes the problem at hand (i.e. designing adaptable systems for future unknown requirements) and its evolution in the minds of its creators. In addition, this chapter describes the early attempt to fund an initiative aiming at resolving this problem.

Chapter 5.3: Gaining deeper insights. This chapter describes how understanding of the problem evolved over time and the emergence of a creative solution. In addition, this chapter encapsulates clear and measurable project objectives to be achieved, in light of the existing state-of-the-art in the problem/solution domain.

---

1 Portions of Part V of the book appeared previously:
  ● in several Wiley's publications. Reproduced with permission of the John Wiley & Sons, Inc.
  ● in the AMISA project public website: http://amisa.eu/. All AMISA products are in the public domain.

Practical Creativity and Innovation in Systems Engineering, First Edition. Avner Engel.
© 2018 John Wiley & Sons, Inc. Published 2018 by John Wiley & Sons, Inc.

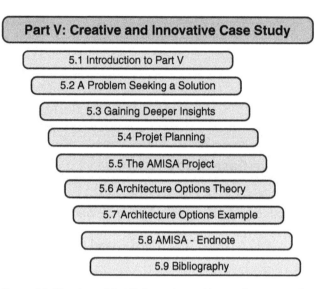

Figure 5.1 Structure of Part V: A creative and innovative case study

Chapter 5.4: Project planning. This chapter describes in details the planning of an innovation project in preparation for submitting a funding request to the European Commission (EC). Project planning includes detailed descriptions of individual work packages, risks and mitigation plan, management structure and procedures, project participants, and an analysis of needed resources.

Chapter 5.5: The AMISA project. This chapter describes the actual AMISA (Architecting Manufacturing Industries and Systems for Adaptability) project key ingredients. This includes the encapsulation of Design for Adaptability (DFA) state of the art, establishment of project requirements, development and expansion of architecture options theory, implementation of a software support tool, development of six pilot projects, generation of technical and management deliverables, exploitation and dissemination of project results, assessment of the overall project results, and consortium meetings throughout the project. An EC final assessment of the project is also provided.

Chapter 5.6: Architecture options theory. This chapter describes architecture options (AO), the theoretical background for this case study. It explains the concepts of financial and engineering options, transaction cost and interface cost, architecture adaptability value (AAV) and Design Structure Matrix (DSM) as well as static and dynamic system value modeling and optimization.

Chapter 5.7: Architecture options example. This chapter provides a comprehensive architecture options example. It includes detailed depiction of a general architecture option process model and a practical AO example using a solid state power amplifier (SSPA) system.

Chapter 5.8: AMISA – Endnote. This chapter provides a closure to this case study and considers the overall success and failure of the AMISA project.
Chapter 5.9: Bibliography. This chapter provides bibliography related to Part V topics.

Note

Readers mostly interested in the creative aspects of this case study are advised to focus on Chapters 5.2, 5.3, 5.6 and 5.7. Readers primarily interested in the innovative aspects of this case study are advised to read Chapters 5.4 and 5.5.

5.2 A Problem Seeking a Solution

5.2.1 The Problem and Its Inception

During the 2002–2005 time frame, the author led the international research project SysTest.[2] In September 2003 the project team traveled to Bergen, Norway (Figure 5.2) in order to discuss ongoing project issues.

At that time, Shalom Shachar, from the Israel Aerospace Industries (IAI) suggested to the group that, at the end of the SysTest project, the current consortium or parts thereof undertake a new applied research aimed at improving the methodology of systems' upgrade during use and maintenance. Over their lifetimes, systems provide value to stakeholders; however, as time goes by, systems tend to lose value due to technological opportunities, growth in stakeholder wants, increased system maintenance costs, and changes in the environment. The ability to maintain systems and also upgrade them quickly and at low cost is an important way to maintain and increase systems' value to stakeholders.

A small lead team, including Shalom Shachar, Tyson Browning from the Texas Christian University (TCU), and the author, set out to investigate the matter and present results to the SysTest project team. The lead team investigated conventional approaches such as open systems, standard interfaces, and

---

2 SysTest attempted to respond to a fundamental Systems Verification, Validation, and Testing (VVT) problem, namely, "What is the optimal yet practical amount of testing appropriate for systems?" The aim of SysTest was to develop a generic VVT methodology and processes as well as evolve a VVT economic model to optimize the VVT process and implement it in software. These products must be tailorable by users in order to fully respond to different risk and quality objectives, different stages of systems life cycles and support different industrial sectors. The project was funded by the European Union, Research, Technological Development & Demonstration (RTD), Fifth Framework Program (FP5) under contract G1RD-CT-2002-00683. See: http://cordis.europa.eu/project/rcn/61935_en.html. Last accessed: September 2017.

Figure 5.2 View from Mount Floyen, Bergen, Norway

modular designs, as well as more advanced approaches like real options and real options "in" projects. As a result, the team concluded that the existing approaches are constrained to deal with expected future requirements (e.g. designing and building a one- or two-story building with the strength to support additional floors to be added in the future as needed).

The lead team concluded that a desired solution to future maintenance and upgrade is a generic systems design that supports a wide variety of unexpected future requirements. In other words, systems designers should consider how adaptability can be designed into systems so that they will provide maximum value to stakeholders throughout their lifetime.

The lead team presented their finding to the SysTest project team during a consortium meeting in Toulouse, France, on June 2004. The findings were evaluated in a classical brainstorming session (Section 3.3.1, Classic Brainstorming) and adopted by the consortium, which recognized the problem as an important one for most members. Several persons indicated ways and means to exploit the results of such research in their respective companies. For example, the head of the SysTest project team from DaimlerChrysler, in Germany, raised the idea that automobile manufacturers could create in-car information and entertainment subsystems that will be designed to be periodically upgraded upon driver request.

5.2.2 Initial Funding Effort

It was also agreed that the task at hand requires the participation of several partners from industry, small and medium-sized enterprises (SMEs), and universities. In addition, external funding will be required, such as from the European Commission (EC) of the European Union. Over the next year the lead team assembled a group of participants and, together, developed a formal research proposal.

At the time, the only relevant requests for proposals (*calls* in EC lingo) emanated from the Information Society Technologies (IST), an arm of the European Union's 6th Framework Program. Accordingly, the proposal was restructured and submitted on September 2005 under the name "Developing Adaptable Embedded Systems for Economic Opportunities (DAESEO)." However, the IST is more interested in computers and software and less in systems engineering and even less so in economic matters. Thus, under the intense competition for funding, the DAESEO proposal failed to attract EC financial support.

5.2.3 Further Reading

- Engel, 2010.

5.3 Gaining Deeper Insights

Following the DAESEO failure, it was clear that (1) the right proposal must be submitted to the right call and (2) a deeper understanding of issues related to Design for Adaptability (DFA) must be achieved in order to increase the likelihood that a future proposal could attract funding for an extensive research project.

First, the concepts and objectives of such research must be expanded. This included precise definition of the project's concepts, the main ideas of the proposed work, the specific objectives of the project stated in a measurable and verifiable form as well as the basis for predicting that these objectives are actually attainable during the lifetime of the proposed research. Finally, it was critical to show that a proposal to be submitted is indeed relevant to the specific EC call.

Second, the expected progress beyond the state-of-the-art must be identified. This included specifically state-of-the-art related to architecting manufacturing systems for adaptability, project's partners' historical adaptability experience, expected advantages associated with using quantitative methods for architecting adaptable systems, and the particular assumptions made within the proposal.

5.3.1 The Problem and the Approach

The purpose of this section is to describe the concept of the project and the main ideas of the proposed work. In addition, its purpose is to define the objectives of the project stated in a measurable and verifiable form and justify the basis for the expected attainability of these objectives.

5.3.1.1 The Problem

Manufacturing industries, system products, and customer services (hereinafter *systems*) provide value through their ability to fulfill stakeholders' needs and wants. These needs evolve over time and may diverge from an original system's capabilities. Thus, a system's value to its stakeholders diminishes over time. Some reasons for this decrease include technological opportunities and growth in stakeholder wants, which make an existing system seem inadequate. Other reasons are growth in a system's maintenance costs, due to effects such as depreciation and component obsolescence. Still other reasons are changes in the environment (e.g. new rules and regulations). As a result, systems have to be periodically upgraded at substantial cost and disruption to normal operations. Since complete replacement costs are often prohibitive, system adaptability is a valuable characteristic.

Current concepts, methods, and tools for architecting systems (emanating from engineering disciplines) lack vital business and economic considerations. As a result, most system architectures are not easily adaptable to evolving manufacturing needs and product variants. This gap hinders industry from delivering updated products/services quickly and cost-effectively and prevents optimal manufacturing performance. In summary, increasing a system's lifetime value requires improved methods of architecting it. The problem, therefore, is:

> How can adaptability[3] be designed into systems so that they will provide maximum lifetime value to stakeholders?

5.3.1.2 The Approach

The leader of each pilot project used lateral thinking methodology (Section 3.2.1, Lateral Thinking) in order to generate a new and out-of-the-box approach to embed adaptability into product and systems. More specifically, the project sought to develop a next-generation technology for model-based architecting

---

3 According to the Merriam-Webster dictionary, to adapt means "to make fit, often by modification" (...from the outside). Adaptability is distinguished from "Flexibility," which is derived from the Latin word flexus and literally refers to what is capable of withstanding stress without injury and figuratively to what may naturally adjust itself as needed.

of adaptable systems. Such systems will increase cost-efficiencies, reduce production and development cycle times, and ensure longer, more valuable lifetimes. This technology was also to be harmonized with relevant intelligent manufacturing system (IMS) projects and existing standards.

The creation of artifacts adaptable for a succession of upgrades and variations is of great importance to all systems since technology and business environments are likely to change over their lifespan. Figure 5.3 shows one example of manufacturing enterprise architecture. Several raw materials are used to build components, several components are used to build subsystems, and several subsystems are used to build the final system. Sometimes, products are sold to end customers. At other times a manufacturer retains ownership of their products in order to create a revenue-generating service. Most often, manufacturers combine these two approaches.

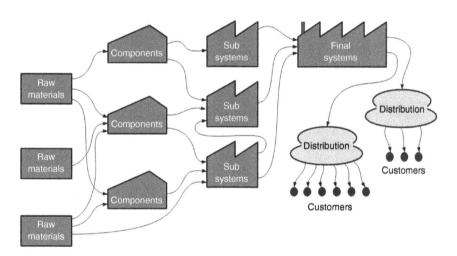

Figure 5.3 Example: Architecture of a manufacturing industry

This conglomerate represents a particular and unique architecture composed of various elements and interfaces distributed along geographical areas, production philosophy, management structure, etc. In addition, this conglomerate is very dynamic. Different raw materials replace old ones, components are replaced or combined together, subsystems' manufacturers change or go out of business and the final system or service itself evolves over time, as well as the distribution network and customer base.

Similarly, Figure 5.4 depicts the automotive electronics architecture of a Scania truck (circa 2010). This specific architecture was essentially created by way of historical experience and engineering gut-feeling, not specifically by optimizing for adaptability to future, unknown needs. Although various

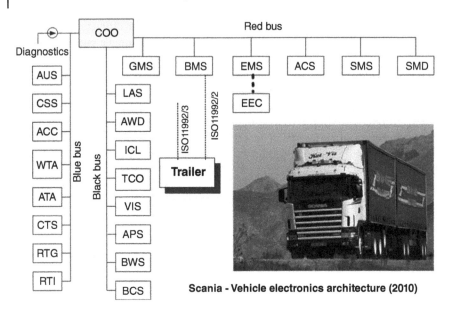

Figure 5.4 Example: Architecture of vehicle electronics system

> The project claim is that a more adaptable architecture could increase the life-time value of systems.

qualitative methods exist to increase the adaptability of systems (e.g. modularity, open systems, interface standardization, etc.), a methodology was lacking to quantify the value and achievable benefits of deliberately incorporating adaptability into system architectures.

The challenge, therefore, was to create theoretical methods to architect systems (i.e. manufacturing systems as well as products and services) for optimal adaptability. The corresponding challenge was to prove that such a method can be implemented in industry that will, in fact, lead to substantial lifelong savings.

5.3.2 Main Ideas of the Proposed Work

The project was to create a quantitative architecture option theory for designing systems for optimal adaptability fusing two well-known theories,[4] transaction cost theory (Coase, 1937) and financial options theory (Black and Scholes, 1973).

---

4 The creators of both theories were awarded the Nobel Prize in economics for their work.

5.3.2.1 Financial/Engineering Option Theory

A financial option is a contract that gives the buyer the right, but not the obligation, to buy or sell an underlying asset at a specific price on a specified date. Black and Scholes (1973) formulated the first reliable model (equation) to compute the cost of financial options given various parameters associated with such transactions. In engineering, financial options had evolved into real options and then real "in" options, culminating in architecture options. This analogous engineering concept expresses the "right, but not the obligation, to undertake some future engineering project at a specific price on a specified date."

Engineering options capture the value of managerial flexibility to adapt decisions in response to unexpected circumstances. This method represents the state-of-the-art technique for the valuation and management of future flexibility (i.e. system adaptability).

5.3.2.2 Transaction Cost Theory

In economics, transaction costs are costs associated with the exchange of goods or services. They include, among other things, expenditures related to communication, legal, enforcement, information, quality, and durability, transportation etc. incurred in overcoming market imperfections. Coase (1937) discovered that internal transaction costs within an organization are generally substantially lower than similar external transaction costs. As a result, organizations tend to expand or shrink as a function of their transaction costs. In engineering, transaction costs are associated with interface costs among system's elements as well as between the system and its environment. The key concept here is that interfaces between components are subject to various transaction costs whereas interfaces buried within modules represent minimal or no transaction costs, thus may be ignored altogether.

5.3.2.3 Static Value Paradigm

Fundamentally, the architecture of an engineered system defines its components and the interfaces between them. The project was to develop a new quantitative method to evaluate different ways of packaging a system in order to optimize its adaptability relative to its interface cost. The term *packaging* was used to indicate the "degree of distributiveness" of the system. In an extreme architecture, the entire system is embodied as a one large module. In the other extreme architecture, the system is composed of multitude of decentralized, individual components connected by means of multiple interfaces. Decentralization promotes adaptability as it is relatively easy to upgrade a component without disturbing the rest of the system. However, decentralization necessitates appropriate interfaces, which represent burdensome transaction costs. It was hypothesized that, for a given system, there exists a specific architecture that optimally balances the adaptability of a given system with its interface costs.

5.3.2.4 Design Structure Matrix (DSM)

A Design Structure Matrix (DSM), also referred to as an N^2 matrix, is a compact visual model of a system in the form of a square matrix. It shows objects and their relationships in a similar manner to adjacency matrix in graph theory, and is used in systems engineering to model the structure of complex systems or processes. The fusion of transaction costs and financial or engineering options is achieved by way of a square DSM in which its diagonal represents engineering options associated with systems components and the areas above and below the diagonal represents interface costs associated with interfaces between individual components or between the system and its environment.

5.3.2.5 Dynamic Value Paradigm

As mentioned before, a system's value to its stakeholders diminishes over time. As a result, systems have to be periodically upgraded at substantial cost and disruption. The project scientists proposed a quantitative model that attempts to minimize the value loss accumulating over the lifetime of systems. The model used estimates like economic growth, technology advances, wear-out costs, obsolescence costs, and system upgrade costs in order to provide quantitative estimates, indicating the optimal expected time until the next system upgrade.

5.3.3 Measurable Project Objectives

The project expected to deliver a step-change in the performance of industry characterized by a higher reactivity to customer needs and more economical production lines, product systems and customer services. The project conducted a plus-minus-interesting (PMI) analysis (Section 3.4.1, PMI Analysis) in order to identify the project's objectives and rate them in accordance with their importance. By the end of this process, the project objectives were as follows:

Objective 1: Develop a generic (widely applicable) and tailorable quantitative and usable method for architecting systems for optimal adaptability to unforeseen future changes in stakeholder needs and technology development. Such systems will exhibit better cost-efficiency and longer lifetime as well as reduced upgrade cycle time,[5] thus providing more value to stakeholders.

Objective 2: Validate and prove the methodology by means of real-life pilot projects in order to provide concrete evidence that it is (1) generic and tailorable, (2) scalable, (3) usable, and (4) cost effective.

Objective 3: Show by the end of the project that reconfiguring manufacturing systems or products/services designed for adaptability yields savings of 20% either in the cost or upgrade cycle time or a combination thereof. Confirmation will be based on measuring cost/time parameters during the pilot projects and simulations of these systems using future projection techniques.

---

5 The elapsed time between successive system's upgrades.

Objective 4: Show, by the end of the project, that the lifespan of manufacturing systems or products/services designed for adaptability increases by 25%. Confirmation will be based on analysis of the pilot projects data and simulations of these systems using future projection techniques.

Objective 5: Show that systems yielding more service for a longer duration will exhibit the following qualitative benefits: (1) during the manufacturing process, the overall usage of natural resources and energy consumption as well as the overall pollution and byproduct waste will be reduced and (2) adaptable systems will be more amenable to sustained evolving regulatory framework (i.e. environmental, health, safety, etc.).

5.3.4 Basis for Predicting the Objectives

The project coordinator together with the WP leaders utilized the SCAMPER (Substitute, Combine, Adapt, Modify, Put, Eliminate and Reverse) analysis (Section 3.3.5, SCAMPER Analysis) in order to assess whether the objectives of the project are attainable. The expected measurable objectives were based on two elements: (1) existing researches on cost efficient designs and (2) historical information provided by the project industrial and SME partners regarding losses due to lack of adaptability in their respective manufacturing lines, system products and customers services.

5.3.4.1 Existing Research

Many researches showed that, over time, design of various systems tended to evolve and improve again and again due to market pressure, combined with technology advances and engineering ingenuity. Figure 5.5 exemplifies this phenomenon. A company manufacturing transmission systems with hydraulic torque converters was able to continually reduce their manufacturing cost by about 70% (inflation considered) over a period of three decades. This was achieved by re-architecting the system and reducing the number of parts by 70% (Hundal et al., 2007).

5.3.4.2 Partners' Adaptability Experience

The six industrial and SME partners evaluated their own historical experiences vis-à-vis systems adaptability. More specifically, the partners estimated the savings that could have been realized had their systems been architected for adaptability. Table 5.1 depicts a condensed summary of this analysis. As can be seen, lack of manufacturing as well as products and customer services adaptability is a common phenomenon, permeating through most organizations.

The following is a detailed historical adaptability analysis, which illustrates the importance of adaptability to the industrial and SME project's partners.

TPPS. Tetra Pak is a global supplier of food technology processes, packaging, distribution lines, stand-alone equipment and services. In the past, our

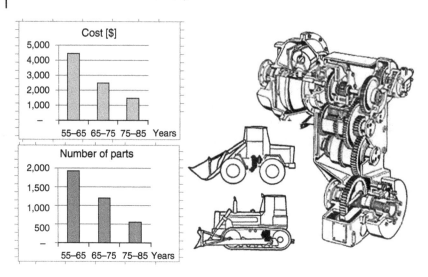

Figure 5.5 Cyclical architecture improvements

Table 5.1 Condensed partners' estimates of potential historical cost savings

Company	Business	Type	Category	Cost savings
TPPS	Food packaging systems	Industry	Manufacturing lines	20%
MAG	Machine tools	Industry	Manufacturing lines	20–30%
MAN	Trucks and buses	Industry	Customer service / System products	30%
IAI	Aircraft avionics / Vehicle electronics	Industry	System products	20–30%
TTI	Communication systems	SME	System products	30–35%
OPT	Optoelectronics systems	SME	Manufacturing lines	20%

liquid food packaging production lines were designed rigidly to meet existing and specific production requirements. As a result, our product development process used to be focuses on a single and isolated aspect of the production processes. However, we often experienced significant delays in upgrading our production line. In addition, we found that often, the improvements failed to achieve our expected environmental and value benefits. At a later stage, Tetra Pak implemented a product development strategy consisting of a systematic and holistic approach to the entire packaging solution. This was done by adopting an efficient combination of project management and systems engineering techniques. It was observed that under the new, holistic approach, a typical production line development project lasted about three years and cost about 55,000 person-hours. TPPS estimated that if the production line was originally architected for adaptability, we could have saved about 11,000 person-hours for each development project. This stems from an average reduction of around 20% in developments costs related to minimization of rework (15%) and eliminating "gold-plated" activities (5%).

MAG. MAG is a leading machine tool and systems company serving the durable-goods industry worldwide. A customer producing solar panels using MAG production equipment wanted to reduce the cycle time after the initial ramp-up to increase the throughput of the production line. However, products' dimensions as well as production equipment had to be adapted for each customer. The modification required 800 engineering hours plus about 8€ million in materials. Within one year, MAG was involved in two upgrade operations for solar industry. MAG estimated that, at the current rate of technology transformation and strong market demand, the company is required to perform about four to five such upgrades per year. If the system would have been originally designed for flexibility and adaptability about 20–30% of upgrading costs could have been saved.

MAN. MAN develops and produces trucks and buses suitable for multiple purposes, especially for the German and European markets. With the internationalization of our markets through cooperation agreements with Latin America, India, and China, however, the range of necessary products has considerably widened. More specifically, the number of variant models at a component level basically doubled. As a result, the organization suffers from deviating too much from economically viable set of standardized components and modules. MAN estimated that if they were architecting each range of products (e.g. busses for international markets) for adaptability with regards to different applications (e.g. reuse of different axles in vehicles of varying length and overall weight) they could have reduced the number of standardized modules from 350 to about 60. Such reduction could, on average, be translated into savings of up to 30% for a product range in both development and production costs with the added benefit of improved quality through targeted reuse of already available modules and subsystems.

IAI. IAI/Lahav is an avionics system house designing and fielding avionic systems for aircrafts and helicopters with a wide range of applications. In the past, the requirements for the internal communication among avionic units required relatively limited volume and therefore, the system's architecture was based on a central control and distribution of data by means of a standard MIL-STD-1553 Mux-Bus. New aircrafts have become "Systems of Systems" necessitating three orders of magnitude increase in the volume of data distribution. In addition, control had to be distributed among different avionic units. As a result the architecture of the avionics system had to be redesigned and many avionic units had to be upgraded to handle an Ethernet technology and protocols. The system re-architecture effort, including individual avionic units and interface redesign, required about 50,000 person-hours and lasted for about 2.5 years. IAI estimated that if the system was originally designed for adaptability we could have saved about 20–30% of the above cost.

TTI. TTI is an SME working in the space, defense, science, and telecommunications industries. Currently, TTI is developing and fielding Ka-band solid state power amplifiers (SSPA) systems for the Asian market. New applications requiring different SSPA systems are emerging continuously, and the time-to-market requirements decrease dramatically. In addition, each world region has a different frequency band allocation for the same application coupled with specific requirements for power and gain levels. As a result, our core systems must be upgraded periodically at a substantial cost in order to adapt them to the new technological and market opportunities. Typically, TTI needed to customize each SSPA system every two years at a cost of 1,800 person-hours. However, we manufacture a large number of communication components (i.e. Advanced radio and microwave components, RF transceivers and outdoor VSAT units, Satcom mobile terminals, antennas, etc.) so we estimated that if our products were originally designed for adaptability, TTI could save about 30–35% of the total design and integration cost for each customer.

OPT. OPT is an SME that owns and operates a production line for holograms. Originally, the facility was designed for analog instrumentation yielding limited adaptability. This included strict requirements on environmental noise and vibration, long exposure time of the holographic master, and an ongoing need for multiple chemicals used during the development of holographic plates. After intensive efforts, we upgraded our production line using digital exposure system, which significantly increased the efficiency of the holographic process as well as its quality. The new system reduced production time of holographic masters by over 70% and the overall cost for the consumption of raw materials was reduced by 50–60%. At the time, the system was able to reach a resolution of 10,000 dots per inch (DPI) and the minimum height of text could reach 25 μm, thus allowing the technicians to design more complex holograms that open our facility to a larger market segment. OPT estimated that if our system was originally designed for flexibility OPT could have saved up to 20% of the direct investment and reduced the total time used for system installation by a similar amount.

5.3.5 Systems Adaptability: State-of-the-Art

The purpose of this section is to describe the state-of-the-art[6] in architecting manufacturing and systems for adaptability. In addition, this section discusses the advantages of using quantitative methods for architecting

---

6 This case study is depicted in chronological order thus, this section refers to State-Of-The-Art up to 2010.

adaptable systems and the assumptions made by the project's designers and their justifications.

5.3.5.1 Architecting Systems for Adaptability

By and large, systems are typically architected solely to meet stated requirements at a point in time. Many designers do not consider the fact that systems are repeatedly upgraded as they evolve. Meanwhile, there is ample literature (e.g. Fricke and Schulz, 2005) indicating that systems undergo major upgrades every few years due to:

Market demand. Users or customers desire increased capabilities.
Technology improvements. Opportunities are provided by new technologies.
Maintenance costs. Aging causes increased maintenance costs and down times.
Component obsolescence. Partial systems must be redesigned due to obsolete components.

The earliest, formal, and public Design for Adaptability (DFA) reference appeared in 1986 as applied to computer hardware and software architecting (Alexandridis, 1986). Such philosophy eventually led to the development of computer hardware and software packages possessing open systems architectures (e.g. object-oriented). An alternative DFA methodology has been developed by the Software Engineering Institute at Carnegie Mellon University in Pittsburgh, Pennsylvania. The Product Line Practice Initiative (PLPI) guides organizations away from traditional, one-at-a-time system development and towards the systematic, large-scale reuse paradigm of product lines. PLPI is limited, however, to architecting software components and for reusability as the system evolves (CMU/SEI, 2003). There were several other budding research centers interested in various aspects of software DFA. For example, the Distributed Systems Research Group at Carnegie Mellon University was interested in identifying, understanding, and constructing technology that facilitates adaptable software systems. However, these efforts were oriented toward a narrow band of existing systems within the software domain.

Open systems was another (limited) DFA approach emphasizing standard interfaces and modularity of subsystems. This was both a technical approach to systems engineering and a preferred business strategy that was applied by the US Department of Defense (DoD) for large and complex systems (Hanratty, 1999). Of course, the issue of DFA is much wider than the scope of open systems as is defined by the US DoD. Researchers at the Tel-Aviv University (TAU) devised a method to calculate the total design cost of product platforms if components or subsystems would undergo standardization or modernization. Simulating different alternatives, and comparing the total design cost of each option, a design team could be directed to exploit systems upgrade opportunities. In other words, this method used an estimation of design effort as the basis for architecture decision in contrast with commonly used static measures of components' interactions (Sered and Reich, 2006).

Design Structure Matrix (DSM) attracts a worldwide research effort (see, e.g., Maurer et al., 2005; Karniel and Reich, 2009). In general, a DSM depicts some key constituents of a system and the corresponding information exchange and dependency patterns. In other words, it details systems' elements as well as information flow and other dependencies within and outside a system. In this way, one can quickly recognize which component depends on information generated by other components. The main strength of DSM is that it can represent a large number of system elements and their relationships in a compact way that highlights important patterns in the data (such as feedback loops and modules). DSM analysis provides insights into how to manage complex systems or projects, highlighting information flows, task sequences and iteration. It can help teams to streamline their processes based on the optimal flow of information between different interdependent activities.

Researchers at the Massachusetts Institute of Technology (MIT) have been developing a theoretical approach to the value of flexibility (de Neufville et al., 2004). These concepts, defined as real options "in" projects, are options created by changing the architecture of the technical system. Real options in systems can be very effective (Wang, 2005). For example, the use of real options in satellite communication systems can increase the value of satellite communications systems by 25% or more. In that case, the real options "in" the satellite constellation involve additional positioning rockets and fuel in order to achieve a flexible architecture that could adjust capacity according to need (de Weck et al., 2004). In addition, much research on flexibility, particularly related to the oil and gas exploration (Lin et al., 2009) and manufacturing sector in the United States (Suh et al., 2008) and Europe, is conducted (see, e.g., Maurer et al., 2005; Lindemann et al., 2008).

Another state-of-the-art technology, relevant to the project involved quantitative research on virtual manufacturing (VM). This was the process of architecting a model of a manufacturing enterprise, a product system or a customer service and conducting experiments with this model for the purpose of capturing its future behavior. Typically, such a model represents the structure and interconnectivity of a virtual system which may be simulated in order to understand its physical and logical operations. Based on current research and several case studies (see, e.g., Ali and Seifoddini, 2006), it was found that simulated models of business processes could predict the manufacturing dynamics and business impact of a proposed architecture of manufacturing industries by evaluating the process performance of a number of measures such as resource constraints, manpower utilization, departmental interactions, lead times, productivity under different condition and so on. A business process simulation (BPS) provides the opportunity to test for unexpected future interactions with the system or check the robustness of the design of business scenarios quantitatively and visually.

5.3.5.2 Using Quantitative Methods

The project's researchers committed to developing the next generation of quantitative method for architecting adaptable systems using the concept of architecture options (AO). Table 5.2 summarizes the project technical approach as opposed to the prevailing state of the art practices, up to 2010.

Table 5.2 Technical approach versus prevailing state of the art

Current practice	The project technical approach
Narrow systems architecting concepts / philosophy	The project sought to expand the systems architecting concepts and philosophy by demonstrating to the business and engineering communities, the advantage of reframing the architecting problem from meeting specifications at a point in time to providing best value over a range of scenarios throughout the system's lifetime.
Lack of systematic DFA methodology	The project planned to develop an approach enabling systems engineers to make informed judgments about architecting adaptability into their manufacturing systems and products/services. It defined features affecting the adaptability of systems as well as procedures to harmonize them with current and emerging architecting methodologies.
No optimization of DFA strategy	The project planned to develop quantitative procedures and a software tool for translating future uncertainties, risks, and opportunities into quantified value scenarios. These scenarios should help determine the costs of adapting manufacturing systems and products/services to new opportunities or circumstances. The project goal was to realize 20% improvement in systems lifetime value and extension of 25% in systems lifetime.
No DFA assessment within industrial projects	The project planned to perform formal assessments of the effectiveness of using AO methodology and tools within industrial setting by way of conducting six real-life pilot projects in diverse industries (i.e. food packaging, machine tools, trucks and buses, aerospace, communication, and optoelectronic). The results were to be used to enhance the project research products/services as well as to measure the overall performance of the pilot projects against the stated original requirements.

5.3.5.3 Project Assumptions and Justifications

The project proposal was based on a number of assumptions. These are delineated in the following table together with their justifications (Table 5.3).

Table 5.3 Project assumptions and justifications

Assumptions	Justifications
Manufacturing industries and system products and services provide value to stakeholders. This value diminishes over time.	These has been shown by many researches (e.g. Browning and Honour, 2005, 2008; Hundal et al., 2007, etc.)
Systems have to be periodically upgraded since complete replacement costs are often prohibitive. Therefore, system adaptability is a valuable characteristic.	• These have been shown by many researches (see above).
By and large, current architecting and design concepts address present requirements and ignore future unanticipated customer needs.	• See section 5.3.4.2 Partners' Adaptability Experience.
Adaptability can be designed into systems so that they will provide maximum value to stakeholders throughout their lifetime.	This has been shown in numerous studies (see references).
The project is based on two economic theories: transaction cost theory and financial options theory.	These theories are well established, extensively used in business and academic environments, and sufficiently proven to be useful. See References.
The project is based on architecture options (AO) theory.	AO theory has been published in the past and demonstrated mathematically.
The project is based on Design Structure Matrix (DSM).	DSM methodology is a well-established field used in many business and academic domains (e.g. Maurer et al., 2005; Karniel and Reich, 2009).
Both AO theory and DSM will be expanded in order to utilize them effectively within industrial settings.	These activities are planned to be done in WP2. Methodology development.
It is possible to develop a generic quantitative method for architecting systems for optimal adaptability to unforeseen future changes.	• Soundness of underlying economic and engineering theories • Project strategy based on fusing scientific research and pilot projects
Systems architected for future adaptability will exhibit better cost-efficiency, longer lifetime, as well as reduced cycle time, thus, provide more value to stakeholders.	• The generic and tailorable attributes of the methodology is sufficiently "proven" by selecting the six pilot project domains.

(Continued)

Table 5.3 (Continued)

Assumptions	Justifications
It is possible to provide concrete evidence that the method is generic, scalable, usable and cost effective.	• See Section 5.3.4, Basis for Predicting the Objectives. • Positive experience is gained in the EC supported SysTest project (contract number G1RD-CT-2002-00683) using comparable overall strategy • Scientific knowledge, industrial diversity and leadership of the project consortium
Show by the end of the project, that re-architecting manufacturing system or products/services for adaptability yields: • Savings of 20% either in cost or cycle time or a combination thereof. • Increase by 25% the lifespan of manufacturing system or products/services	• See Section 5.3.4.2, Partners' Adaptability Experience. • Scientific research appears in the literature, especially (Hundal et al., 2007; de Weck et al., 2004; Lin et al., 2009; Suh et al., 2008; Lindemann et al., 2008) • Measuring cost/time parameters during the pilot projects • Simulating future projection of pilot projects data (see Section 5.4.1.3, Extrapolating future objectives based on early data sets).

5.3.6 Further Reading

• Alexandridis, 1986.
• Ali and Seifoddini, 2006.
• Black and Scholes, 1973.
• Browning and Honour, 2005, 2008
• CMU/SEI, 2003.
• Coase, 1937.
• de Neufville et al., 2004.
• de Weck et al., 2004.
• Engel and Browning, 2006, 2008.
• Fricke and Schulz, 2005.

• Hanratty, 1999.
• Hundal et al., 2007.
• Karniel and Reich, 2009.
• Lin et al., 2009.
• Lindemann et al., 2008.
• Maurer et al., 2005.
• Sered and Reich, 2006.
• Suh et al., 2008.
• Wang, 2005.

5.4 Project Planning

Meticulous project planning is a prerequisite for successful engineering projects. The careful preparations for what became the AMISA project was instrumental in achieving most of the project's objective, as well as creating an excellent team spirit that prevailed throughout the three-year project and beyond.

The problems of managing R&D projects funded by international, national, or regional agencies are often compounded because many of these projects are operated by way of a consortium arrangement. A consortium is generally composed of independent companies and organizations where each one has its own institutional as well as personal understanding and interests. More often than not, these points of views are not quit aligned and even opposed. Exasperating the management problems is the fact that in a consortium the lines of command between the coordinator and the different organizations are blurred or nonexistent. A coordinator can explain, ask, beg, or cajole – but not much more. Thus, comprehensive project planning goes a long way to alleviating the shortcoming inherent in consortium structures.

Accordingly, the purpose of this chapter is to provide an example, a template, for engineers undertaking the difficult task of planning a complex scientific and engineering R&D project, especially, under a consortium structure.

5.4.1 Project Planned Activities

5.4.1.1 Planning the Logical Progression and Information Flow of the Project

The project coordinator utilized the Contradictions resolution methodology (Section 3.2.2, Resolving Contradictions) in order to lead the detailed planning of the project. The planned logical progression of the project is described below (Figure 5.6).

Capture relevant information. The project is to start by capturing the state of the art in DFA as well as other information related to relevant IMS projects and European standards (WP1).

Define requirements. Next, all partners are to identify their DFA methodology and tool requirements (WP1).

Develop methodology and tool. Next, a DFA generic and tailorable methodology and tools are to be created. The methodology and tools include built-in means for tailoring them to different users' needs. The tailoring is to be done along the nature of DFA business domain (i.e. manufacturing lines, product systems or service business) as well as system size/scope and other relevant parameters (WP2, WP3).

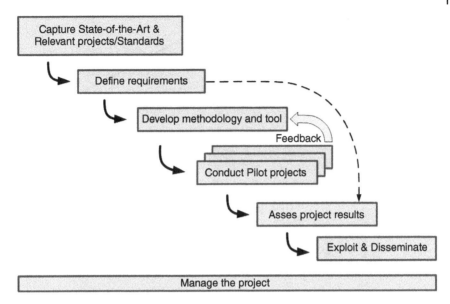

Figure 5.6 Project's logical progression

Conduct pilot projects. Next, four industries and two SMEs are to tailor the methodology to their specific needs and then conduct real-life pilot projects in order to verify the viability of the methodology and tools. In addition, and if needed, all partners will engage in adjusting the DFA methodology and tool for real-life implementation within different industrial settings (WP4).

Assess project results. The pilot projects results are to be assessed against the original objectives and requirements (WP5).

Exploit and disseminate results. The results of the project are to be exploited by the partners and disseminated to the public (WP6).

Manage the project. The project is to be managed and monitored throughout its duration (WP7). The project coordinator, together with the work package (WP) leaders created a process map (Section 3.6.1, Process Map Analysis). The planned internal flow of information as well as the planned flow of information between the project and the outside world is depicted in Figure 5.7. This includes general input and output information as well as project deliverables.

5.4.1.2 Pilot Projects

The pilot projects are considered a vital component of the project. Here, each pilot project is to assess the viability and effectiveness of the DFA methodology

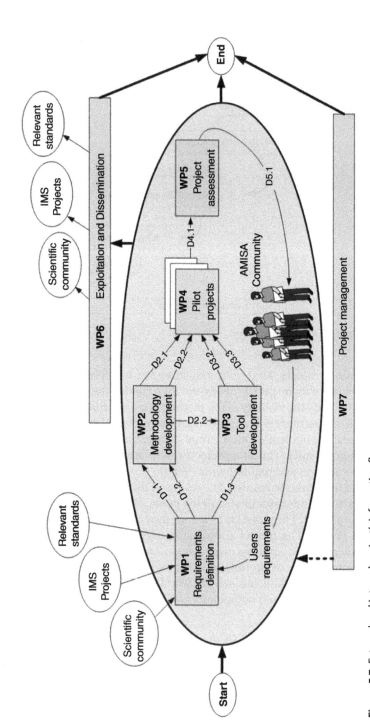

Figure 5.7 External and internal project's information flow

and economic model under true, real-life industrial conditions. Table 5.4 summarizes the originally planned six pilot projects.

Table 5.4 Summary of six pilot projects

Ind.	SME	Business	**Architecting adaptability in:** 1. Product systems 2. Service business 3. Manufacturing lines		
TPPS		Food packaging systems	Biodegradable products and reduced energy / wastes	X	
MAG		Machine tools	Turnkey solar modules production systems	X	
MAN		Trucks, buses, large engines	Customer services and environmental truck engines / exhaust architecture	X	X
IAI		Aircraft avionics / Vehicle electronics	Partially known requirements of unmanned ground vehicle		X
	TTI	Communication systems	Large and changing variety of solid-state power amplifiers		X
	OPT	Optoelectronics systems	Digitally controlled holographic production process	X	

A more specific description of the planned pilot projects follows:

TPPS. Tetra Pak's pilot project was to include the development of an overall family of packages supported by the same manufacturing system of systems. The scope of the pilot project included the updating of several industrial production plants and packaging lines of small, medium and large dimensions for adaptability as well as acquiring relevant process data.

MAG. MAG pilot project was to be involved in infusing adaptability into MAG turnkey production systems, mainly for manufacturing solar modules.

MAN. MAN was planning to undertake two pilot projects: (1) a project involves in the development of an adaptable truck or bus customers service organization and, (2) a project focused on adapting truck engines and exhaust architectures for yet unspecified environmental needs such as the Euro 6 norm

IAI. IAI/Lahav selected its new unmanned ground vehicle (UGV) as a candidate for evaluating the effectiveness of the project's proposed technologies. The pilot project was to include the design of a robotic vehicle control system. The scope of the pilot project included mission planning and operations under autonomous and tele-operation modes. In addition, the system was expected to support a "system of systems" concept requiring coordination among a number of UGVs.

TTI. TTI sought to achieve a flexible design of its solid state power amplifiers (SSPA) so it will be adaptable to new technological developments and to the new allocated transmission bands depending on market demand. In this way, the SSPA design could support systems ranging from high power transmitters to small distribution and portable systems. TTI considered the SSPA system a candidate for evaluating the effectiveness of the project's proposed methodology. The pilot project was an attempt to design highly adaptable SSPAs.

OPT. OPT sought to conduct a pilot project aiming at infusing maximum adaptability into its holographic products/services production line architecture.

5.4.1.3 Extrapolating Future Objectives Based on Early Data Sets

Normally, a production line or a significant system is designed for a lifetime duration of 5, 10, or 20 years. Such systems may be updated or renovated a few times during their lifetime so, at those times it is easy to find the cost and other parameters associated with such upgrade. However, the pilot projects were to be carried out within about 20 months whereas the intention is to determine the success or failure of the project in meeting future objectives. In other words, the intent is to use relatively short duration pilot projects in order to determine whether a system designed for adaptability may be upgraded to unplanned specifications at less cost or after longer duration. That is, whether the lifetime value of a system designed for adaptability is, in fact, enhanced.

Originally, the project planned to exploit modern quantitative research on Virtual Manufacturing and use a selected commercially available tool (e.g. Arena, Automode, Delmia-quest, etc.) to perform discrete event factory simulation (DEFS). The intent is to use business process simulation (BPS) to predict the behavior of a virtual manufacturing enterprise (or system) designed in an "as-is" manner (only to meet stated requirements at a point in time) and compare the results with BPS of a system that is built to meet future unknown requirements (designed for adaptability).

5.4.1.4 Project Schedule

A graphical presentation of the project schedule master plan is depicted in Figure 5.8. The formal deliverables that were to be generated by the different WPs is superimposed on the image.

5.4.1.5 Work Packages and Tasks

A graphical presentation of the work breakdown structure (WBS) depicting the seven work packages (WPs) and the tasks of the project are depicted in Figure 5.9.

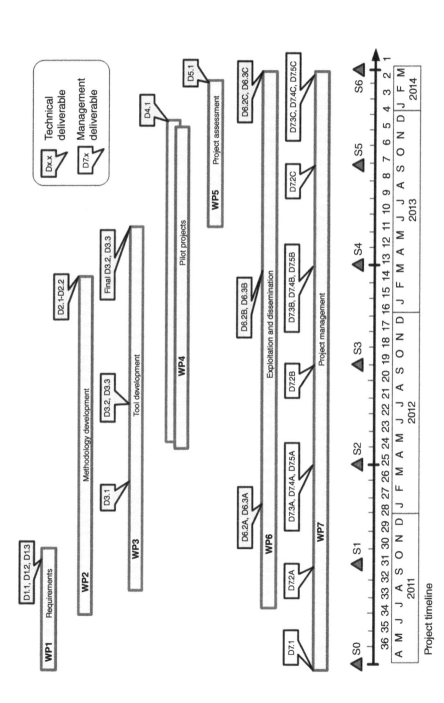

Figure 5.8 Project schedule master plan

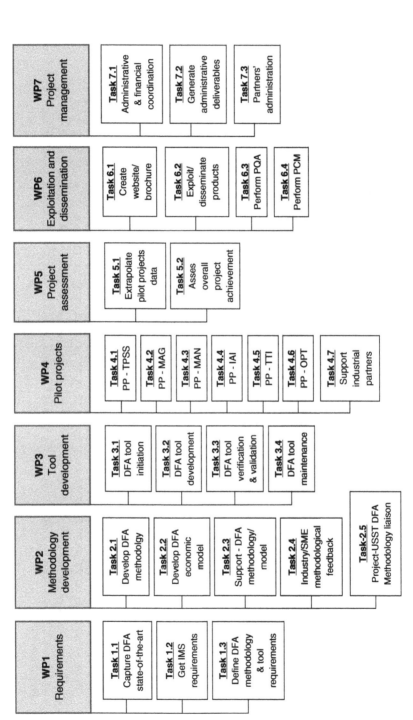

Figure 5.9 Project tasks and work packages

5.4.1.6 Work Package List

The planned project's work packages are depicted in Table 5.5. The table defines the work packages numbers and titles, the types of activities, the WP leaders as well as the scope of each WP, and the start and end months of each WP (The type of activity is: RTD = Research and Technology Development; MANAG = Management).

Table 5.5 Work package list

WP number	Work package title	Type of activity	Lead participants	Person-months	Start month	End month
1	Requirements definition	RTD	MAN	38	1	8
2	Methodology development	RTD	TUM	56	4	24
3	Tool development	RTD	TAU	41	5	27
4	Pilot projects	RTD	IAI	153	13	34
5	Project assessment	RTD	MAG	43	27	36
6	Exploitation & dissemination	RTD	TPPS	30	6	36
7	Project management	MANAGE	TAU	27	1	36
	TOTAL			388		

5.4.1.7 Deliverables List

Table 5.6 shows the planned deliverables. This includes the deliverable ID and name, WP number responsible for producing the deliverable, the nature of the deliverable as well as the dissemination level and the deliverable planned delivery month. The nature of deliverables is: R = Report, P = Prototype, O = Other. The dissemination levels of each deliverable were: PU = Public, PP = Restricted to other program participants as well as the customer.

Table 5.6 Deliverable list

Deliverable ID	Deliverable name	WP number	Nature of deliverable	Dissemination level	Delivery month
D1.1	DFA state-of-the-art report	1	R	RP	6
D1.2	Requirements for a DFA Methodology	1	R	RP	6
D1.3	Requirements for a DFA Economic Model support tool	1	R	RP	6
D2.1	DFA Methodology Guide (DFA-MG)	2	R	PU	24, 34
D2.2	DFA Economic Model (DFA-EM)	2	R	PU	24, 34
D3.1	DFA-Tool specifications and design document	3	R	PP	10
D3.2	Design For Adaptability-Software Tool (DFA-ST),	3	P	PU	15, 34
D3.3	DFA-Tool training and assimilation kit	3	O	PU	15, 34
D4.1	Report on six pilot projects and their results	4	R	PP	33
D5.1	Assessment of the project overall achievements	5	R	PP	35
D6.1	Brochure and internet site promoting the project	6	O	PU	8
D6.2-A	Exploiting and disseminating of knowledge year-1	6	R	PP	12
D6.2-B	Exploiting and disseminating of knowledge year-2	6	R	PP	24
D6.2-C	Exploiting and disseminating of knowledge year-3	6	R	PP	36

(Continued)

Table 5.6 (Continued)

Deliverable ID	Deliverable name	WP number	Nature of deliverable	Dissemination level	Delivery month
D6.3-A	Report on PQA and PCM end year 1	6	R	PP	12
D6.3-B	Report on PQA and PCM end year 2	6	R	PP	24
D6.3-C	Report on PQA and PCM end year 3	6	R	PP	36
D7.1	Consortium agreement	7	R	PP	0
D7.2-A	Periodic interim activity reports midyear 1	7	R	PP	6
D7.2-B	Periodic interim activity reports midyear 2	7	R	PP	18
D7.2-C	Periodic interim activity reports midyear 3	7	R	PP	30
D7.3-A	Periodic activity reports end year 1	7	R	PP	12
D7.3-B	Periodic activity reports end year 2	7	R	PP	24
D7.3-C	Periodic activity reports end year 3	7	R	PP	36
D7.4-A	Financial follow-up end year 1	7	R	PP	12
D7.4-B	Financial follow-up end year 2	7	R	PP	24
D7.4-C	Financial follow-up end year 3	7	R	PP	36
D7.5-A	Periodic management report end year 1	7	R	PP	12
D7.5-B	Periodic management report end year 2	7	R	PP	24
D7.5-C	Periodic management report end year 3	7	R	PP	36

5.4.1.8 List of Milestones

Table 5.7 depicts a list of planned project's milestones. Each milestone is identified by its ID, its name and description, the work package responsible to execute the milestone, the expected month of occurrence and the ways to prove each milestone verification.

Table 5.7 List of project milestones

Milestone ID	Milestone name/description	Responsible WP	Expected month	Means of verification
M0.1	Kickoff meeting	7	0	N/A Relevant deliverables: D7.1
M1.1	Initiate Workshop-0, open to representatives from two IMS projects: 1. MyCar-FUTURA and 2. VFF-MTP	1	1	Workshop is dedicated to sharing the understanding regarding project research: Theoretical concepts in DFA, Options (Financial, Real, Architectural) and DSM
M1.2	Synthesize DFA methodology requirements versus the state-of-the-art and current practices.	1	6	Reach a consensus on requirements for the DFA methodology and economic model. Relevant deliverables: D1.1, D1.2, D1.3
M2.1	Complete first phase development of DFA methodology and economic model.	2	15	Reach a consensus on a generic and tailorable DFA methodology. Agree on DFA tailoring rules for different environments and industries. Reach a consensus on a generic and tailorable DFA economic model.
M2.2	Finalize DFA methodology and economic model development.	2	30	Successful evaluation of the DFA methodology and economic model based on usage in different industrial pilot projects. Relevant deliverables: D2.1, D2.2

(Continued)

Table 5.7 (Continued)

Milestone ID	Milestone name/description	Responsible WP	Expected month	Means of verification
M3.1	Complete first phase development of the DFA software tool.	3	15	Agree on the design and initial interfaces and operations of the DFA software tool. Relevant deliverables: D3.1, D3.2, D3.3
M3.2	Finalize the development of the DFA software tool.	3	30	Successfully operate the DFA software tool within various industrial pilot projects.
M4.1	Verify DFA assimilation activities.	4	18	DFA assimilation kit containing training material for using DFA methodology and economic model in the various pilot projects is ready. Ongoing DFA support to pilot projects is available. Industrial pilot projects have begun running.
M5.1	Project users' group conference on assessment of DFA methodology utilizing real-life pilot projects data and COTS tool to perform Discrete Event Factory Simulation (DEFS).	5	34	Availability of the following reports: • D1.2 Requirements for a DFA methodology • D1.3 Requirements for a DFA economic model support tool • D5.1 Assessment of the project overall achievements
M6.1	Project Workshop-1 (open to the public).	6	24	Workshop devoted to project progress and achieved results. Opinions about project topics and the methods of solving the technical issues will be discussed, which should help in the next phase of the work. Relevant deliverables: D6.1, D6.2, D6.3
M6.2	Project Workshop-2 (open to the public).	6	36	Workshop will be conducted in order to present project results and for discussing further implementation activities. Relevant deliverable: D4.1, D5.1
M7.1	Project closure.	7	36	Relevant deliverables: Final D7.2, D7.3, D7.4, D7.5

5.4.1.9 Summary of Staff Effort

Table 5.8 depicts a summary of staff effort distributed by project participant and work package.

Table 5.8 Summary of staff effort

Participant no.	Short name	WP1	WP2	WP3	WP4	WP5	WP6	WP7	Total person Months	%
1	TAU	5	12	24	6	6	6	20	79	20%
2	TUM	8	22	4	6	12	6	1	59	15%
3	TPPS	3	2	1	19	3	3	1	32	8%
4	MAG	4	4	4	24	6	3	1	46	12%
5	MAN	6	4	2	24	4	3	1	44	11%
6	IAI	4	4	2	26	4	3	1	44	11%
7	TTI	4	4	2	24	4	3	1	42	11%
8	OPT	4	4	2	24	4	3	1	42	11%
Total		38	56	41	153	43	30	27	388	100%

5.4.2 Detailed Work Package Descriptions

Table 5.9 to Table 5.15 describe each of the planned work packages.

Table 5.9 WP1: Requirements definition

Start date:	Month 1							
Activity type	RTD							
Participant number	1	2	3	4	5 (WP leader)	6	7	8	Total
Participant short name	TAU	TUM	TPPS	MAG	MAN	IAI	TTI	OPT	
Person-months per participant	5	8	3	4	6	4	4	4	38
WP objectives	The objective of this WP is to analyze the current state-of-the-art in Design for Adaptability (DFA) process and practices in industry and academia and to define the requirements for the (1) DFA Methodology, (2) DFA Economic Model of the project and (3) software DFA-Tool needed by all industries / SME participating in the project.							
Description of work									Part.
Task 1.1 Capture DFA state-of-the-art	Identify and gather all relevant current research representing state-of-the-art in DFA methodologies. This will cover methodical review of: (1) advances in US, European, and other academic institutions, (2) Governments organizations (e.g. NASA, ESA), (3) Government / Industries collaborative research (e.g. EC research programs) and (4) Industries (e.g. The IMS organization).								TAU, TUM, MAN, USST
Task 1.2 Get IMS requirements	Agree on collaboration between the project and two emerging IMS projects (MyCar-FUTURA and VFF-MTP) and get DFA methodology requirements relevant to the above IMS projects.								TAU, TUM, MAN

(Continued)

Table 5.9 (Continued)

Description of work		Part.
Task 1.3 Define DFA methodology & tool requirements	Define requirements from the Systems Engineering / TAU perspectives.	TAU
	Define requirements from the Mechanical Engineering / TUM perspectives.	TUM
	Define requirements from the Manufacturing / food industry perspectives.	TPPS
	Define requirements from the Manufacturing / Machine tools perspectives.	MAG
	Define requirements from the customers' service of truck and buses perspectives.	MAN
	Define requirements from the avionics and aerospace systems products perspectives.	IAI
	Define requirements from the Information and Communication systems products perspectives.	TTI
	Define requirements from the Optoelectronics equipment perspectives.	OPT
	Define requirements from the US academic perspectives.	USST

WP deliverables		Res.
D1.1	DFA State-of-the-art report	MAN
D1.2	Requirements for a DFA methodology	TUM
D1.3	Requirements for a DFA economic model support tool	TAU

Table 5.10 WP2: Methodology development

Start date:	Month 4								
Activity type	RTD								
Participant number	1	2 (WP leader)	3	4	5	6	7	8	Total
Participant short name	TAU	TUM	TPPS	MAG	MAN	IAI	TTI	OPT	
Person-months per participant	12	22	2	4	4	4	4	4	56

WP objectives	The objective of this WP is to develop a generic and tailorable DFA Methodology and a quantitative Economic Model. Ultimately, the objective is to generate a DFA Methodology Guide (DFA-MG) and a formal DFA Economic Model (DFA-EM) for architecting manufacturing enterprises, product systems and customer services to be optimally adaptable to unforeseen future changes. The DFA methodology and model were to incorporate a built-in tailoring facility which is considered the nature of DFA improvement (i.e. manufacturing lines, product systems, or service business) and system class as well as other relevant parameters.	
Description of work		Part.
Task 2.1 Develop DFA methodology	Create the project Design for Adaptability-Methodology Guide (DFA-MG). This involves the following: (1) Propose the most appropriate set of procedures, methods and guidelines needed in order to foster practical DFA thinking and DFA implementation within industrial setting, (2) define a set of heuristic rules for DFA methodology tailoring for different industrial domains and systems classes and (3) support the six pilot projects and adjust the DFA-MG in accordance with ongoing encountered issues.	TAU, TUM

(*Continued*)

Table 5.10 (Continued)

Description of work		Part.
Task 2.2 Develop DFA economic model	Create the project Design for Adaptability-Economic Model (DFA-EM). This involves the following: (1) Propose the most appropriate and practical models to quantify DFA parameters and their mathematical relationships, (2) Identify practical means to gather such information in an industrial settings, (3) Devise means to find and optimize architecture under realistic constraints, (4) support the six pilot projects and adjust the DFA-EM in accordance with ongoing encountered issues.	TAU, TUM
Task 2.3 Support DFA methodology/model	Maintain the project DFA methodology. This involves the following: (1) Provide DFA methodological support to the six pilot projects in terms of explanation of the methodology and responding to any problem detected in the field and (2) Adjust and upgrade the DFA methodology and model in accordance with the practical needs within the pilot projects.	TAU, TUM
Task 2.4 Industry/MSE methodological feedback	All four industries and two SMEs will provide feedback concerning the effectiveness and problems associated with the DFA methodology and DFA economic model.	TPPS, MAG, MAN, IAI, TTI, OPT
Task 2.5 Project-USST DFA methodology liaison	Participate in ongoing technical meetings. Provide DFA methodology inputs based on current US research. Review reports and presentation materials created by the project's partners.	USST
WP deliverables		Res.
D2.1	DFA methodology guide (DFA-MG)	TUM
D2.2	DFA economic model (DFA-EM)	TAU

Table 5.11 WP3: Tool development

Start date:	Month 5							
Activity type	RTD							
Participant number	1 (WP leader)	2	3	4	5	6	7	8	Total
Participant short name	TAU	TUM	TPPS	MAG	MAN	IAI	TTI	OPT	
Person-months per participant	24	4	1	4	2	2	2	2	41

WP objectives: The objective of this WP is to build a prototype software tool embodying the DFA-EM as well as tailoring heuristics and rules. This DFA-Tool is to be evaluated by the academic and the industry/SME partners and calibrated within six different industrial sectors. The different industrial partners were to guarantee the practical applicability of the DFA-EM theory in industrial setting and its harmonization with company best practices.

Description of work	Part.
Task 3.1 DFA tool initiation	This task will initiate development of the DFA-Tool software package. This shall include using the existing DFA-Tool requirements in order to create tool specifications and design. The specifications / design will be presented to the project's partners for evaluation and approval.	TAU
Task 3.2 DFA tool development	Develop the DFA-Tool software according to the available requirements and specifications / design. The development will be carried out iteratively where new functionalities and re-testing will be done on a cyclical basis. In addition, the DFA-Tool will be demonstrated to the consortium members in order to validate the general concepts of the implementation.	TAU

(*Continued*)

Table 5.11 (Continued)

Description of work		Part.
Task 3.3 DFA tool verification & validation	Test the DFA-Tool software package. Two project's partners (one academic and one industrial) will formally evaluate the DFA-Tool.	TUM, MAG
Task 3.4 DFA tool maintenance	Maintain the project DFA-Tool. This involves the following: (1) Demonstrate the DFA-Tool to all users, (2) train the users on the operation of the DFA-Tool, (3) respond to any user queries regarding the utilization of the software package, (4) provide maintenance facilities to the consortium members for the duration of the project and (5) adjust and upgrade the DFA software tool in accordance with the practical needs occurring within the pilot projects.	TAU, TUM, TPPS, MAG, MAN, IAI, TTI, OPT

WP deliverables		Res.
D3.1	DFA-Tool specifications and design document	TAU
D3.2	Design for Adaptability-Software Tool (DFA-ST),	TAU
D3.3	DFA-Tool training and assimilation kit	TAU

Table 5.12 WP4: Pilot projects

Start date	Month 13								
Activity type	RTD								
Participant number	1	2	3	4	5	6 (WP leader)	7	8	Total
Participant short name	TAU	TUM	TPPS	MAG	MAN	IAI	TTI	OPT	
Person-months per participant	6	6	19	24	24	26	24	24	153
WP objectives	The objective of this WP is to support the formal assessments of the DFA methodology and economic Model by conducting real-life pilot projects in different industrial sectors. Specifically, this work package is to include the selection of DFA pilot projects and the tailoring of the DFA methodology to the specific manufacturing industries, product system and customer service classes. The results are to be used to enhance the project research products as well as to measure the overall performance of the projects against the stated original requirements.								
Description of work									Part.
Task 4.1 Pilot project TPPS	Select a real-life pilot project within the domain of the food packaging industry for optimizing its manufacturing architecture for future adaptability. In addition, tailor the DFA methodology to suite the specific needs of the project. In addition, conduct a real-life pilot project and collect appropriate process data related the specific domain / industry.								TPPS
Task 4.2 Pilot project MAG	Select a real-life pilot project within the domain of the machine tools industry for optimizing its manufacturing architecture for future adaptability. In addition, tailor the DFA methodology to suite the specific needs of the project. In addition, conduct a real-life pilot project and collect appropriate process data related the specific domain / industry.								MAG

(Continued)

Table 5.12 (Continued)

Description of work		Part.
Task 4.3 Pilot project MAN	Select two real-life pilot projects within the domain of the trucks and buses industry for (1) optimizing its customer service architecture for future adaptability and (2) optimize engines/exhaust architecture adaptability for future environmental requirements. In addition, tailor the DFA methodology to suite the specific needs of the project. In addition, conduct a real-life pilot project and collect appropriate process data related the specific domain / industry.	MAN
Task 4.4 Pilot project IAI	Select a real-life pilot project within the domain of vehicle electronics associated with UGV products / services for optimizing the system architecture for future adaptability. In addition, tailor the DFA methodology to suite the specific needs of the project. In addition, conduct a real-life pilot project and collect appropriate process data related the specific domain / industry.	IAI
Task 4.5 Pilot project TTI	Select a real-life pilot project within the domain of SSPAs for communications systems products for optimizing the system architecture for future adaptability. In addition, tailor the DFA methodology to suite the specific needs of the project. In addition, conduct a real-life pilot project and collect appropriate process data related the specific domain / industry.	TTI
Task 4.6 Pilot project OPT	Select a real-life pilot project within the domain of manufacturing holographic products for optimizing its manufacturing architecture for future adaptability. In addition, tailor the DFA methodology to suite the specific needs of the project. Conduct a real-life pilot project and collect appropriate process data related the specific domain / industry.	OPT
Task 4.7 Support industrial partners	Support all the industrial partners and provide methodological help in order to: (1) eliminate problems while conducting the pilot projects and collecting relevant date, (2) methodological support in extrapolating the collected data and verifying whether and to what extant each pilot project has met its original objectives.	TAU, TUM
WP deliverables		Res.
D4.1	Report on six pilot projects and their results.	IAI

Table 5.13 WP5: Project assessment

Start date	Month 27							
Activity type	RTD							
Participant number	1	2	3	4 (WP leader)	5	6	7	8	Total
Participant short name	TAU	TUM	TPPS	MAG	MAN	IAI	TTI	OPT	
Person-months per participant	6	12	3	6	4	4	4	4	43
WP objectives	The objective of this WP is to conduct formal assessments of the DFA methodology and Economic Model using data obtained from the real-life pilot projects. Business process simulation (BPS) will be used to predict the long term impact of implementing DFA methodology within each pilot project. The results were to be used to enhance the project research products as well as to measure the overall performance of the projects against the stated original requirements.								
Description of work								Part.
Task 5.1 Extrapolate pilot projects data	Identify method and tools to extrapolate the results of the pilot projects (by means of simulation of virtual manufacturing processes). In addition, establish criteria to determine the level of success vis-à-vis meeting the quantitative objectives of the project.								TAU, TUM, MAG
Task 5.2 Asses overall project achievements	Perform virtual manufacturing simulation, unique for each industrial / SME partner and generate a report on the assessment of the project overall achievements.								TPPS, MAG, MAN, IAI, TTI, OPT
WP deliverables									Res.
D5.1	Assess the project's overall achievements.								MAG

Table 5.14 WP6: Exploitation and dissemination

Start date	Month 6								
Activity type:	RTD								
Participant number	1	2	3 (WP leader)	4	5	6	7	8	Total
Participant short name	TAU	TUM	**TPPS**	MAG	MAN	IAI	TTI	OPT	
Person-months per participant	6	6	**3**	3	3	3	3	3	30
WP objectives	The objective of this WP is to facilitate exploitation of the project products within the consortium and disseminate the project products to the affiliated IMS projects (MyCar-FUTURA and VFF-MTP), manufacturing and the product systems design community as well as the public at large. In addition this WP is to conduct products quality assurance (PQA) and products configuration management (PCM) services for the project.								

Description of work		Part.
Task 6.1 Create website / brochure	Create a project website and brochure.	TUM
Task 6.2 Exploit/disseminate products	Exploit the project's products within the consortium and disseminate them to the manufacturing, the product systems design community and the public at large.	TAU, TUM, TPPS, MAG, MAN, IAI, TTI, OPT, USST
	Offer the project products to the collaborating IMS projects (MyCar-FUTURA and VFF-MTP) and provide implementation assistance.	
	Conduct two project workshops, open to the public:	
	1) At the end of two years. The workshop will be devoted to project progress and achieved results. Opinions about project topics and the methods of solving the technical issues will be discussed, which should help in the next phase of the work.	
	2) At the end of the project. The workshop will be conducted in order to present project results and for discussing further implementation activities.	

(Continued)

Table 5.14 (Continued)

Description of work		Part.
Task 6.3 Perform PQA	Perform PQA on all the project deliverables.	TAU, TUM
Task 6.4 Perform PCM	Perform PCM on all the project deliverables.	IAI
WP deliverables		Res.
D6.1	Brochure & Internet site promoting the project	TUM
D6.2-A	Exploiting and Disseminating of Knowledge year 1	TPPS
D6.2-B	Exploiting and Disseminating of Knowledge year 2	TPPS
D6.2-C	Exploiting and Disseminating of Knowledge year 3	TPPS
D6.3-A	Report on PQA and PCM end year 1	IAI
D6.3-B	Report on PQA and PCM end year 2	IAI
D6.3-C	Report on PQA and PCM end year 3	IAI

Table 5.15 WP7: Project management

Start date	1								
Activity type	MANAGE.								
Participant number	1 (WP leader)	2	3	4	5	6	7	8	Total
Participant short name	TAU	TUM	TPPS	MAG	MAN	IAI	TTI	OPT	
Person-months per participant	20	1	1	1	1	1	1	1	27

WP objectives	The objective of this WP is to perform the administrative and financial activities and coordination for both the partners and the project levels.
Description of work	Part.
Task 7.1 Administrative & financial coordination	Perform administrative and financial coordination at the project level. This will include (1) administrative, (2) planning, (3) coordination, (4) monitoring, and (5) day-to-day running of the project. This task will ensure effective administrative processes and financial performance and accountability of the project.	TAU
Task 7.2 Generate administrative deliverables	Generate all formal administrative deliverable required by the contract throughout the project	TAU
Task 7.3 Partners' administration	Perform individual partner's administrative activities throughout the duration of the project.	TAU, TUM, TPPS, MAG, MAN, IAI, TTI, OPT

(Continued)

Table 5.15 (Continued)

WP deliverables	Description of work	Part. Res.
D7.1	Consortium agreement	TAU
D7.2-A	Periodic Interim Activity Reports midyear 1	TAU
D7.2-B	Periodic Interim Activity Reports midyear 2	TAU
D7.2-C	Periodic Interim Activity Reports midyear 3	TAU
D7.3-A	Periodic Activity Reports end year 1	TAU
D7.3-B	Periodic Activity Reports end year 2	TAU
D7.3-C	Periodic Activity Reports end year 3	TAU
D7.4-A	Financial follow-up end year 1	TAU
D7.4-B	Financial follow-up end year 2	TAU
D7.4-C	Financial follow-up end year 3	TAU
D7.5-A	Periodic Management Report end year 1	TAU
D7.5-B	Periodic Management Report end year 2	TAU
D7.5-C	Periodic Management Report end year 3	TAU

5.4.3 Risks and Contingency Plans

The overall technical approach of the project – namely, to develop a generic and tailorable industry-wide DFA methodology and a quantitative DFA economic model, as well as a practical software tool embodying the economic model – are key innovations that were beyond the current state-of-the-art and were accompanied by some risks. The consortium used a cause-and-effect diagram (Section 3.5.3, Cause-and-Effect Diagram) in order to identify potential project risks.

5.4.3.1 Identified Risks

Five specific risks have been identified in the project. These risks were to be addressed at the following early milestones (M2.1, M3.1) and late milestones (M2.2, M3.2), with possible appropriate corrective action exerted. The identified risks and their corresponding mitigations strategies are provided in Table 5.16 to Table 5.20.

Table 5.16 Risk 1: Scalability of DFA methodology and economic model

Risk	Mitigation strategy
The chosen approach implies that a generic standard DFA methodology and a quantitative DFA economic model can be developed. The risk is that the above elements are: • Too general and vague, therefore not practical, or too detailed, therefore not generic enough for the variety of users. • Not scalable to different industry sectors, product systems' classes and lifetimes as well as project types. • Not practicable due to difficulties in accurately forecasting future human needs and technology evolution.	• Include diverse industries and domain experts from different fields like engineering, economics, etc. in the consortium. • Develop tailoring rules, as an integral part of the DFA methodology and quantitative DFA economic model for different industrial users. • Assess and improve the DFA methodology and a quantitative DFA economic model by conducting pilot projects in different industrial sectors, design environments, and project types.

Table 5.17 Risk 2: Convergence and accuracy of the DFA-EM

Risk	Mitigation strategy
The chosen approach assumes that a quantitative DFA-EM could be developed and issues like human desire, product costs, risks and quality as well as opportunities parameters could be identified and quantified. The risk is that	• Investigate existing solutions in other fields (e.g. economics, engineering, social science, etc.) where advanced domain theory exists and such parameters are routinely quantified. In particular, consider the two collaborating IMS projects as a source of diverse inputs.

(Continued)

Table 5.17 (Continued)

Risk	Mitigation strategy
the DFA-EM may not converge into stable solution for different environments therefore optimal design of different architectures could only be marginally achieved.	• Implement DFA-EM in software using, as much as possible, COTS tools and database and tailor it to specific environments and industries. Thereafter, assess and calibrate the DFA methodology and quantitative DFA-EM based on the six real-life, pilot projects within the different industries and SMEs domains.

Table 5.18 Risk 3: Extending current theory of Design Structure Matrix (DSM)

Risk	Mitigation strategy
The practicality of the project is dependent on several extensions to the current Design Structure Matrix (DSM) theory. In particular: • Provisioning of hierarchical attributes into DSM theory • Provisioning of restriction and dependency attributes into DSM theory • Selecting the most appropriate model to compute combined option value of a model • Considering different stakeholders	DSM theory is an active area of current research. It is used in divergent fields of science, technology, economics, medicine, and the like. • Scientists from all academic institutes within the consortium (i.e. TAU, TUM, and USST/TCU) have been working extensively and published their work on DFA- and DSM-related issues. Initial solutions have been discussed within the consortium. • A worldwide body of scientists is interested in such problems and active research related to these is underway.

Table 5.19 Risk 4: Pilot projects assessment uncertainty

Risk	Mitigation strategy
The chosen approach assumes that the assessments of the DFA methodology and quantitative DFA economic model would proceed properly in the various industrial sectors and different environments.	• As part of WP2 and WP3, specific attention will be given to the tailoring of the generic DFA methodology and quantitative DFA economic model to different industrial sectors and project environments. Representatives from all the project partners' organizations will be involved in this process. • As part of WP3, an assimilation kit will be developed. All participants in the assessment process will be trained in DFA methodology and economic model as well as in the tailoring process needed to fit the above to the unique environment of each assessor.

(Continued)

Table 5.19 (Continued)

Risk	Mitigation strategy
• One risk is that some, or all, pilot projects would fail to conduct the DFA assessments due to lack of sufficient understanding regarding the complexity of the DFA methodology and the economic model. • Another risk is that it would not be possible to extrapolate long term project objectives from pilot projects lasting only 20 months.	• Presentations and discussions of DFA methodology and economic model assessment via pilot projects will be the core of the first user group conference, defined as milestone M5.1. • Mitigating the second risk is discussed in Section 5.4.1.3, Extrapolating future objectives based on early data sets. • TAU and TUM will provide methodological support for this aspect of the assessment process so the project will implement a business process simulation (BPS) for each pilot project in order to predict the long term behavior of a virtual manufacturing enterprise. The project will use existing methodologies and commercial off-the-shelf software to carry out this process. Due to the criticality of this issue, a fair number of person-months are allocated in the program to prove this very point.

Table 5.20 Risk 5: Software DFA-Tool development

Risk	Mitigation strategy
The success of the pilot projects is vitally dependent on the proper and timely construction, operation and maintenance of the DFA-Tool. The risk is that the software package might not meet users' needs regarding the following: • It fully embodies the economic model. • It is available on time. • Operation is not cumbersome. • Users of the pilot projects can understand it. • It is fully maintained throughout the project, causing users to abandon it altogether.	• As part of WP3, the software-based DFA-Tool will be developed in an iterative manner, under a conventional development procedure, which shall include requirements definition, system design, coding, integration, and formal testing. • The DFA-Tool will be demonstrated to the project partners and, especially to potential users throughout its development cycle and comments / criticism will be solicited in order to make the tool as user-friendly and robust as possible. • As part of WP3, an assimilation kit will be developed. All users of the DFA-Tool will undertake training sessions in order to learn how to operate it. • Sufficient funds are allocated to support individual users on a day-to-day operations as well as upgrading to the tool as needed over the course of the project. • The programmer group assigned to create the DFA-Tool is highly experienced, having built several specialized tools similar to this one (e.g. VVT-Tool created during the EC-supported SysTest project; contract number G1RD-CT-2002-00683).

5.4.3.2 Scaling the Risks

The levels of individual risks have been heuristically estimated as follows:

Effect. Each risk was classified on a scale of 1 (low) to 5 (high) in relation to its seriousness/effects.

Likelihood. Each risk was classified on a scale of 1 (low) to 5 (high) in relation to its likelihood of occurrence.

Overall. The overall risk level was computed by multiplying each risk's seriousness/effects by its likelihood of occurrence (Table 5.21).

The AMISA coordinator, together with the work package (WP) leaders, used decision tree analysis (Section 3.4.3, Decision Tree Analysis) in order to plan contingencies in case one or more of the identified risk will actually materialized.

Table 5.21 Scaling the risks

#	Identified risk	Seriousness/ effects	Likelihood of occurrence	Overall risk
1.	Scalability of DFA methodology and economic model	3	4	12
2.	Convergence and accuracy of DFA economic model	5	3	15
3.	Extending current theory of Design Structure Matrix (DSM)	4	1	4
4.	Pilot projects assessment uncertainty	2	5	10
5.	Software DFA-Tool development	1	2	2

5.4.4 Management Structure and Procedures

5.4.4.1 General Description

TAU had been identified as the coordinator of the project. project. TAU is to be assisted by the work-package leaders and by the steering committee (SC), which is empowered to make high-level decisions on every aspect of the project. The consortium agreement described the organizational structure and the rights and duties of the operational bodies responsible for the decision-making. Figure 5.10 depicts the overall structure of the management organization.

5.4.4.2 Steering Committee

A steering committee (SC) was expected to be created at the beginning of the project. It is to be composed of one senior representative from each partner organization, and chaired by the project coordinator. All committee members were to have many years of experience in technology development and participation in previous European Commission funded programs. Their role is to

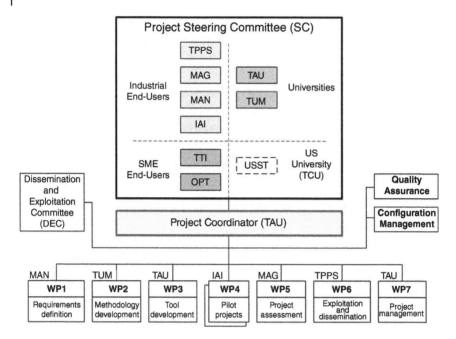

Figure 5.10 The project management structure

make high-level decisions concerning every aspect of the project life: technical, financial, schedule, partnership, dissemination, and exploitation. The steering committee planned to meet at least twice a year on an ordinary basis. Extraordinary meetings could be called for, should the need arise. More specifically, the SC's role will be as follows:

Propose contract amendments. Make proposals to the participants for the review and/or amendment of terms of the contract and/or the consortium agreement, by unanimous vote.

Make project decisions. Decide to suspend all or part of the project, or to terminate all or part of the contract, or to propose to exclude one or more participants.

Authorize participant's withdrawal. Authorize the withdrawal of a participant after the signature of the EC contract.

Act against participants. Take actions against a defaulting participant, and making proposals to dispatch remaining tasks among the consortium members, and, if appropriate, agreeing on a new participant to join the consortium.

Make WP changes. Decide on changes in work packages.

Set technical roadmap. Decide on a technical roadmap for the project.

Approve disseminations. Agree to press releases, publications, and any dissemination actions.

Support the coordinator. Support the coordinator in preparing meetings with the Commission and related data and deliverables.

5.4.4.3 Dissemination and Exploitation Committee

A dissemination and exploitation committee (DEC) is to be established immediately after the beginning of the project. The consortium agreement described the organizational structure and the rights and duties of the operational bodies responsible for the decision-making. The DEC role undertook to (1) manage all new knowledge generated in the project as well as to (2) insure proper and optimal exploitation and disseminated of this knowledge. Publications will be encouraged but will be refereed internally to ensure that knowledge is protected. Several international conferences together with national or international colloquia will offer opportunities to publish results achieved during the project. The day-to-day knowledge management is to be performed by the DEC manager (with approval by the DEC members). In particular, the DEC must:

Ensure consistent dissemination. Ensure a consistent approach in dissemination activities by reviewing materials before their release in order to remove sensitive information and protect intellectual property rights (IPR) through appropriate measures (copyrights, patents, etc.).

Manage disseminated. Monitor, maintain, and update all disseminated materials and the links with major dissemination channels (including IMS ongoing collaboration and standardization bodies).

5.4.4.4 Project Coordinator

The project coordinator is to be Dr. Avner Engel. He is expected to assume responsibilities for the overall progress of the project. He has followed the project throughout its whole lifetime, on a day-to-day basis. His main interactions within the consortium were to be with the WP leaders, from whom he gathered the necessary information for efficient communication with the EC project officer (activities, financial and final reports, audits, etc.) and within the consortium (meeting reports, progress reports, etc.) The project coordinator had been defined as the sole interface between the customer and the consortium, and the contact point for communication with other projects in the program. Should any major problem arise, the project coordinator had the possibility to call for an extraordinary meeting of the SC or for a meeting of the partners. The project coordinator has specific roles:

Manage consortium agreement. Produce and maintain the consortium agreement.

Receive and dispatch funds. Receive funds from the Commission and dispatch them to the partners, and to keep accounts of allocated funds.

Manage the project. Ensure the project proceeds at the scheduled pace, and meet its scientific and technological objectives through innovation-related activities.

Submit deliverables. Submit reports, audit certificates, and other deliverables to the Commission.

Chair project meetings. Organize and chair the steering committee meetings, as well as progress meetings, to draft their minutes, and to follow-up the implementation of the decisions taken.

Disseminate internal information. Ensure that information circulates between participants.

Manage project objectives and targets. Manage possible redefinition of objectives and targets during project progress and obtain the approval of the SC and the Commission on this matter.

5.4.4.5 Role of Work Package Leaders

The work package (WP) leaders were to be selected by the organizations responsible for the various work packages. Each WP leader managed the activities within his/her work package as well as coordinated the work with other WPs in the project. More specifically, the WP leaders also have defined roles:

Manage the WP. Supervise the task(s), the schedule, events, and budget of the WP.

Disseminate WP status. Provide a progress status report to the project coordinator and the steering committee, on a monthly basis. The progress status report should include items like results obtained and problems encountered, work scheduled, decision and questions.

Control WP schedule. Control the progress of the scheduled work within the work package in terms of technical achievement, planned deliverables and expenses.

Collect WP data. Collect the information (technical, programmatic) needed to prepare periodic progress reports and to transmit them to the project coordinator.

Apprise WP partners. Transmit information from the project coordinator to the partners involved in the WP.

Lead and report WP issues. Chair work package's meetings, and to report to the project coordinator on all relevant matters.

Support information flow. Allow a smooth upstream and downstream exchange of information, regular contact will be kept between project coordinator and WP leaders.

5.4.4.6 Decision-Making Mechanisms and Conflicts Resolution

The decision-making mechanism chosen for the project is illustrated in Table 5.22.

Specific decisions (such as the repartition of funds and the management of the flow of project money) could be taken following a different scheme (specific voting rights). However, usual or standard ways were preferred.

Table 5.22 Decision-making mechanisms

	Financial	Technical	Strategic	IPR
WP Leaders	Mediation	Mediation		
Coordinator	Mediation			
Steering Committee	Vote			

Concerning the conflict resolution mechanisms, the experience of such projects shows that a certain number of criteria must be fulfilled, especially before the project start:

Management commitment. Commitment, at the highest level within the companies, to the project and its objectives.

Fair balance. Balance between the partners regarding the contribution, the interests and the budget.

Consortium agreement. Signature of a consortium agreement accepted by all the parties.

All these criteria are anticipated within the project. However, two conflict scenarios were specifically addressed: (1) one partner versus the entire project and (2) one partner versus one or a few partners. The following steps were planned to be taken on any of these situations:

Mediation by the coordinator. Mediation via the coordinator or its representatives within 1 month after being officially informed by letter.

Vote by the Steering Committee. Vote by the SC (if an extraordinary session is needed, at the expense of the parties).

5.4.4.7 Communication Strategy

The communication strategy aimed at keeping all the partners fully informed of the project status, the planning and all other issues, which are important to the partners in order to obtain maximum transparency for all involved and to increase the synergy of the cooperation. It includes:

Written communications. It is the responsibility of the project coordinator together with the work-package leaders that all the relevant written documents coming from the partners (reports, tests results, publications, meeting minutes, etc.) be circulated to interested partners.

Oral and personal communications. This process is critical to ensure proper interpersonal communication between the partners. To facilitate this, meetings were organized on a frequent basis hosted by all partners in sequence. During these meetings, time is allowed for extensive, informal, technical discussions, as well as plant and laboratory visits. In practice, two full days per meeting seems a minimum.

External communication. This process is managed by the SC. The idea is to effectively communicate with parties outside the consortium, such as other European project consortia and other potential users of DFA concepts. The communication strategy included a plan for producing publications, giving presentations and planning conferences to be attended on behalf of the consortium. External communication is to be a topic at each SC. However the technical information resulting from the project is to be considered as confidential. The publication of those results must be agreed by the partners.

5.4.4.8 Monitoring and Reporting Status

Each partner and each work-package leader is expected to formally report every three months the status of their work to the project coordinator using a regularly updated detailed planning. Reporting included information about the technical progress, results obtained (e.g. deliverables) and compliance with the work program. The progress status of each task is also to be reported in terms of percentage of completion, estimated time to completion and actual person-months needed to complete the task.

After 12 and 24 months as well as at the end of the project, the coordinator is expected to prepare a consolidated overview of the budgetary situation of the project, on the basis of the costs received from the partners for submission to the Commission and of the payments that have been made. The budgetary situation is also to be compared with the initial costs-per-year planning, which is to be made at the kick-off phase of the project.

The planning is to have a meeting every three months with all the partners directed by the project coordinator. Of course, technical meeting between those meetings may take place if the work program requested it. All deliverables to the customer were to be written in English and given to all partners. At the first meeting the work plan would be updated with respect to progress, which has occurred in the meantime. This updated work plan is to be the reference point for the assessment of progress, consumption of resources, milestones, and meeting of project objectives.

5.4.4.9 Product Quality Assurance

Product quality assurance (PQA) is expected to support the project coordinator in ensuring the high level of the products / deliverables. PQA is to be conducted throughout the project lifetime. The PQA involved in the incorporation of quality into the products of the project. Quality means adherence to external and internal requirements of users. The criteria for insuring the quality of products will be:

Internal consistency. Internal consistency meant that (1) no two statements in a document contradict one another, (2) a given acronym or abbreviation means the same thing throughout the document, and (3) a given item or concept is referred to by the same name or description throughout the document.

External consistency. External consistency meant that two or more documents are free from contradictions with one another. Elements of external consistency are (1) no two statements contradict one another, (2) a given acronym or abbreviation means the same thing in all documents, and (3) a given item or concept is referred to by the same name or description in all documents.

Understandability. Understandability meant that (1) the document uses rules of capitalization, punctuation, symbols, and notation consistent with agreed standards or English dictionary, (2) all items not contained in agreed standards or English dictionary are defined, (3) all acronyms and abbreviations are preceded by the word or term spelled out in full the first time they are used in the document, and (4) all tables, figures, and illustrations are called out in the text before they appear, in order in which they appear in the document.

Traceability. Traceability meant that a document in question is in agreement with a predecessor document to which it has a hierarchical relationship. Traceability has these elements: (1) the document in question contains or implements all applicable stipulations of the predecessor document, (2) all the material in the successor document has its basis in the predecessor document.

5.4.4.10 Product Configuration Management

Product configuration management (PCM) is expected to support the project coordinator in ensuring that the products / deliverables are stored in safe and retrievable place. PCM is to be conducted throughout the project lifetime. The purpose of PCM is to establish and maintain the integrity of the products of the projects throughout the project's life cycle. PCM involves:

Identification. Identifying the configuration of the products (e.g. identifying and naming all products).

Repository and dissemination. Creating a central repository and dissemination of all products. The project consortium members will provide project products to the central repository. The project consortium members as well as all other external users of the project products were to receive products from this central repository.

Control changes. Systematically controlling changes to the configuration, and maintaining the integrity and the traceability of the configuration throughout the project life cycle. The work products placed under configuration management included all items delivered to either consortium members, or outside users of the project's products. A product baseline library is established containing the products as they are developed and evolved.

Change control board. Changes to the product baseline library were systematically controlled via a Change Control Board (CCB) that had been established.

5.4.4.11 Timely Corrective Actions

The following elements constituted management structure feedback loops facilitating timely corrective actions before reaching final results:

Technical manager. The coordinator is to act as a technical manager, ensuring the coherence of work products of the various work packages carried out in the project. He also provided feedback and identified required corrective actions.

Technical face-to-face meetings. Regular technical meetings, every two to three months in the early stages of the project and every three to four months in the late stages of the project were scheduled in order to review the technical status of the project and facilitate timely corrective actions.

Technical tele-conferences. Monthly technical tele-conferences were scheduled in order to review the technical status of the project and facilitate timely corrective actions.

Product quality assurance (PQA). Product quality assurance has been conducted throughout the project lifetime. The PQA team reviewed all technical and management deliverables to verify meeting requirements and, if necessary, timely corrective actions.

Risk manager. The coordinator also is to act as a risk manager for the project. He reviewed the status of all risk components and presented it at steering meetings as well as technical meetings. If needed, an appropriate corrective action is formulated and the project coordinator is be responsible for any corrective action or mitigating process.

5.4.5 Project Participants

The project coordinator performed management morphological analysis (Section 3.4.2, Morphological Analysis) in order to select the most suitable partners for the AMISA project. The following describes each participant and its role in the project.

5.4.5.1 TAU

Background. The Tel Aviv University (TAU) was created in 1956. It is the largest university in Israel. Tel Aviv University is a major center of teaching and research, offering an extensive range of study programs within its faculties of Engineering, Exact Sciences, Life Sciences, Medicine, Humanities, Law, Social Sciences, Arts, and Management. More specifically, it comprises 9 faculties, 106 departments, and 90 research institutes. In addition, the University maintains academic supervision over the Center for Technological Design in Holon, the New Academic College of Tel Aviv-Yaffo, and the Tel Aviv Engineering College. With close to 100 faculty members, 2000 undergraduates, and 1000 graduate students, the Faculty of Engineering teaches and conducts research in fields at the forefront of science and technology, including electrical, mechanical, biomedical, software, environmental, and industrial engineering, and materials science. It trains students to participate in the economy and industry of Israel. It provides them with the tools for professional receptiveness and flexibility.

Role in the project. TAU is designated as the coordinator of the project. This includes managing and administrating the project. TAU assumes a leadership role in orchestrating the entire project to ensure effective technical and management of the project as well as providing effective communication between EC representatives, affiliated IMS projects, and the project consortium. As a coordinator, TAU focuses on team leadership, risk management, financial auditing, and meeting schedules with special consideration on gender equality. Initially, TAU participates in generating DFA and tool requirements. In addition, TAU assumes key position in the development of the DFA methodology so it may be used in any particular real-life project. Also, TAU is responsible for implementing, fielding, and maintaining the prototype software tool embodying the DFA economic model. Finally, TAU is the leader of WP3, DFA-Tool development, and, naturally TAU is the leader of WP7, Project Management, utilizing its administrative expertise in order to organizing various technical and management meetings as well as generating administrative documents.

5.4.5.2 TUM

Background. The Technical University of Munich (TUM) is one of Europe's top universities. It is committed to excellence in research and teaching, interdisciplinary education and the active promotion of promising young scientists. The university also forges strong links with companies and scientific institutions across the world. Within TUM, the Institute of Product development supports industry through competent and practical education of our students, the development of effective methods and tools for product development, and the transfer of knowledge into industry.

Role in the project. Initially, TUM is assigned to participate in the generating DFA and tool requirements. In addition, TUM leads the development of the DFA methodology so it may be used in any particular real-life project. Furthermore, TUM is responsible for gathering pilot project data, analyzing it, and assessing the DFA methodology implemented within the pilot projects. Ultimately, TUM's role is to ascertain the level of success achieved by the partners, vis-à-vis the original objectives of the project. Finally, TUM is the leader of WP2, Methodology Development.

5.4.5.3 TPPS

Background. Tetra Pak (TPPS) supplies completely integrated food processing, packaging and distribution lines, and stand-alone equipment, carefully tested to make sure it gives to the final customer optimal performance. Tetra Pak Packaging Solutions, is active in all the EU countries, automates entire processing and packaging lines, trains staff to operate them, and assists in getting operations up, running, with proper maintenance and disposal. Our Systems Engineering department, based in Modena, Italy, was recently created by Tetra Pak according to the full rollout of systems engineering within the D&E organization. Our mission is focused on managing the development of increasingly complex systems to meet cost, schedule and technical performance objectives. Our specific tasks include assisting D&E organization in translating diverse stakeholder needs into system-level technical requirements and targets, which are then cascaded to the subsystems, defining and managing the interfaces between subsystems, taking responsibility for the virtual and physical integration of the complete technical solution including full requirements validation. Within these tasks TPPS are responsible for the fulfillment of all the legal, environmental, quality, functional and value-related requirements through the overall system life cycle.

Role in the project. Initially, TPPS is assigned to participate in generating the DFA and tool requirements. In addition, TPPS introduces DFA methodology into a real-life project involving the development of an adaptable production line for packaging liquid food. In this process, TPPS collects relevant process data and evaluates the benefits of using the project concepts. Finally, TPPS is the leader of WP6, Exploitation and Dissemination.

5.4.5.4 MAG

Background. MAG is a leading machine tool and systems company serving the durable-goods industry worldwide with complete manufacturing solutions. With over 4,300 employees at 26 production sites the company offers a comprehensive line of equipment and technologies including process development, automated assembly, turning, milling, automotive powertrain production, composite processing, maintenance, automation and controls, and core components. Key industrial markets served by these technologies include aerospace, automotive and truck, heavy equipment, oil and gas, rail, solar energy, wind turbine production, and general machining. With manufacturing and support operations strategically located worldwide, MAG ranks as a leader in the capital equipment market. A growing number of leading international companies are relying on the impressive innovation power of MAG to assure their technological leadership and prepare for future challenges. The MAG Switzerland AG in Schaffhausen, Switzerland, is the technology and engineering center of MAG. Here, novel innovative solutions for machine tools and manufacturing are being developed by a professional team of engineers and technicians.

Role in the project. Initially, MAG is assigned to participate in generating the DFA and tool requirements. In addition, MAG introduces DFA methodology into a real-life project involving the development of an adaptable machine tools production line. In this process, MAG collects relevant process data and evaluates the benefits of using the project concepts. In addition, MAG is the leader of WP5, Project Assessment.

5.4.5.5 MAN

Background. The MAN Nutzfahrzeuge Group is the largest subsidiary of the MAN Group and one of the leading international suppliers of commercial vehicles and transport solutions. Its brands include MAN trucks and buses, NEOPLAN buses and bus chassis. The MAN Group, active in 150 countries, is one of Europe's leading industrial players in transport-related engineering, with revenue of approximately €12 billion in 2009. As a supplier of trucks, buses, diesel engines, turbo machines, and special gear units, MAN employs approximately 47,700 people worldwide (31,519 in the MAN Nutzfahrzeuge Group). MAN's product portfolio is highly adapted to provide an apt basis for customization; as such, MAN is particularly suited to cater for particular needs of customers that go beyond the basic vehicle architectures. This, however, comes at the price of a high internal variance that is particularly growing now with the recent internationalization of the company.

Role in the project. Initially, MAN is assigned to participate in generating the DFA methodology and tool requirements. In addition, MAN introduces DFA methodology into two real-life projects: (1) a project involves in the development of an adaptable truck or bus customers service organization, and (2) a project focused on adapting truck engines and exhaust architectures for yet-unspecified environmental needs such as the Euro 6 norm. In this process, MAN collects relevant process data and evaluates the benefits of using the project concepts. In addition, MAN is the leader of WP1, Requirements definition.

5.4.5.6 IAI

Background. Israel Aerospace Industries, Ltd. (IAI), is Israel's largest industrial producer and exporter of systems and subsystems. It is a globally recognized leader in the commercial and defense markets. IAI provides unique and cost-effective technological solutions for a broad spectrum of needs in space, air, land, and sea. This includes maintenance, conversion and upgrading of aircrafts, unmanned air and ground vehicles, missiles, satellites, and launchers. IAI develops produces and export many types of subsystems including radars, communications, optoelectronic payloads, navigation, and more. In addition, IAI is involved in many other core technologies, products, and services.

Role in the project. Initially, IAI is assigned to participate in generating the DFA and tool requirements. In addition, IAI introduces DFA methodology into a real-life project involving the development of unmanned ground vehicle. In this process, IAI collects relevant process data and evaluates the benefits of using the project concepts. In addition, IAI is the leader of WP4, Pilot projects.

5.4.5.7 TTI

Background. Communication Technology (TTI) is a Spanish SME founded in 1996 that comprises a team of 100 highly qualified engineers well supported by key lab and fabrication assets. TTI works in the technological forefronts of space, military, telecommunications, science, and information technology sectors. Its main expertise is in communications, in areas like microwave and RF technologies (active RF front ends elements, specific modules like SSPA&LNA, etc. as well as integrated equipment and systems), active antennas (mobile and fix applications, fully flat, electronic scanning from L to Ka bands), and satellite systems. TTI provides expertise in system design, prototyping, integration, and testing of advanced RF and microwave components, providing added features like high miniaturization electronics and low power consumption. Main RF areas of activity are indoors and outdoors mobile communications systems, large RF integrated systems, transceiver subsystems in RF, and microwave bands, microwave hybrid integrated customized circuits, etc., and has background in radio software multi-band RF front-ends (UMTS, Wifi, Wimax, and Galileo applications), Antennas Integration with Vehicles structural bodies, etc.

Role in the project. Initially, TTI is assigned to participate in generating the DFA and tool requirements. In addition, TTI introduces DFA methodology into a real-life project involving the development of solid state power amplifiers. In this process, TTI collects relevant process data and evaluates the benefits of using the project concepts.

5.4.5.8 OPT

Background. S.C. Optoelectronica 2001 S.A (OPT) is a private Romanian SME company. The main activity is research, technological development, and innovation in applied physics and advanced technological developments in the optoelectronics. This domain includes UV, VIS, IR radiation physics, laser devices, and laser technologies, optoelectronic equipment for industrial, medical and military applications, data acquisition systems, and image processing, as well as security equipment (i.e. false document analyzer, biometry, night vision, and thermal vision).

Role in the project. Initially, OPT is assigned to participate in generating the DFA and tool requirements. In addition, OPT introduces DFA methodology into a real-life project involving the development of an adaptable optoelectronic production line. In this process, OPT collects relevant process data and evaluates the benefits of using the project concepts.

5.4.5.9 USST / TCU

Background. The United States Scientific Team (USST) is composed of economists involved with Design for Adaptability (DFA) and Design Structure Matrix (DSM). They are located at the Texas Christian University (TCU) and budgeted by their own US research agencies.

Role in the project. Initially, USST is assigned to participate in generating the DFA and tool requirements. In addition, USST participates in the development of the DFA methodology. Due to its advance knowledge in the key science aspects of the project, it is expected that USST will assume a leading role during the initiation workshop-0 (M1.1), which is dedicated to sharing understanding regarding the research: theoretical concepts in DFA, options, DSM, and architecture options.

5.4.6 Resources Needed

The project is to develop DFA methodology and a support tool for architecting manufacturing lines, product systems, and customer services for future adaptability. In addition, the project evaluates the fulfillment of its objectives by conducting six real-life pilot projects. The financial resources requested in the proposal were to be used for covering research and technology development (RTD) costs as well as management cost plus travel, and costs associated with hardware and software purchasing.

5.4.6.1 Project-Wide Funding Statistics

The overall project planned cost is 3,911,152€; and the total planned requested EC funding is 2,430,000€) (62%). Of the requested EC funding, 2,114,248€ (87%) is earmarked for RTD work, 201,565€ (8%) is earmarked for management, and 114,187€ (5%) is earmarked for expenditures like travel and purchasing of hardware and software. Considering the number of partners in the project, the expenditures for each category seemed reasonable, maximizing science and engineering value. Figure 5.11 depicts the requested EC funding (€) in graphical form divided into the three types of expenditures.

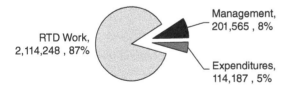

Figure 5.11 Requested EC funding (€) distributed by funding type

5.4.6.2 Work Effort Distributed among Partners

The project is expected to require 388 person-months (PM). Of which, 361 (93%) PM is to be dedicated to RTD and 27 (7%) PM is to be devoted to project management. Figure 5.12 depicts the planned work effort distribution among the partners.

5.4.6.3 Work Effort Distributed over the Project Duration

It is planned to have about 30–40 skilled and experienced scientists, engineers, and managers working on the project (most on a part-time basis). These individuals have been involved in past EC funded projects, therefore, and are

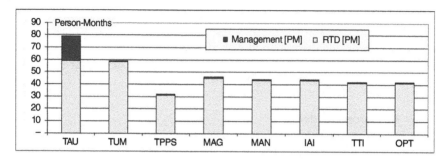

Figure 5.12 Work effort distributed among partners

familiar with the dynamics and the management process of such projects. Figure 5.13 depicts the overall planned work effort (in synthetic full time positions) distribution over the project's duration.

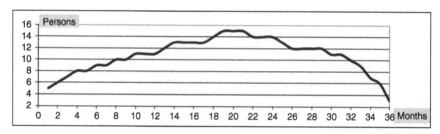

Figure 5.13 Work effort distributed over the project's duration

5.5 The AMISA Project

5.5.1 AMISA Initiation

It became clear that in order to implement the ambitious ideas discussed in previous chapters, a suitable organization had to be established and an appropriate source of funding had to be secured. A consortium of eight participants was established under the name AMISA. It proceeded to prepare a skeleton, three-year, 4,000,000 euro research proposal on the basis of the deeper insight gained over the preceding few years. Then, on July 30, 2009, a relevant, EC-sponsored call had been unannounced.[7]

The AMISA consortium finalized its research proposal, now renamed "Architecting Manufacturing Industries and Systems for Adaptability (AMISA)" and submitted it to the EC on December 8, 2009. Then, the AMISA project (Number: 262907) was approved on July 1, 2010, and actually started on

---

7 A *call* is a request for proposal (RFP) in European commission (EC) lingo. The call details: (1) Call identifier: FP7-NMP-2010-SMALL-4, (2) Call Title: Theme 4 – NMP Nanosciences, Nanotechnologies, Materials and new Production Technologies – SMALL 2010, (3) Call area: NMP.2010.3.1-1 New industrial models for a sustainable and efficient production.

April 1, 2011 (Figure 5.14). The AMISA project ended on time, three years later on March 31, 2014. Internet references to the AMISA website as well as the EC description of the AMISA project are available in Section 5.5.12, Further Reading.

Figure 5.14 AMISA kickoff meeting at TUM, Munich, Germany, April 2011

Table 5.23 depicts the AMISA actual workforce statistics. Note that several persons are identified in more than one category. Also note that most individuals supported the AMISA project on a part-time basis.

Table 5.23 AMISA workforce statistics

Position categories	Women	Men	Total
Scientific and management leaders		7	7
Work package leaders	1	5	6
Experienced researchers (i.e. PhD holders)	1	11	12
PhD Students	1	5	6
Other (Employees)	3	13	16
Total persons involved in AMISA (approximate number)	**6**	**41**	**47**

5.5.2 Identifying the DFA State-of-the-Art

The consortium created a Design for Adaptability (DFA) state-of-the-art report viewed from academic perspective (Contributors: TAU and TUM) as well as industrial perspective (Contributors: TPPS, MAG, MAN and IAI) and SME perspective (Contributors: TTI and OPT).

5.5.3   Establishing Requirements for AMISA

The requirements for the AMISA DFA methodology, the DFA economic model, and the DFA software tool were established by way of a multistep process. More specifically, in accordance with the methodology described in Section 3.5.6.4, Quantitative Approach: Multiple Criteria Decision-Making. Due to the size and broad scope of the AMISA consortium, transparency of the process and consensus decisions supported by all involved stakeholders were essential elements during the requirement acquisition process.

Initially, all partners documented the requirements from their own perspectives. Those initial set of requirements were consolidated into one document that was made available to the entire consortium as a basis for further work. During two requirement workshops with representatives from each partner present, the requirements were reviewed and elaborated further. Redundancies were removed; requirements were challenged, discussed and sharpened. In a final step each requirement within this document had to pass a consensus decision in order to be taken into further account. The results of the requirement analysis were consolidated and divided into the following eight groups:

1) System architecture/modeling
2) Input data
3) Output data
4) Cost calculation

5) Computational specifications
6) Usability
7) Uncertainties/Risks
8) Transparency of calculation

5.5.4   Implementing a Software Support Tool

A software package (DFA-Tool) implementing all relevant AO functionalities along with a comprehensive user guide have been developed and used by all the industrial/SME partners throughout the AMISA project time frame (Figure 5.15). The DFA-Tool as well as the user guide and other information are available to anyone, courtesy of the AMISA team, free of charge at the AMISA project public website.

The DFA-Tool includes the following key capabilities:

Definitions. The DFA-Tool supports the definition of: (1) all target systems components as well as the target system's environment, (2) all internal and external target system interfaces, (3) all relevant parameters needed to compute option value (OV) of each component (4) all parameters needed to compute interface cost (IC) associated with each interface (5) the content of each components' exclusion set and (6) the sets of target system variants.

Computations. The DFA-Tool supports the computation of: (1) the option value of each component, (2) the interface cost of each interface, (3) the architecture adaptability value of any given system architecture and (4) recommended time-frame and cost for performing target system upgrade.

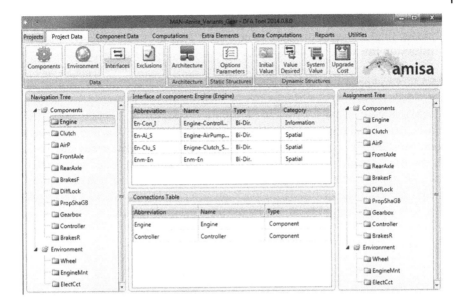

Figure 5.15 Example: A DFA-Tool screen shot

Modeling. The DFA-Tool supports the modeling of the target system by way of a design structure Matrix (DSM).

Optimization. The DFA-Tool supports the optimization of the architecture adaptability value (AAV), a measure to optimize a target architecture for adaptability.

Sensitivity analyses. The DFA-Tool supports sensitivity analysis of the system architecture to variations in option values and/or interface cost. Thereby, users may evaluate the robustness of the optimal architecture as well as identify near-optimal alternative system's architectures.

Display. The DFA-Tool supports the display of (1) all components and interfaces, as well as data associated with them. (2) target system's architecture and its equivalent DSM structure associated with different types of interfaces (material, spatial, energy and information), (3) target system's upgrade information, (4) target system's sensitivity analysis information as well as (5) variant system's data.

5.5.5 Developing Six Pilot Projects

Leaders of the pilot projects used a SWOT (strengths, weaknesses, opportunities and threats) analysis (Section 3.3.3, SWOT Analysis) in order to select suitable pilot projects. This section provides a brief description of the six pilot projects conducted under the AMISA project. Four of the pilot projects involved re-architecting product systems supplied to customers, and two pilot projects involved re-architecting portions of the companies' production lines.

5.5.5.1 Cap Applicator Machine (CAM) at TPPS

Tetra Pak Packaging Solutions (TPPS) selected the cap applicator machine (CAM), shown in Figure 5.16 as its pilot project system. The CAM is one of the most critical stations in a liquid food-packaging production line. It brings semi-finished packages and plastic caps together at very high speeds and binds them using special, fast-reacting glue. TPPS's central project issue was to design the cap applicator machine to be adaptable to future, unspecified requirements, for example, manipulating package size containers, different cap sizes, different cap placement locations on the physical package, etc.

Figure 5.16 Cap applicator machine (CAM)

TPPS acknowledged three main insights from its pilot project. First, managers realized that their development engineers should add the concept of DFA to their current design skills. Second, applying the DFA methodology requires sufficient organizational integration across disciplines. More specifically, estimating components' OA parameters is a team effort requiring diverse competencies from finance, marketing, and engineering. Third, a DFA approach requires appropriate support tools as well as new engineering mindsets. TPPS's management also consider the improved architecture as an important marketing advantage for the company with several key customers.

As of 2016, TPPS had officially adopted the optimized CAM design as a standard product. They claimed that it provided significant economic advantage because of quicker and less expensive accommodations to new customer requirements. In fact, since 2014, the CAM production capability had doubled in terms of cap/package configurations, cap positioning issues, and optimized glue applications, all while requiring limited modifications to the optimized CAM design. TPPS found that customers using the new CAMs expressed satisfaction and remarked that the systems had maintained their original value two years after installation. TPPS had also confirmed that the new CAM design reduced waste generation and the consumption of natural resources (mainly the glue and cartons needed to manufacture the liquid food containers). Therefore, TPPS had also begun to use the AAV framework in another project.

5.5.5.2 Fiber-Placement System (FPS) at MAG

During the early phases of AMISA, MAG sold its solar panels production subsidiary and a new subsidiary, MAG-IAS, technology center for composite materials processing, located near Stuttgart, Germany, established new contracts with the European Commission (EC) and the AMISA consortium. MAG IAS selected its fiber placement system (FPS), shown in Figure 5.17 as its pilot project system. The FPS is a machine used for orienting different fiber materials with variable tow widths used in the buildup of composite parts. MAG-IAS's central issue was to design its FPS to be adaptable to unspecified future requirements related to its ability to fabricate different types of products, to use a variety of raw fiber materials and shapes, etc. In addition, MAG-IAS was seeking to reduce the building and operation costs of its FPS.

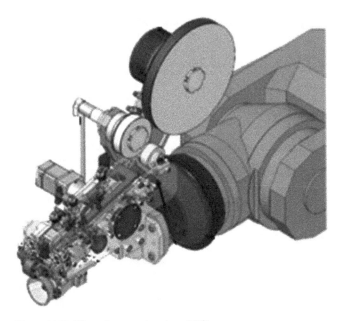

Figure 5.17 Fiber placement system (FPS)

The pilot project findings prompted MAG to apply the AAV model further within selected subsystems. For example, within the robot arm, MAG developed a deeper model concentrating on smaller components and their interfaces. Insights from this model prompted MAG to develop a different fiber-placement head that would keep the wedge component from depending on the tow width.

About six months after the end of the AMISA project, MAG sold its fiber composites division to the Brötje Automation Company in southern Germany. Although most of the staff involved with this study did not relocate to the new company, it stated that by early 2016 the AAV results had been used in the development of the "Staxx Compact," a new, automated, fiber-placement machining center.

5.5.5.3 Power Train (PT) at MAN

MAN Nutzfahrzeuge Group selected for its pilot project the power train (PT) of a standard MAN TGM truck with all-wheel drive shown in Figure 5.18.[8] The PT included elements that generates mechanical energy and transfer it to the wheels in order to push the vehicle. MAN's central project issue was to design the power train of its trucks to be adaptable to future requirements, mostly related to converting trucks to their "second life" in specialized markets in Asia, Africa, and South America. In addition, MAN was seeking to reduce the quantity and variety of spare parts in its trucks' inventory.

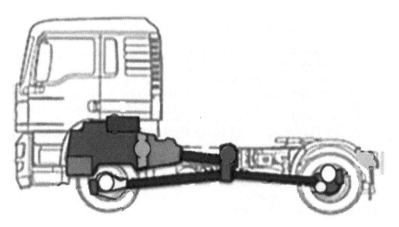

Figure 5.18 Power train (PT) of a standard MAN TGM truck

---

8 During the AMISA project, MAN selected its trucks' power train system as a pilot project instead of its originally planned trucks' engines and exhaust system. In addition, MAN did not fulfill its original commitment to explore a customer services platform as a pilot project.

5.5.5.4 Vehicle Localization System (VLS) at IAI

Israel Aerospace Industries (IAI) selected its vehicle localization system (VLS) for a robotic, unmanned ground vehicle (UGV), shown in Figure 5.19 as its pilot project system. The VLS provides state vector (i.e. position, velocity, acceleration and attitude data) to the robotic, UGV by fusing data from various space-based, terrestrial and intravehicular sensors in real time. IAI has recently entered the field of unmanned ground vehicles and therefore, its central project issue was to design the vehicle localization system to be adaptable to continually evolving requirements. In particular, the VLS must provide accurate information needed to navigate the UGV within minimal special location margins.

Figure 5.19 Vehicle localization system (VLS) for an unmanned ground vehicle

IAI had already considered hardware and software modularity to be an important aspect of its product design philosophy. For example, many IAI products use a standard computer system with modular electronic boards and standardized interfaces. The importance of design flexibility and adaptability was also ingrained in IAI's engineering culture. This often translated into designs based on relatively large numbers of separate components. However, IAI learned from this research that having too many modules has detrimental effects on interface cost and complexity. As a result, IAI asked its designers to maximize AAV value rather than maximize modularity.

As of early 2016, IAI had officially adopted the optimized VLS. They had also conducted two, two-day DFA methodology and tool orientation courses, which some 60 engineers and first-line managers had already attended. IAI had also applied the DFA approach to another project where, again, unexpected results results emerged: Two subcontractors, who had previously provided two separate subsystems for integration by IAI, were prompted to collaborate on a joint (single module) subsystem that is superior to and less expensive than the original two modules.

5.5.5.5 Solid State Power Amplifier (SSPA) at TTI

TTI Norte (TTI) selected its SSPA, a transmitter that uses semiconductor devices to amplify signals, shown in Figure 5.20, as its pilot project system. TTI aim was to design a family of SSPAs that could provide quick response to changing markets while reducing the development time and cost. TTI selected its SSPA system because user requirements, especially related to output power and frequency bands, are changing over time within a variety of markets.

Figure 5.20 Solid state power amplifier (SSPA)

By early 2016, the improved SSPA had become a standard product for TTI, and they had also applied the overall DFA methodology and optimization tool to several other products, claiming that the approach is well understood by the stakeholders and currently applied by their engineering team. The SSPA's new architecture allowed TTI to develop several product variants and upgrades more quickly and inexpensively, which, in turn, increased customers' options and TTI's responsiveness. TTI noted that, in 2014–2016, customers' new requirements would have mandated five major redesigns of the old SSPA, whereas using the optimized architecture necessitated only a single, minor redesign. During this time span, TTI realized a 20% reduction in the SSPA's lifetime cost (due to savings in development and manufacturing costs) and a 20% reduction in its upgrade lead-time.

5.5.5.6 Hologram Production Line (HPL) at OPT

Optoelectronica 2001 (OPT) selected the Hologram Production Line (HPL), shown in Figure 5.21 as its pilot project. Holography is a technology used for creation of 3D images using laser beam that engraves the image on a sensitive media. A colored image is created by light diffraction and the sophistication of the technology used in the production process limits counterfeiting. Accordingly, holograms labels are used for authentication of documents and products. OPT central project issue was to design the Hologram production line to be adaptable to changing customer requirements as well as rapidly evolving Holography technology.

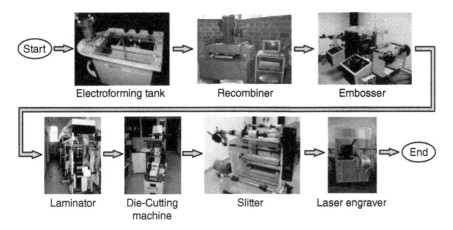

Figure 5.21 Hologram Production Line (HPL)

5.5.5.7 Overall Pilot Projects Results

Table 5.24 summarizes the key results emanating from the six AMISA pilot projects. The AAV comparisons are between a base (original) architecture and the improved design.

5.5.6 Generating Deliverables

A total of 38 technical and administrative deliverables have been generated during the AMISA project time. All deliverables have been checked and approved for quality (met quality assurance requirements), formally released and archived (placed under configuration management) and delivered to the EC on time.

Table 5.24 Architecture data from the six pilot projects

System	Primary functions	Number of modules (M)	Number of inter-module interfaces	AAV
Cap applicator machine (CAM)	• Bind plastic cups to semi-finished packages with fast-reacting glue at very high speed.	• Base: 19 • Improved: 12	• Base: 90 • Improved: 79	• Base: −8,863 • Improved: −5,453 • In percent: 39%
Fiber placement system (FPS)	• Control the manufacturing process of large, composite parts using fiber placement technology.	• Base: 17 • Improved: 12	• Base: 18 • Improved: 13	• Base: 95 • Improved: 98 • In percent: 3%
Truck power train (PT) system	• Generate, distribute and control the power in MAN TGM truck with all-wheel drive.	• Base: 11 • Improved: 5	• Base: 34 • Improved: 21	• Base: 5276 • Improved: 6242 • In percent: 18%
Vehicle localization systems (VLS)	• Receive data from space-based, terrestrial, and intravehicular sources. • Analyze input data and compute vehicle location, orientation and dynamics.	• Base: 10 • Improved: 6	• Base: 20 • Improved: 15	• Base: −19,361 • Improved: −17,142 • In percent: 11%
Solid state power amplifier (SSPA)	• Amplify low-level electrical signals. • Transmit /high-frequency data. • Provide power and cooling to internal components.	• Base: 25 • Improved: 18	• Base: 32 • Improved: 25	• Base: 967 • Improved: 3,586 • In percent: 371%
Hologram Production Line (HPL)	• Manufacture hologram labels.	• Base: 32 • Improved: 22	• Base: 82 • Improved: 66	• Base: 9,770 • Improved: 10281 • In percent: 5%

5.5.7 Planning Exploitation beyond AMISA

By the end of the AMISA project, each industrial/SME partner had adopted the DFA methodology and economic model to its pilot projects. Each of these organizations committed to exploit the AMISA results beyond the end of the project as summarized below.

TPPS. The company planned to continue the insertion of DFA technology into a new development project in the future.

MAG. The company planned to use the methodology and tool developed under AMISA in future projects.

MAN. The company, being a large company, planned to coordinate the introduction of DFA methodology with several engineering departments. This is critical now, that the company has been acquired by the VW Group, and standard product development processes (PDP) must be agreed by the entire VW concern. Nevertheless, organized process of sensitizing the development engineers to the importance of DFA mindset were planned to start within the near future.

IAI. The company planned to introduce the DFA-Tool and methodology to the systems engineering (SE) group at IAI. In addition, IAI planned to include the AMISA results in the internal SE-Intranet site of IAI, create special trainings in DFA methodology and tools, and use it in future projects.

TTI. TTI reported that a final SSPA prototype, designed using AMISA methodology is ready. TTI expects to identify a potential customer for this new generation SSPA systems next year. TTI considered the mental change on the part of development TTI staff, crucial. Therefore, the company planned to conduct an internal seminar to guide TTI staff as to the importance of developing adaptable solutions.

OPT. The company planned to use the methodology and tool developed within AMISA in future projects. This could help OPT to expand the capabilities of its production lines while reducing the cost of production, in particular, with respect to enhancing the hologram product quality as well as enhancing the laser-based system performances.

5.5.8 Disseminating Project Results

Table 5.25 summarizes the dissemination activities performed by the AMISA partners. The list contains the following categories: (1) major submissions – This set includes papers published by refereed journal as well as MSc and PhD dissertations, (2) conference submissions – This set includes papers presented at conferences, and (3) exhibitions and other activities – This set includes theoretical presentations and demonstrations of products develop under AMISA to third parties.

Table 5.25 Summary of AMISA disseminations by categories

	Major submissions	Conference submissions	Exhibitions and other activities	Total
TAU	4	4	2	10
TUM	12	13	2	27
TPPS			4	4
MAG				
MAN	4	4		8
IAI				
TTI			3	3
OPT	1		3	4
Total	**16**	**17**	**14**	**47**

5.5.9 Assessing the AMISA Project

5.5.9.1 Assessment Problem

The AMISA project was committed to perform formal assessments of the effectiveness of using AO methodology and tools within industrial setting by way of conducting six real-life pilot projects in diverse industries. The result was expected to be used to enhance the AMISA research products as well as to measure the overall performance of the projects against the originally stated requirements. The fundamental assessment problem facing the AMISA researchers was that normally, a production line or a large/complex system is designed for lifetime duration of 5, 10, or 20 years. However, the pilot projects within AMISA are carried out within about 20 months. The difficulty therefore was: How to use relatively short-duration pilot projects in order to determine whether a system designed for adaptability may be upgraded to unplanned specifications at less cost or after longer duration. That is, whether the lifetime value of a system designed for adaptability is, in fact, enhanced.

5.5.9.2 Assessment Approach

After careful analysis, the academic participants came to the conclusion that the original strategy of using business process simulation (BPS) is unreliable due to the uncertainty associated with the input data needed for the BPS model.

The AMISA consortium used the group decisions methodology (Section 3.5.5, Group Decisions: Theoretical Background and Section 3.5.6, Group Decisions: Practical Methods) in order to assess the effectiveness of using architecture option approach. An online questionnaire was created to collect the opinions of the industrial/SME partners. The opinions of the participants were based on their experience of using the methodology and the tool for developing designs and products in their respective organizations. The questionnaire consisted of 26 questions divided into three sections. Most of the questions were multiple-objective type with only one option to choose. The three sections were: (1) background information, (2) status before AMISA project, and (3) effect of knowledge outcomes of the AMISA project. The objective of the first section was to understand the background information of the respondents and their organizations. The objective of the second section was to understand the state of adaptability and its effects on various factors related to the AMISA objectives. The objective of the third section was to understand how using the methodology and the tool helped in meeting the objectives of the AMISA project.

5.5.9.3 Summary of Project Results

An assessment questionnaire was produced, evaluated, and approved by all the AMISA partners. Thereafter, each industrial and SME partner provided one or more sets of responses. The results were collated and are presented in a summary format (Table 5.26).[9]

Table 5.26 Summary of project results

Objectives	Subobjective	Meet AMISA objectives?
Objective-1	Better cost-efficiency	Yes
	Longer lifetime	Yes
	Reduced cycle time	Yes
	Provide more value to stakeholders	Yes
Objective-2	Genericity	Yes
	Tailorability	Yes
	Scalability	Yes
	Usability	Yes
	Cost-effectiveness	Yes
Objective-3	20% cost savings	Partially. Assessment identified 15% cost saving
	20% cycle-time savings	Partially. Assessment identified 17% cycle-time savings
Objective-4	25% lifespan increase	Partially. Assessment identified 17% lifespan increase
Objective-5	Decrease in usage of natural resources and energy consumption	Yes
	Decrease in pollution and by-product waste	Partially

---

9 The reader may refer to the AMISA project objectives stated in section 5.3.3 Measurable Project Objectives.

5.5.10 Consortium Meetings

As already mentioned, there were over 40 persons involved in the AMISA project – some on a full-time basis, many on a part-time basis, and a few just drifted in and out of the project during its three-year duration. The need to communicate and know people on a personal basis was deemed vital to the success of the project. As a result, AMISA practiced a frequent physical meetings policy. That is, the consortium met every two to three months, totaling 16 meetings, twice at each partner's home base (Figure 5.22). Each consortium meeting used a focus group technique (Section 3.3.5, Focus Groups) where individual participants bring diverse points of view regarding relevant issues. Usually, these meeting lasted three days and stressed technical, management, and social issues.

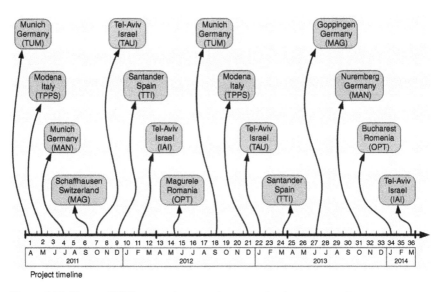

Figure 5.22 Sixteen AMISA consortium meetings over the three-year project

Technical and management discussions centered on: (1) resolving ongoing matters as well as (2) work performed by the partners during the previous months, and (3) work to be done in the near future (Figure 5.23).

Industrial and academic excursions were undertaken in order to enrich the technical knowhow of the participants in the specific work environment of the hosting partners (Figure 5.24).

Figure 5.23 Technical / management discussions at OPT, Bucharest, Romania, January 2014

Figure 5.24 Industrial excursion at MAN's facility, Nuremberg, Germany, October 2013

Similarly, historical and cultural tours were undertaken in order to acquaint visitors in the rich history and culture of the hosting partners' country as well as develop personal bonding within the team (Figure 5.25).

Finally, shared lunches and dinners were a great chance for individuals to meet, get to know one another, and discuss technical and other issues, not necessarily, on the official agenda (Figure 5.26).

Figure 5.25 Touring Jerusalem, Israel (IAI hosting), February 2012

Figure 5.26 AMISA courtesy dinner in Modena, Italy (TPPS hosting), May 2011

5.5.11 EC Summary of the Project

The following is a summary of the project written by the EC program officer Mr. Andrea Gentili.

EUROPEAN COMMISSION DIRECTORATE GENERAL FOR RESEARCH & INNOVATION Directorate D – Key Enabling Technologies

D.2 – Advanced Manufacturing Systems and Biotechnologies

Brussels, 06th May 2014
G2/AG/aa I (2014)[10]

NOTE TO THE FILE

Subject: Final assessment of the project AMISA Contract N° NMP2-SL-2011-262907 AMISA (IP) "Architecting Manufacturing Industries and Systems for Adaptability"

1. MAIN OBJECTIVES OF THE PROJECT

The project aimed at developing and validating a generic methodology with supporting IT tools to define system architectures allowing future adaptations in an optimal way with respect to economics criteria related to stakeholders' interest (e.g. maximizing shareholders profit or minimizing product life cycle cost). Targeted systems were manufacturing processes, complex products or service organizations that are likely to need adaptation for handling new requirements due to changing environments. The main idea of the project was to provide a suitable trade-off between a "rigid system" featuring lower initial cost and higher adaptation cost and an "adaptable system" featuring higher initial cost and lower adaptation cost.

During the project, a Design for Adaptability (DFA) methodology was developed together with a stand-alone tool supporting practical implementation of the methodology. The methodology and tool were used in six real-life pilot projects from several industrial sectors with the objective to provide concrete evidence that it is generic, tailorable, scalable, usable and cost effective.

The project finished on 31 March 2014, a final revue meeting was held on 25–27 March 2014. The project was followed by a PTA (Pierre Mereau) until October 2013.

---

10 Rue de la Loi 200, B-1049 Bruxelles / Wetstraat 200, B-1049 Brussels Belgium.

2. TECHNICAL/ADMINISTRATIVE ASPECTS

The AMISA consortium is composed of four large manufacturers and two SMEs representing the Food, Machinery, Aerospace, Automotive, Communication and Optronics sectors, giving a strong industrial focus to the project. The consortium also includes four research centers with experience in fusing engineering and economic theories with practical applications in industry and government.

All partners have been active during the project and participated to the different meetings (sometimes with more than one representative) showing satisfaction for the work done and strong commitment for the additional work to be done after the end of the project to ensure a proper exploitations of the project results. Since AMISA focused on systemic issues, its expected impact is vastly wider than just the industries that are directly involved in the project. Accordingly, the project operated within an international framework, including US scientists and collaboration with relevant Intelligent Manufacturing System (IMS) projects.

Project work was performed by following closely the AMISA Description of Work. Progress went as planned, work schedule was respected, milestones were achieved as planned and all due deliverables were submitted on time. This remarkable achievement is due in particular to efficient project management and good collaboration of all the partners who showed to have performed their work and delivered results as expected.

The topic addressed has a solid theoretical basis and a project challenge was to use theoretical concepts to make an industrially exploitable method giving relevant results and to solve some sort of dilemma between theory/generic feature and practicality/relevance. This was done under the guidance of the academic partners TUM (focusing on the methodology) and TAU (focusing on software tool). Collaboration with the so-called "United States Scientific Team" was a good initiative allowing consolidating of the theoretical basis.

AMISA has delivered a step-change in the performance of European industry, characterized by a higher reactivity to needs and more economically compatible products and services. Manufacturing systems or products/services designed for adaptability will save 20% either in cost or cycle time and increase their valuable lifespan by 25%. During manufacturing these systems will consume less energy and natural resources and produce less pollution and waste. Adaptable systems will also be more amenable to adjustments in regulatory frameworks (i.e. environmental, health, safety, etc.).

The administrative management of the project was carried out by the coordinator with care and devotion. It should be noticed that the coordinator has proven to manage very well the meetings, distributing responsibilities to the work package leaders and assuring a good coordination of all of them and the activities interconnected.

3. OUTPUT AND RESULTS

All the project objectives were achieved, namely:

- A generic, customable, and quantitative method, based on a DFA methodology and an economic model, for architecting systems in view of optimal adaptability to unforeseen changes
- Easy-to-use software tool supporting implementation of the method
- Validation of the method with real life projects
- Demonstration that the method is generic, tailorable, scalable, usable and cost effective

The industrial partners had a key role in contributing to progressively validate and improve the method and the tool by implementing several versions and supplying feedback through a cycle approach. Demonstration of the sought features (i.e. generic, tailorable, usable, and cost effective) is specific to the particular industrial projects. Very promising results were obtained which can be seen as a first step toward a more general demonstration of the features within a significant application scope.

4. FUTURE PERSPECTIVES

Further theoretical and development work could be done in the DFA area in view of obtaining robust results (i.e. with respect to forecasted data uncertainty) and increasing the application scope. The academic partners intend to continue research work and to create some momentum through working groups and introduction of DFA concepts into design courses.

The project results have strong potential to be applied in many industrial cases. The project applications brought benefit by allowing better understanding of industrial needs with respect to architecting systems/manufacturing and the issues at stake. All the industrial partners intend to exploit the method within their business activities by applying it on an increasing number of cases. They expect to get very positive impact.

Commercial exploitation of the project results on open markets is not foreseen at the moment. The partners' intention is to make the results available. Business opportunities are expected to emerge, like through consulting, around efficiently using the method and tool developed in the project, as well as with future methods coming from further research. The AMISA partners attended an Exploitation Strategy Seminar, which provided very good awareness on results exploitation aspects. The report which was produced after the seminar, could make a good basis for eventual commercial exploitation.

5. CONCLUSIONS

AMISA is a successful project that delivered good results opening possible ways for further progress at theoretical and industrial exploitation levels.

The project coordinator has done a very good job during the entire duration of the project and has been shown to have good communication and management capabilities.

It should be noted the extremely nice atmosphere among the partners, which stimulated a good collaboration, even beyond the boundaries of the project itself and promoted bilateral cooperation among the consortium. The partners are encouraged to build on the project outputs by further developing results and/or exploiting results in industrial cases.

In conclusion, AMISA project will be proposed to be flagged as a "Success Story."

Andrea Gentili
Programme Officer

5.5.12 Further Reading

- The AMISA project public website: http://amisa.eu/. Accessed: May 2017.
- EC description of the AMISA project: http://cordis.europa.eu/project/rcn/ 98517_en.html. Accessed: May, 2017.

5.6 Architecture Options Theory

The purpose of this chapter is to describe and elaborate on the theory of architecture options (AO). More specifically, this chapter will expand on the following AO components: (1) financial and engineering options, (2) transaction cost and interface cost, (3) architecture adaptability value, (4) Design Structure Matrix, and (5) dynamic system value modeling.

5.6.1 Financial and Engineering Options

Options, whether financial or engineering, are mechanisms to infuse flexibility into future operations. Financial options theory (Black and Scholes, 1973) established quantitative relations between options costs and their parameters.

Here, an option is "a contract giving an owner the right, but not the obligation, to buy or sell an underlying asset at a specified strike price on or before a specified date," Real option is an analogous engineering concept. It expresses the "Right, but not the obligation, to undertake some future engineering project or business decision at a specified price on or before a specified date." Over the years, the Black-Scholes model has been subject to criticism in both the financial domain and beyond (e.g. Mathews et al., 2007). For example, the model assumes that an option will not be exercised until its expiration date (T), whereas many options can be exercised at any future date; furthermore, it assumes that the interest rate (r) is constant, which often, is not the case.

Engineering options mirror flexibility needs in the engineering domain. Real options capture the value of managerial flexibility to carry out decisions in response to evolving circumstances (de Neufville and Scholtes, 2011; Hommes and Renzi, 2014). Real "in" options is an extension to real options that categorizes options as either "on" or "in" projects. Real options "on" projects are financial options taken on technical things, treating the particular system as a "black box," while real options "in" projects are options created by changing the system design. A simple example of a real option "in" a system is a spare tire on a car: It gives the driver the right (flexibility) without the obligation to change a tire at any time (Wang and de Neufville, 2005).

The AMISA project scientists adopted some of the above concepts and proposed a new quantitative method to design adaptability into systems in order to maximize their lifetime value to its stakeholders (Engel and Browning, 2006, 2008). They coined the term *architecture options* (AO) to indicate the next generation method for the architecting of adaptable systems (Figure 5.27).

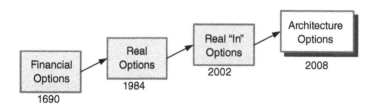

Figure 5.27 Ladder of financial and engineering options

Architecture options use a variant of the Black-Scholes model in order to apply it to the engineering domain, especially with regards to systems' future value estimation. This is depicted in equation (1) where the option value (*OV*) is the difference between the component's gain in value (if the component is to be upgraded) and the expected present cost of future upgrading.

$$OV = \left(S' - S\right) X e^{-rT} N \left(\frac{\ln\left(\dfrac{S}{X}\right) + T\left(r + \dfrac{\sigma^2}{2}\right)}{\sigma\sqrt{T}} - \sigma\sqrt{T} \right) \tag{1}$$

Here, S is the component's current value, S' is its estimated future value[11] (e.g. after an upgrade), ($S' - S$) is therefore the expected benefit of the upgrade. The large right-side term of equation (1) represents the expected present cost of future upgrading. X is the estimated cost to upgrade (redesign, remanufacture, reinstall, etc.) the component at a future time T, $N(x)$ is the cumulative, standardized normal distribution of the upgrade cost, σ is the estimated annual volatility of the upgrade cost (i.e. a measure of its likelihood of changing) expressed as a fraction of that cost, and r is the estimated risk-free interest rate over the change time horizon, T. In accordance with option theory, OV increases with uncertainty (σ) and time horizon (T) because this increases the chances that the option will expire at an advantageous moment. According to Schilling (2000), components that exhibit a fast rate of technological change tend to increase the value of modularity. Hence, such components would have high OV compared to others within the same system.

Another approach, proposed by Schrieverhoff (2014), is to use Monte Carlo simulations in order to identify a large number of possible future scenarios of system's behavior. This provides means to assess the adaptability characteristics of a system under uncertain economic environment. Another line of thought (Yassine, 2012) took adaptability one step forward by suggesting a framework and a corresponding computational model that supports adaptability optimization in a multigenerational environment. Yassine demonstrates this concept using historical automotive industry data related to two design parameters: engine power and vehicle weight.

5.6.2 Transaction Costs and Interface Costs

Transaction costs (Coase, 1937) represent additional expenditures (e.g. search costs, bargain costs, enforcement costs, etc.) beyond the direct cost incurs in making an economic exchange. The main thesis of this economic theory is that transaction costs between two entities, for example, two commercial enterprises, can be reduced significantly if the two companies merge into one. Therefore, transaction costs could be employs as markers for contracting or

---

11 The term *current value* often embodies the present cost of a component. Whereas, the term *Future value* represents value (but not necessarily cost) to a future's stakeholder. For example, a PC may cost one $600 (i.e. current value). However, in five years a new PC, having many improved features, may cost one the same but its value may exceed $1,000.

expanding the firm. That is, when certain economic tasks should be performed within a firm (i.e. by expanding the firm), or when they should be performed by the open market (i.e. by divesting portions of the firm).

In engineering, the concept of interface[12] and, in particular, interface costs are quite analogous to the concept of transaction costs in economics. Interface costs are additional expenditures, beyond the cost of designing, manufacturing, maintaining and disposing systems or subsystems. Their costs may also be divided into three categories:

Search costs. Costs associated with finding the most cost-effective means to connect different, sometimes unrelated, systems or subsystems.

Bargain costs. Costs associated with negotiating, designing, and documenting an agreeable interface control document (ICD), as well as creating and maintaining physical interfaces amongst systems or subsystems.

Enforcement costs. Costs associated with testing interfaces and rejecting the ones that fail to meet ICD requirements.

As in other aspects of systems engineering, the perception of an interface in a given system is often elastic and dependent on the particular system engineer. More specifically, it depends on the size of the system and the position of the engineer within the system hierarchy. Referring to Figure 5.28(a), one engineer may consider the bolts and the gaskets between the flanges as the interface. Another one, may consider the six flanges as the interfaces and yet another engineer may consider the entire apparatus (i.e. the pipe tee with its six flanges) to be the interface. Analogously, as depicted in Figure 5.28(b), most of the interface costs will be reduced or nearly eliminated if two or more components are designed and manufactured as a single module (of course, the degree of adaptability of this solution is being reduced dramatically).

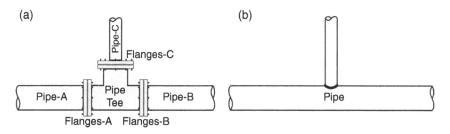

Figure 5.28 Pipe interface example: (a) adaptable but expensive; (b) rigid but cheap

---

12 The noun *interface* may be defined as: "The place at which independent and often unrelated systems meet and act on or communicate with each other." See: http://www.merriam-webster.com/dictionary/interface. Accessed: August 2017.

5.6.3 Architecture Adaptability Value

Architecture option (AO) theory seeks to identify an optimal system design between a monolithic, nonadaptive, but less expensive system and a fully adaptive but expensive one. The tradeoff is found by combining some components into modules and thus reducing or eliminating their intra-module interface costs. The particular set of modules depends on the mix of each component's option value and interface cost derived from the component's interconnections. Architecture adaptability value (AAV) is the core of AO theory. It is the sum of component option values less the sum of the intermodule interface costs:

$$AAV = \sum_{m=1}^{M} \left[\sqrt{\sum_{i=1}^{N_m} OV_i^2} - \left(\sum_k I_{mk} + \sum_l E_{ml} \right) \right] \quad (2)$$

where M is the number of modules $(M \le N)$, N_m is the number of components in the mth module, I_{mk} are the outgoing interface costs of the components in module m to other modules, $k \ne m$ (i.e. only intermodule interface costs), and E_{ml} are external interface costs between the components in module m and the environment. AAV, OV_i, I_{mk} and E_{ml} are expressed in monetary units (e.g. dollars, euro, etc.). Within the large brackets in equation (2), the first (square root) term is the option value of the mth module, modeled as the geometric sum of the option values OV_i of its constituent components. As the number of components merged into a module increases, the option value of the module also increases, albeit at lesser rate than the algebraic sum of the components' option values. Therefore, the sum of the option value terms is maximized when none of the components are merged together into any larger modules (i.e. when $M = N$) and minimized when $M = 1$.

The second set of terms (in round parentheses) in equation (2) sums the intermodule interface costs associated with each module. Because each (internal) interface is outgoing from one component and incoming to another, the measure covers only the outgoing interface costs. As more components are merged into modules, intermodule interface costs decrease (becoming intramodule interface costs, which are not included in equation (2)). Intermodule interface costs are maximized when $M = N$ and minimized when $M = 1$. Thus, each module's AAV is measured as the difference between its option value and its intermodule interface costs, and a product's overall AAV is the sum of the AAVs of its modules. $AAV > 0$ indicates that the option value term dominates in a particular architecture (i.e. it exceeds the total interface costs), whereas $AAV < 0$ implies domination by the interface costs. The primary concern is not the sign of AAV but its amount of positive change as a result of architectural adjustments such as splitting and merging of components and modules.

This *AAV* measure captures the benefits of component mergers from transaction cost theory by disregarding interface costs among any components within a module. That is, I_m includes only inter-module, not intramodule, interface costs. Hence, in the extreme case where each component is its own module ($M = N$), the *AAV* will merely be the algebraic sum of all component option values less the algebraic sum of all interface costs (all of which will be intermodular). This extreme architecture provides maximum adaptability, often however at maximum interface cost. In the other extreme case where all components are merged into one module ($M = 1$), the *AAV* will be the square root of the sum of the squares of all component option values, less the external interface costs, because all internal interface costs will be intramodular ($I_m = 0$, $\forall m$). This other extreme architecture provides minimal option benefits at zero internal interface cost. The optimum (maximum) *AAV** would normally exist between these two extremes (i.e. $1 < M^* < N$). Thus, this *AAV* measure recognizes the trade-off between the benefits of having many small options (Merton's theorem, 1973) and the costs of the interfaces to maintain them. We use it because of its grounding in the theories of options and transaction costs, and because it was developed through interaction with the engineers conducting the diverse pilot projects.

5.6.4 Design Structure Matrix

The fusion of transaction cost theory and financial options theory is achieved by way of a Design Structure Matrix (DSM). A DSM (Section 3.6.4, Design Structure Matrix Analysis) represents various types of systems (Eppinger and Browning, 2012). For example, Figure 5.29 depicts an architecture and its equivalent DSM representation. A DSM is composed of a square matrix identifying each component of the system (i.e. AA, BB, CC). In addition, the environment of the system is also defined as part of this DSM's layout (Env). The option value of each component is placed along the diagonal of the matrix (e.g. the option value of component AA is 200.4). The interface costs between components are also depicted (e.g. the internal interface cost from AA to BB is 250). In addition, the interface cost between each component and the environment is depicted (e.g. the external interface cost from BB to the environment is 40).

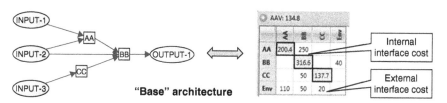

Figure 5.29 Design structure matrix (DSM): Base architecture

Optimizing the AAV of the system (Figure 5.30) may entail combining several components into a module. For example, combining component AA and component BB eliminate the interface cost between them yielding a combined option value for the AA–BB module of $OV_{AA-BB} = \sqrt{200.4^2 + 316.6^2} = 374.8$. Note that the overall option value of a module is always smaller than the algebraic sum of its component's option values.[13] As can be seen in this example, the architecture adaptability value (AAV) associated with the base system increased dramatically from 134.8 to 242.5 in the optimized system.

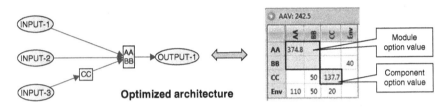

Figure 5.30 Design Structure Matrix (DSM): Optimized architecture

In general, the more modules in a system, the more options there are (representing adaptability). However, the more modules, the more interfaces there are (representing transaction costs). Therefore, the emerging methodology follows these steps: (1) identify needed functionalities / components, (2) associate each component with future option value, (3) identify each functional interface among modules and between modules and the environment then determine their costs, and (4) use an optimization technique (e.g. genetic algorithm), to identify optimal architectures.

5.6.5 Dynamic System Value Modeling

Stakeholders expect that the value of systems will increase due to technological opportunities and realization of what is possible and desirable. At the same time, systems' value diminishes due to increased maintenance costs, parts obsolescence and products/services that are no longer in demand. This difference, defined as "value loss," may be partially offset by upgrading systems to stakeholders' evolving needs and expectations (Figure 5.31a). As can be seen, architecting a system for adaptability increases the system's "lifetime value" in two ways: (1) frequent, small, and inexpensive upgrades of the system keep its value high, and (2) adaptable systems tend to exhibit longer life span because it

---

13 This statement is based on Merton theorem (1973), which states that: "A portfolio of options is more valuable than an option on a portfolio." That is, many small options are more desirable than a few large ones because they provide greater flexibility in exercising these options.

takes longer for the gap between provided and desired values to grow large enough to merit the development of entirely new systems (Figure 5.31b).

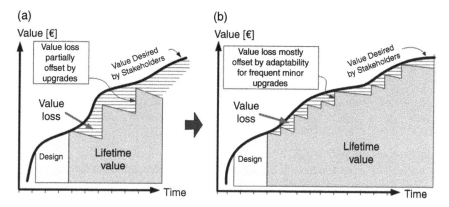

Figure 5.31 Two types of architectures: (a) Traditional; (b) Adaptable[14]

Figure 5.32 depicts in more details the various parameters associated with a single system's loss of value to its stakeholders and its upgrade. The text below

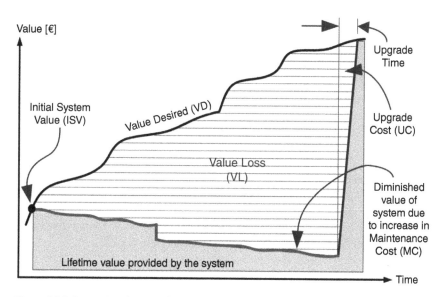

Figure 5.32 Dynamic value paradigm

---

14 Figure adapted from Browning and Honour, 2005, 2008.

describes a model proposed by the AMISA scientists to estimate the optimal time to upgrade a system.

Initial System Value (ISV). The initial system value is measured in monetary units (e.g. \$, euro, etc.). The assumption is that a system's initial value to its stakeholders (upon delivery) is equal to the sum of the costs of developing, manufacturing, and deploying the system.

Value Desired (VD). The value desired by stakeholders is measured in monetary units. This value tends to increase over time due to increase in the expected economic growth (EG) and technological advances (TA).

$$VD_i(t) = ISV + f_{EG_i}(t) + f_{TA_i}(t) \tag{3}$$

Maintenance Cost (MC). The maintenance cost is measured in monetary units. This cost tends to increase because of increased hardware and software wear-out costs (WC), as well as components and infrastructure obsolescence costs (OC).

$$MC_i(t) = f_{WC_i}(t) + f_{OC_i}(t) \tag{4}$$

Value Loss (VL). The value loss to stakeholders is measured in monetary units. The instantaneous value loss during the time period leading up to the ith upgrade $t_{i-1} \to t_i$ equals the accumulated value desired plus maintenance cost of the system (Figure 5.33a).

$$VL_i(t) = \int_{t_{i-1}}^{t_i} \left[VD_i(t) + MC_i(t) \right] dt \tag{5}$$

Upgrade Cost (UC). A system's upgrade cost is equal to its upgrade development and production costs (DPC), plus its suspension of service costs (SSC). In other words, they include costs of any disruption to the existing system while the upgrade occurs.

$$UC_i(t) = DPC_i(t) + SSC_i(t) \tag{6}$$

Optimal Upgrade Strategy. The system should be designed such that the sum of the value loss and upgrade cost for system lifetime upgrades is minimized over n upgrade cycles. Therefore, it makes sense to upgrade only at a time when $UC \le VL$. Note that premature upgrades might serve to increase value desired faster than it might otherwise increase (Figure 5.33b).

$$Min\left(\sum_{i=1,2,\dots}^{n} \left| VL_i(t) - UC_i(t) \right| \right) \tag{7}$$

Conclusions. The dynamic system value model captures the value dynamics of a system over its lifetime. Ultimately, the aim may be to optimize the system architecture and upgrade strategy to maximize the system's lifetime value to stakeholders with dynamic needs and wants.

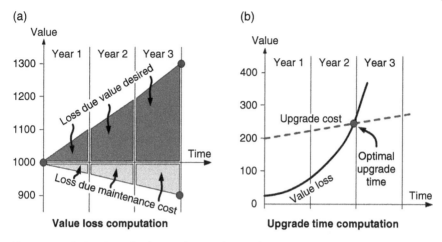

Figure 5.33 Estimating value loss and optimal upgrade time

The dynamic system value model is, of course, a simplification of the various costs that can matter, so it will have to be evaluated during each specific project and tailored to particular contexts. However, its general insights do appear to hold. This method is also susceptible to limitations in forecasting future variables. This vulnerability tends to increase as one project economic, social, and technological trends further and further into the future. Nevertheless, rough predictions are better than none. Also the reader should note that any actual upgrade decision is an "option." It provides "the right but not the obligation" to exercise it once the actual information about costs and values is available.

5.6.6 Further Reading

- Black and Scholes, 1973.
- Browning and Honour, 2005, 2008.
- Coase, 1937.
- de Neufville and Scholtes, 2011.
- Engel and Browning, 2006, 2008.
- Engel and Reich, 2015.
- Engel et al., 2016.
- Eppinger and Browning, 2012.
- Hommes and Renzi, 2014.
- Mathews et al., 2007.
- Merton, 1973.
- Schilling, 2000.
- Schrieverhoff, 2014.
- Wang and de Neufville, 2005.
- Yassine, 2012.

5.7 Architecture Options Example

The purpose of this chapter is to describe an architecture options (AO) implementation example. The example follows a formal 13 steps process designed to support practicing engineers in carrying out their Design for Adaptability

(DFA) using a free software tool (DFA-Tool), which insulates them from the mathematical intricacies of the AO theory (Figure 5.34).

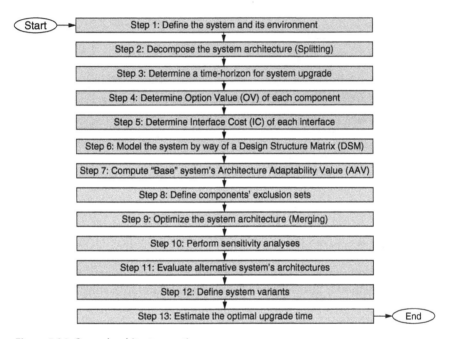

Figure 5.34 General architecture option process

The example depicts the solid state power amplifier (SSPA), a transmitter that uses semiconductor devices to amplify signals, developed by TTI Norte (TTI).

5.7.1 Step 1: Define the System and Its Environment

In this step the system, its boundary and its environment are formally defined. The environment of the SSPA consists of a AC power supply, RF input signal and an amplified RF output signal. The SSPA is also connected to a PC in order to receive commands from an operator and transmit status information for operator's monitoring. A typical SSPA contains six subsystems: a control unit and power supply, a preamplifier stage, an amplifier stage, a protection subsystem, a signal detector, and a cooling subsystem (Figure 5.35). When an RF signal is received from an external source, the preamplifier monitors, controls, and attenuates it to the proper range, and then the amplifier boosts the signal to the output power requirements. The protection subsystem remove spurious and undesired signals, and the signal detector monitors the output via sampling and provides this information to the control unit. The power supply transforms externally sourced alternating current (AC) into direct current

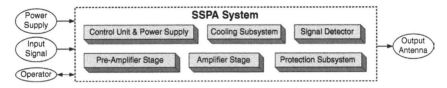

Figure 5.35 Initial SSPA system architecture

(DC) for use by the various subsystems and the cooling subsystem pushes fresh air from the outside in order to cool the SSPA system.

5.7.2 Step 2: Decompose the System Architecture

In this step, the system architecture is decomposed into N components (splitting operation). Here, the system architects, are expected to broaden their attention beyond the subsystems level (their normal focus), and into the components level as well. The system architects determine the interfaces among the N components as well as any external interfaces that cross the system boundary under the expanded granularity of the system. Figure 5.36 depicts a block diagram of the decomposed SSPA architecture including information interfaces among the subsystems.

Up to four types of interfaces may be defined: (1) spatial, (2) material, (3) energy, and (4) information (Pimmler and Eppinger, 1994). For example, Figure 5.37 depicts the detailed decomposed SSPA system architecture and interfaces (henceforth "Base" architecture). However, in the case of the SSPA, the interfaces are composed of three types: (1) energy flow, (2) information flow, and (3) spatial (This model ignores material flow, i.e. air flow throughout the SSPA system).

5.7.3 Step 3: Determine a Time Horizon for System Upgrade

In this step, a rough expected time horizon for system upgrade is determined. Typically, representatives from management, finance, marketing, sales and engineering organizations discuss the issues or preferably, use an established method like Delphi (Section 3.5.1, Delphi Method) to agree on an expected time-horizon to upgrade the system. The selection of appropriate upgraded time horizon is important because the people determining option values of components (Step 4) use the upgraded time horizon as a basis for their appraisal as to future technological and economic momentum. In any case, computationally, the optimized system's architecture (Step 9) is not particularly sensitive to small variations in upgrade time-horizon estimation. In the case of the SSPA, the upgraded time horizon was 5.3 years and the personnel involved concluded that this estimate reflects realistic, historical data.

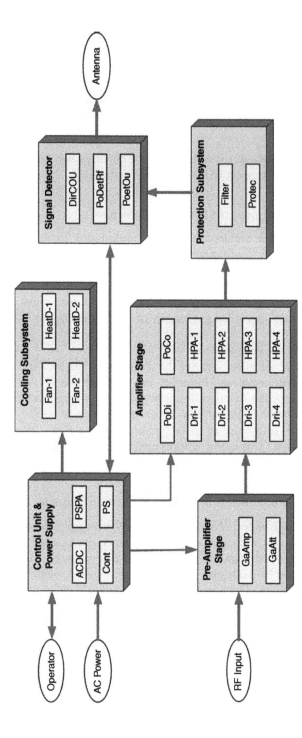

Figure 5.36 Decomposed SSPA system architecture (Information flow)

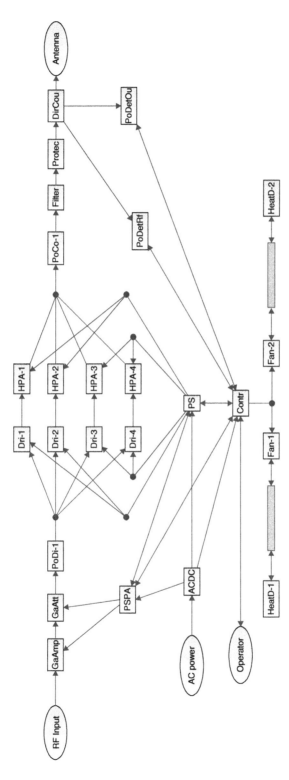

Figure 5.37 "Base" SSPA system architecture and interfaces

5.7.4   Step 4: Determine Option Value (OV) of Each Component

In this step, the option value (OV) of each component in the system is determined. This includes estimating relevant parameters (e.g. technology readiness level) using, for example, methods like TRIZ (Altshuller, 1984) and Delphi as well as utilizing the modified Black-Scholes equation. This step is carried out by following four substeps:

5.7.4.1   Estimating Component's Current Value

In general, component's current value is estimated based on either the market cost of the relevant component or the cost associated with developing and manufacturing of the component. For example, the prevailing value of the SSPA's Control system board was 1,798 Monetary Units (MU).

5.7.4.2   Estimating Component's Future Value

The AMISA scientists extended the existing TRIZ theory and developed a systematic method for estimating component's future value using the Technology forecasting method (Section 3.6.3, Technology Forecasting).[15] Essentially, TRIZ theory defines a set of "Extended laws" governing the evolution of technical systems (see: Appendix B: Extended laws of technical systems evolution). These extended laws are used as a checklist to evaluate each component and forecast its future evolution. The method entails the following elements:

1) Define the S-curve stage appropriate to each component. AMISA's industrial participants used the definition depicted in Table 5.27.

Table 5.27 Generic S-curve stages

Stage	Name	Meaning
1	Initial concept	Basic scientific concepts are observed and reported
2	First implementation	First commercial implementation and use of technology
3	Societal recognition	Technology is recognized by society at large
4	Resources decline	Resources supporting the technology start to decline
5	Technology maximized	Technology maximizes its potentials
6	Technology declines	Usage of technology declines
7	Emerging technology	New and improved technology emerges

---

15 See also: Shishko et al. (2004) and Mann (2003).

2) Examine each TRIZ: Extended[16] laws of technical systems evolution (Table 5.28 and Appendix B: Extended laws of technical systems evolution) to identify relevant technical and/or business parameters likely to evolve and affect the value of the component during the relevant timeframe.

Table 5.28 Extended laws of technical system evolution

Extended TRIZ laws		Extended TRIZ laws (cont.)	
1	System convergence	12	Action coordination
2	Similar systems merging	13	Space coordination
3	Convoluted system merging	14	Parameters coordination
4	Alternative system merging	15	Controllability
5	Integration of inverse systems	16	Dynamization
6	Integration of disparate systems	17	Transition to super system
7	Combining multiple systems	18	Increasing system completeness
8	Flow conductivity	19	Displacement of human
9	Shape/Form coordination	20	Uneven system development
10	Timing and rhythm coordination	21	Technology general progress
11	Materials coordination		

3) Evaluate the technical and business parameters in terms of their initial (I) and future (F) S-curve stages of improvements using an S-curve methodology (Christensen, 1992; Bejan and Lorente, 2011).
4) Estimate the weight of each parameter relative to the other ones, ensuring a sum weight equal to 1.0.
5) Compute the initial and final weighted factors for each parameter and their corresponding totals.
6) Based on the component's current value (S), compute its expected future value (S') and its expected value gain ($S' - S$).

For example, Figure 5.38 depicts the technology forecasting and the value gain expected if the SSPA Control system board options will be exercised: (1) The L1, L2, and L3 columns depict the relevant TRIZ extended forecasting laws, (2) the P column indicates the specific option parameters relevant to upgrading this component, (3) the I and F columns indicate the initial and final S-curve stages associated with each option parameter, (4) the W column identifies the relative weight associated with each option parameter, and (5) the W*I and W*F sum up the weighted value of the initial and future options.

---

16 Users may elect to use the abridged laws of Technical Systems Evolution, described in Chapter 4.2.2 – Laws of Systems Evolution.

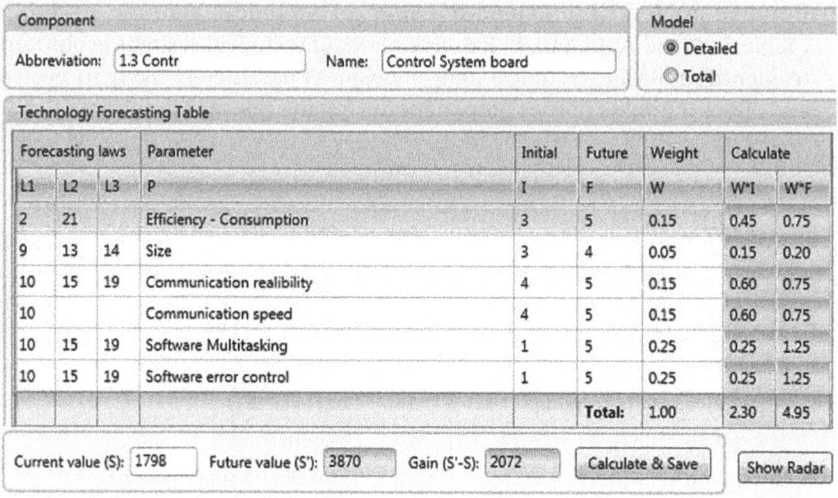

Component								Model
Abbreviation: 1.3 Contr				Name: Control System board				◉ Detailed ○ Total

Technology Forecasting Table

Forecasting laws			Parameter	Initial	Future	Weight	Calculate	
L1	L2	L3	P	I	F	W	W*I	W*F
2	21		Efficiency - Consumption	3	5	0.15	0.45	0.75
9	13	14	Size	3	4	0.05	0.15	0.20
10	15	19	Communication realibility	4	5	0.15	0.60	0.75
10			Communication speed	4	5	0.15	0.60	0.75
10	15	19	Software Multitasking	1	5	0.25	0.25	1.25
10	15	19	Software error control	1	5	0.25	0.25	1.25
					Total:	1.00	2.30	4.95

Current value (S): 1798   Future value (S'): 3870   Gain (S'-S): 2072   Calculate & Save   Show Radar

Figure 5.38 SSPA Control system board technology forecast

Consider the communication speed parameter associated with TRIZ Law 10 (timing and rhythm coordination). This law arises when the rhythm of one component of the system becomes more synchronized with other components so the overall system performance increases. In this case, the electronic part of the system operates at a rather inefficient manner. However, it is expected that the communication bandwidth and speed will increased within the next 5.3 years (the upgrade time-horizon), and this will facilitate a better utilization of the SSPA system. Given the current value of the SSPA Control system board: $S = 1,798MU$, and the weighted value of the upgrade options, the future value and gain of the SSPA Control system board are computed below:

$$\text{Future value} = S' = S\frac{\sum_{i=1,2,\ldots} F_i * W_i}{\sum_{i=1,2,\ldots} I_i * W_i} = 1,798 \times \frac{4.95}{2.30} = 3,870MU;$$

$$\text{Gain} = S' - S = 3.870 - 1,798 = 2,072MU$$

5.7.4.3   Estimating Component's Upgrade Cost

The expected future upgrade cost of each component, given that the option will be exercised, is computed based on a classical project management cost model (Figure 5.39). The overall upgrade cost is estimated by summing up all labor, materials and other expenses associated with developing, testing, and producing each relevant component in the system.

Elements of Cost

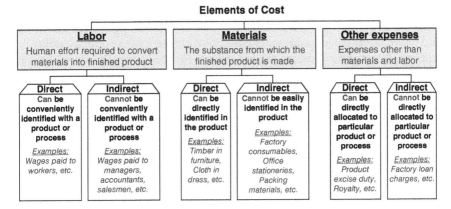

Figure 5.39 Classical project management cost model

For example, Figure 5.40 depicts the estimated SSPA Control system board upgrade cost. The nonrecurring (NRE) material cost for upgrading 10 SSPA units is 246MU (i.e. 24MU per upgrade of each unit). Similarly, the recurring (RE) material cost for upgrading each SSPA unit is 17MU. In this case, the total upgrade cost of the SSPA Control system board is estimated to be 615MU.

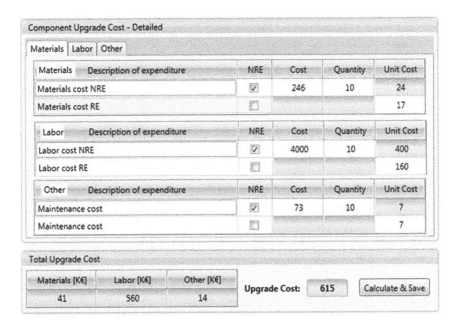

Figure 5.40 SSPA Control system board upgrade cost

5.7.4.4 Computing Component's Option Value

Finally, the expected risk-free interest value for the system upgrade time horizon and the volatility of each component must be estimated. Computing volatility in stock market trading is, pretty much, a straight forward affair. However, in engineering, estimating component's volatility is often an ill-structured problem. AMISA scientists conducted a strategic assumptions, surfacing, and testing (SAST) analysis (Section 3.5.2, SAST Analysis) in order to capture and agree on the meaning of volatility within engineering. Thereafter, option value of each component is computed using the modified Black-Scholes model (Section 5.6.1, Financial and Engineering Options). For the SSPA Control system board, the expected value gain (2,072MU), the upgrade cost (615MU), the volatility (19.7%), the expected upgrade time horizon (5.3 years) and the expected risk-free interest (3.9%) yield an option value of 1,574MU (Figure 5.41).

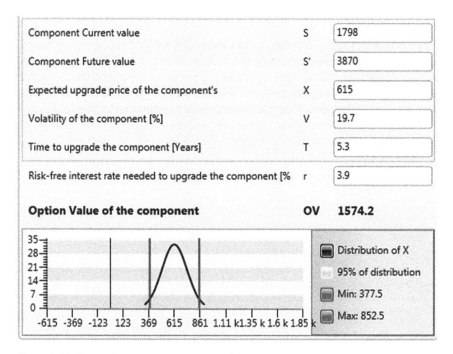

Figure 5.41 Computing component option value

5.7.5 Step 5: Determine Interface Cost (IC) of Each Interface

In this step, the cost of each interface is also computed based on classical project management costing model. More specifically, cost is estimated by summing up all labor, materials, and other expenses over the lifetime of each interface (i.e. during development, production, maintenance, and disposal) using, for example, the Delphi method. Figure 5.42 depicts an interface cost calculation of the

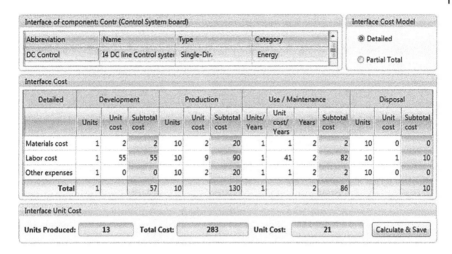

Figure 5.42 Interface cost associated with the SSPA Control system board

SSPA I4-DC line Control system interface. The number of units and their cost is estimated for each phase of the life cycle. In this case, 13 interface units are required for the lifetime of the SSPA (i.e. one during the development phase, 10 during the production phase, and 2 during the use/maintenance phase), for a total cost of 283MU, yielding a cost of 21MU per single interface.

5.7.6 Step 6: Model the System by Way of Design Structure Matrix (DSM)

In this step, the entire SSPA system is modeled by way of a Design Structure Matrix (DSM). The OV of each component is positioned along the diagonal of the DSM[17], and the ICs are placed in the appropriate cells off of the diagonal. Interfaces between two specified system's components are classified as Internal interfaces and interfaces between a system's component and the outside world are classified as external interfaces. The DSM convention is that outgoing interfaces are identified horizontally while Incoming interfaces are identified vertically. Up to five DSM variants may be created reflecting each type of interface as well as a "Combined DSM" showing the overall interface cost emanating from the combined costs of all types of interfaces. For example, Figure 5.43 depicts the base SSPA system. Option value of components ACDC is 47.4MU and it has three outgoing interfaces (i.e. 21MU to the Contr component, 24MU to the PS component and 17MU to the SPSA component). The ACDC component has also one incoming interface (i.e. 55MU) from the environment.

---

17 Note: Some options may or may not be exercised (i.e. not all components will necessarily be upgraded in the future).

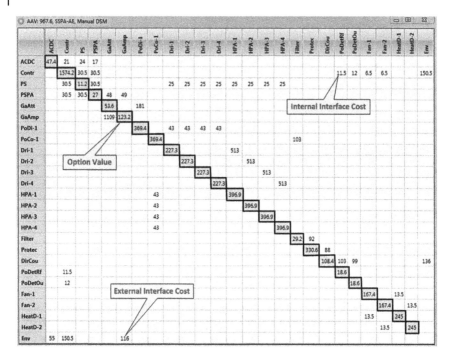

Figure 5.43 The DSM of the "Base" SSPA system

5.7.7 Step 7: Compute Base System's AAV

In this step, the base architecture's adaptability value (AAV) is computed (Equation 2, Section 5.6.3, Architecture Adaptability Value). In this case, the base SSPA architecture yields an *AAV* value of 967.6MU. This architecture provides a great deal of adaptability but requires significant investment in the design, testing, manufacturing, maintenance, and disposal of interfaces.

5.7.8 Step 8: Define Components' Exclusion Sets

In this step, certain groups of components that cannot be combined with other sets (i.e. the exclusion sets) are specified. Team leaders used concept map methodology (Section 3.2.4.1, Concept Map) in order to identify components' exclusion sets for their pilot projects. For example, stationary versus moving components, components that are physically positioned in different locations within the system, components that are produced by different contractors, components that are assembled within different production cells etc. Table 5.29 depicts the SSPA subsystems, components, and the exclusion sets. For example, any component within the AMP set may be combined with any one or more components within the same AMP set but it may not be combined with any component assigned to any other set.

Table 5.29 SSPA subsystems, components, and exclusion sets

Subsystem and environment	Component	Abbreviation	ACDC	AMP	Control	Cooling	Filter	Power	PRE-AMP	Protect
Control Unit & Power Supply	AC/DC convertor	ACDC	X							
	Control System Board	Contr			X					
	Power Supply for HPA & Drivers	PS						X		
	Power Supply for Pre-Amplifier	PSPA						X		
Preamplifier Stage	Variable Gain Attenuator	GaAtt							X	
	Gain Amplifier	GaAmp							X	
Amplifier Stage	Power Divider-1	PoDi-1		X						
	Power Combiner-1	PoCo-1		X						
	Driver-1	Dri-1		X						
	Driver-2	Dri-2		X						
	Driver-3	Dri-3		X						
	Driver-4	Dri-4		X						
High Power Amplifiers	High Power Amp-1	HPA-1		X						
	High Power Amp-2	HPA-2		X						
	High Power Amp-3	HPA-3		X						
	High Power Amp-4	HPA-4		X						

(*Continued*)

Table 5.29 (Continued)

Subsystem and environment	Component	Abbreviation	\<Exclusion sets\> ACDC	AMP	Control	Cooling	Filter	Power	PRE-AMP	Protect
Protection System	SSPA Filter	Filter					X			
	SSPA Protector	Protec								X
Signal Detector	Directional Coupler	DirCou					X			
	Power Detector reflected sig	PoDetRf					X			
	Power Detector out sig	PoDetOu					X			
Cooling System	Fan-1	Fan-1				X				
	Fan-2	Fan-2				X				
	Heat Dissipater-1	HeatD-1				X				
	Heat Dissipater-2	HeatD-2				X				
Environment	External RF input	External RF								
	Antenna	Antenna								
	AC power supply	AC power								
	Control PC	Operator								

5.7.9 Step 9: Optimize the System Architecture (Merging)

In this step, the system's AAV is optimized by searching and evaluating alternative allowable assignments of components to modules. The objective here is to reduce interface costs without losing too much option values in the process. Many architecture clustering algorithm are discussed in the literature, for example, Quan and Kim (2012). In the case of AMISA, the optimization is accomplished by means of a genetic algorithm. This optimization technique is selected because it deals naturally with integer programming and was found by the AMISA team to work well on similar problems (e.g. Sered and Reich, 2006).

As mentioned above, combining components across exclusion sets is not permissible. Obviously, such constrains reduce the effectiveness of the optimization search, yielding substantially fewer potential architectures. Figure 5.44 depicts an optimized architecture DSM of the SSPA, yielding an AAV of 3,586.7MU, a considerable improvement relative to the "Base" AAV of 967.6MU.

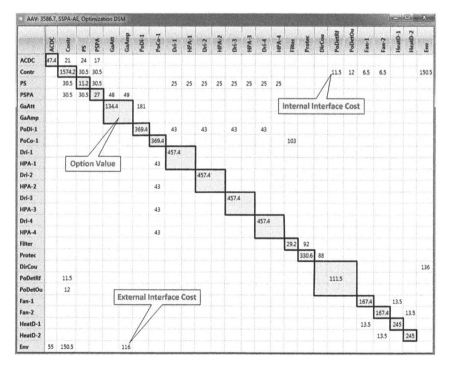

Figure 5.44 DSM of the SSPA optimized architecture B

Similarly, the corresponding block diagram of the optimized SSPA Architecture B is depicted in Figure 5.45.

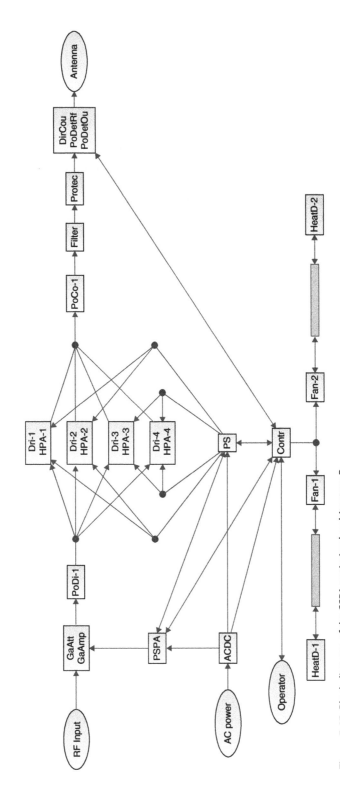

Figure 5.45 Block diagram of the SSPA optimized architecture B

The improved design includes these steps:

1) Within the preamplifier stage, combine the gain attenuator and the gain amplifier, whereby eliminating the interface (a printed circuit board (PCB) and a waveguide-PCB transition).
2) Within the amplifier stage, combine four sets of drivers and high power amplifiers (HPAs). Figure 5.46a shows this part of the system. A wave driver and a high-power amplifier (HPA) initially existed as separate modules with an interface consisting of two transition plugs and a waveguide. The inter-module interface had implied coordination across organizational boundaries, additional assembly in production, and other interface costs. In the new architecture, depicted in Figure 5.46b, the two components merged into a single module (sharing a single printed circuit board). As a result, the interface was internalized and the interface costs eliminated.
3) Within the signal detector, combine the directional coupler with the two power detectors, thereby eliminating the original waveguide-PCB transition and connectors.

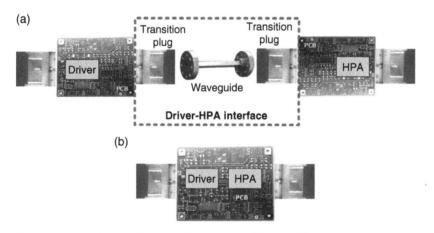

Figure 5.46 Merging the driver and the HPA in the SSPA amplifier stage

Optimizing the AAV model directly motivated these changes. For example, regarding combining the drivers and HPAs (Figure 5.46), the original designers assumed that different combinations of these components would be used in the future. Therefore, they had chosen an elaborate design solution utilizing an expensive waveguide interface. Seeing the maximized AAV architecture caused them to realize that designing this particular adaptability into the product was not economical, because it was unlikely that these options would ever be exercised. Prior to using the AAV model, the product architects had no way to adjudicate various component-level options for DFA from a system perspective. Engineers are often rather conservative in their designs. It is rare that a working design, a system under production, or an ongoing operational procedure is truly reexamined. The optimizer, not being limited by preconceived notions, may propose new architectures that can generate interesting and often valuable insights, as in this case.

5.7.10 Step 10: Perform Sensitivity Analyses

In this step, a sensitivity analyses is performed in order to assess how variations in the estimation of input parameters affect the set of optimized architectures. The AMISA scientists conducted a Pareto analysis (Section 3.4.5, Pareto Analysis), which revealed that: (1) performing sensitivity analysis, where each and every OV and IC is allowed to fluctuate independently, belongs to the NP-Hard class of problems (Fortnow, 2013) and (2) a more limited sensitivity analysis could suffice. Accordingly, a limited sensitivity analysis was performed whereby all the components' option values and all the interface costs are changed simultaneously within certain range. This simplification is also based on the assumption that experts' natural tendencies are to estimate all parameters consistently, either optimistically or pessimistically.

Figure 5.47 depicts a 7×7 matrix where all the OVs and ICs are gradually altered, each within a range of ±60% from their nominal values. Optimized system architecture is computed for each OV and IC set, creating a total of 49 sets of AAVs. For example, the first, top-left square in the optimized sensitivity analysis matrix was computed for OV factor of 1.6 and IC factor of 0.4, yielding $AAV = 8,472.1$MU.

OV factor

	1	2	3	5	7	9	12
1.60	8472.1	7810.6	7396.2	6981.8	6567.4	6153	5738.6
1.40	4	6	8	11	14	16	19
	7206.9	6678.9	6264.5	5850.1	5435.7	5021.3	4606.9
1.20	10	13	15	18	21	23	26
	5961.6	5547.2	5132.8	4718.4	4304	3889.6	3475.2
1.00	17	20	22	25	28	30	33
	4829.9	4415.5	4001.1	3586.7	3172.3	2757.9	2343.5
0.80	24	27	29	32	35	37	40
	3698.1	3283.7	2869.3	2454.9	2040.5	1626.1	1211.7
0.60	31	34	36	39	42	44	46
	2566.4	2152	1737.6	1323.2	908.8	494.4	80
0.40	38	41	43	45	47	48	49
	1434.7	1020.3	605.9	191.5	−222.9	−637.3	−1051.7
	0.40	0.60	0.80	1.00	1.20	1.40	1.60 **factor**

IC

Figure 5.47 Optimized sensitivity analysis of the SSPA

The above process culminated in two unique optimized architectures (A and B): Two set of OV-IC combinations yielded a single architecture A, and 47 sets of OV-IC combinations yielded architecture B (Figure 5.48).

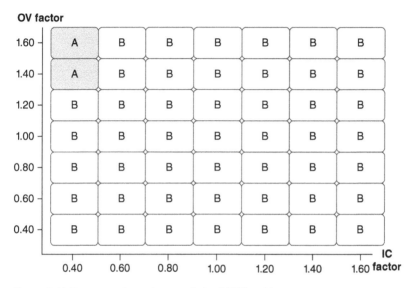

OV factor

	0.40	0.60	0.80	1.00	1.20	1.40	1.60
1.60	A	B	B	B	B	B	B
1.40	A	B	B	B	B	B	B
1.20	B	B	B	B	B	B	B
1.00	B	B	B	B	B	B	B
0.80	B	B	B	B	B	B	B
0.60	B	B	B	B	B	B	B
0.40	B	B	B	B	B	B	B

IC factor

Figure 5.48 Two emerging unique optimized SSPA architectures

The neighborhood around architecture B is depicted in Figure 5.49. This is, in fact, a linear surface, inclined by 67.7 degrees to the horizon, as depicted in Figure 5.50. Note that at the linear surface, inclination is significant. That is, smaller inclination indicates smaller spatial variability of the AAV values (a desired property).

OV factor

	0.40	0.60	0.80	1.00	1.20	1.40	1.60
1.60	229.3%	217.8%	206.2%	194.7%	183.1%	171.6%	160.0%
1.40	197.8%	186.2%	174.7%	163.1%	151.6%	140.0%	128.4%
1.20	166.2%	154.7%	143.1%	131.6%	120.0%	108.4%	96.9%
1.00	134.7%	123.1%	111.6%	100%	88.4%	76.9%	65.3%
0.80	103.1%	91.6%	80.0%	68.4%	56.9%	45.3%	33.8%
0.60	71.6%	60.0%	48.4%	36.9%	25.3%	13.8%	2.2%
0.40	40.0%	28.4%	16.9%	5.3%	−6.2%	−17.8%	−29.3%

IC factor

Figure 5.49 Neighborhood around the SSPA architecture B (numerical presentation)

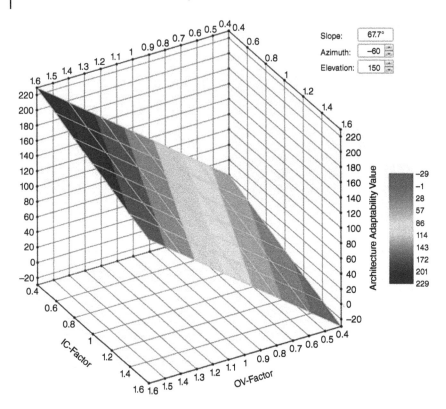

Figure 5.50 Neighborhood around SSPA architecture B (graphical presentation)

A Lattice-graph diagram was used to analyze these results. Lattice-graph diagrams are composed of nodes and edges. Each node in the graph defines a set of elements and each edge defines a relationship between nodes, where moving up along the edges of the graph corresponds to more inclusive nodes. Mathematically, lattices represent multivariate data and algebraic structures, satisfying certain axiomatic identities. In particular, its data is a partially ordered set in which any two or more nodes: (1) Have a supremum (called "Join") and (2) Have an infimum (called "Meet") where each supremum subsume all its infima (Davey and Priestley, 2002). Figure 5.51 illustrates this concept: node X defines a set of one element {a}, node Y defines a set of two elements {a, b} and node Z defines a set of four elements {a, b, c, d}.

Figure 5.52 depicts the base architecture and two unique optimized system architectures that have emerged from the

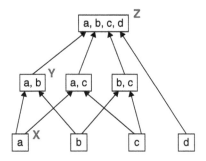

Figure 5.51 Lattice graph illustration

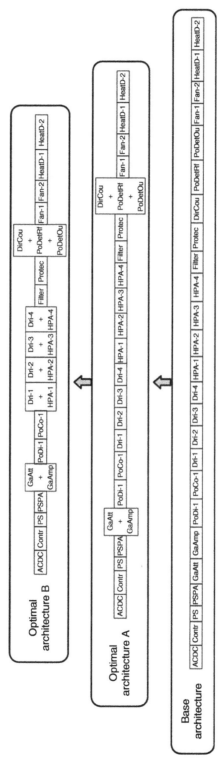

Figure 5.52 Lattice graphs: Two emerging optimal SSPA architectures

sensitivity analysis. Moving up along the edges of the graph corresponds to more inclusive architectures. Each node identifies a module which corresponds either to individual components or to a combined set of components. Architecture A is composed of 22 modules and architecture B is composed of 18 modules. The emerging optimal architectures suggest that the two solutions are quite robust because they do not differ significantly from each other.

5.7.11 Step 11: Evaluate Alternative System Architectures

In this step, the alternative system architectures are evaluated and the most desirable architecture is selected and implemented. This is accomplished by asking the engineering, marketing, sales, and finance organizations to forecast the implications (all costs and benefits) of switching to the optimized architecture over a typical time-horizon. For example, a comparison between the "Base" system and the two alternative architectures is summarized in Table 5.30. As can be seen, each of the two optimized architectures exhibit significantly better AAV relative to the "Base" architecture.

Table 5.30 Comparing three alternative SSPA system architectures

Alternative system architecture	"Base" architecture	Optimized architecture A	Optimized architecture B
Number of components	25	25	25
Number of modules	25	22	18
Equivalent optimal solutions	N/A	2/49	47/49
Surface inclination angle	N/A	79.4	67.7
Architecture adaptability value [MU]	967.6	2,202.0	3,586.7
Architecture adaptability value [Factor]	1.00	2.27	3.70

Among the two optimized architectures, architecture B seems to be the preferred choice. First, 47 of 49 combinations of OV-IC sets lead to this particular architectural solution. This may be interpreted as an indication that this architecture is very robust under a regime of imprecise input data. In addition, architecture B exhibits the highest AAV (3,586.7MU) and the lowest surface inclination angel (67.7 degrees), all indications of robust solution.

The insight here is that the system architect can alter the OV-IC values even more and thus distill several competing architectures. For example, altering the OV-IC set by a range of ±80% from their nominal values creates a total of seven unique optimal architectures. Moving up along the edges of the graph (From A to G) corresponds to designing more inclusive architecture solutions (Figure 5.53). In addition, system architects can make informed architecture decisions incorporating diverse knowledge about production considerations, market competition, customers' preferences, etc.

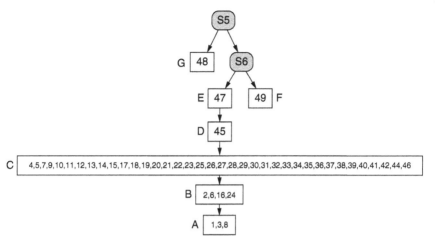

Figure 5.53 Seven near optimal SSPA architectures

Based on historical data, the manufacturer of the SSPA system created a sales and retrofit forecast scenario for the years 2015–2022. The financial savings resulting from transitioning to an adaptable design is depicted in Table 5.31. Some savings stem directly from the elimination of SSPA interfaces (2%). However, most of the saving (70%) occurs as a result of shorter and less expensive upgrading and retrofitting processes required to adapt the SSPA to new requirements, mostly from new customers. Therefore, the overall saving over the stated eight years is 28%.

Table 5.31 Financial savings – eight-year forecast[18]

	Original design [MU]	Adaptable design [MU]	Savings [MU]	Saving [%]
Development cost	750,000	224,000	526,000	70%
Production cost	1,210,000	1,188,000	22,000	2%
Total	1,960,000	1,412,000	548,000	28%

5.7.12 Step 12: Define System Variants

In this step, a variant SSPA may be defined. More specifically, an SSPA variant was designed to feed the antenna with power of 40 Watts instead of 10 Watts. Figure 5.54 depicts this variant system: a larger 40 W power supply was

---

18 The eight years financial saving forecast are specifically related to TTI corporate future business plans. This number is not derived from the upgrade time-horizon of the SSPA system

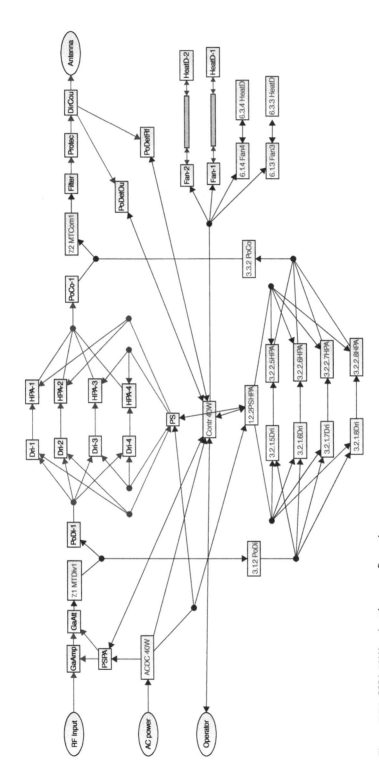

Figure 5.54 SSPA 40W variant base configuration

provided (ACDS 40 W) as well as a controller that can handle 40 W output load (Contr 40 W). In addition, more Drivers (x.DRI) and High Power Amplifiers (x.HPA), as well as more cooling (x.Fan) and heating (x.HeatD) components, have been configured. Once a variant system is defined, its AAV can be optimized in a similar manner, as an altogether new "Base" system.

5.7.13   Step 13: Estimate the Optimal Upgrade Time

Lastly, in this step one may estimate the optimal upgrade time of the system using the dynamic value paradigm (Section 5.6.5, Dynamic System Value Modeling). In the case of the SSPA, the dynamic parameters estimated by the relevant engineers, managers, marketers, and sales representatives are:

1) Initial system value (ISV) ≥ 40,000MU
2) Economic growth (EG) ≥ Yearly slope = 3.0%; Yearly volatility = 1.0%
3) Technology advances (TA) ≥ Yearly slope = 5.0%; Yearly volatility = 1.0%
4) Wear-out costs (WC) ≥ Yearly slope = 4.0%; Yearly volatility = 1.0%
5) Obsolescence costs (OC) ≥ Yearly slope = 3.0%; Yearly volatility = 1.0%

Based on these data, the optimal upgrade strategy is at the end of 30 months of operations at a cost of 17,000MU (Figure 5.55). Note that this time-frame is significantly shorter than the initial time horizon estimated in step 3 by representatives of management, finance, marketing, sales, and engineering.

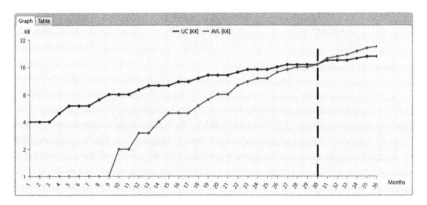

Figure 5.55 SSPA optimal upgrade time

5.7.14 Further Reading

- Altshuller, 1984.
- Bejan and Lorente, 2011.
- Christensen, 1992.
- Cooke, 1991.
- Davey and Priestley, 2002.
- Engel and Reich, 2015.
- Engel et al., 2016.
- Engel et al., 2012.
- Fortnow, 2013.
- Loveridge, 2002.
- Mann, 2003.
- Pimmler and Eppinger, 1994.
- Quan and Kim, 2012.
- Sered and Reich, 2006.
- Shishko et al., 2004.

5.8 AMISA – Endnote

As mentioned before, the story of AMISA was an exemplary creative and innovative undertaking. Together with other engineers and managers, three individuals became the guiding spirits of this effort throughout the ensuing years: Avner Engel from the Tel Aviv University, Tyson Browning from the Texas Christian University, and Shalom Shachar from Israel Aerospace Industries (Figure 5.56).

| Avner Engel | Tyson Browning | Shalom Shachar |

Figure 5.56 Key players in the AMISA project

The idea of maximizing modularity is deeply ingrained in many design and management communities. When the AMISA participants presented their concepts of engineering options, interface costs, and architecture adaptability values, firms' designers and managers were initially skeptical. A formal method was required to guide and support their decisions regarding assignments of components to modules. In addition, it was necessary to convince management as to the viability of the approach. The process of building and using architecture options methodology resolved three challenges: it convinced the designers that the approach is viable, it provided actual design solutions (some intuitive but others quite remarkable and unexpected), and it gained management approval for the new designs.

End-project reports from the AMISA participants confirmed that optimized architecture options approaches were indeed helpful, allowing them to optimize product adaptability, improved cost-efficiency, extend lifespan, and increase overall value. They reported that upgrading products via the DFA methodology provided shorter development lead-times and reduce costs. The AMISA partners considered the approach to be generic, scalable, tailorable, and useful in practice. According to a post-project review by the European Commission, this research delivered a "step-change" in the performance of European industry, characterized by a higher reactivity to market needs and more economically compatible products and services.

However, as this book goes to print, the facts on the ground are as follows:

1) An exemplary creative and innovative research project, AMISA involving academia, industry, and SMEs was conducted.
2) AMISA demonstrated the effectiveness of the Design for Adaptability methodology using AO.
3) The complexity of the problem necessitates the use of a computerized support tool (i.e. DFA-Tool) in order to identify one or more optimal system architectures.
4) Both a software tool and a comprehensive user guide are available to anyone, free of charge.
5) The project and its results have been published in various trade journals, conferences, exhibitions and the internet.
6) Some of the AMISA partners seem to understand the importance of designing systems for future upgrades, yet, no one – not any of the AMISA partners, nor anyone else – have truly incorporated this approach into their day-to-day systems design operations.

The author struggles with this issue. One plausible reason for this outcome could be that institutions, as well as many engineers, are very conservatives by nature and will do their utmost to remain within their comfort zone regardless of any facts and figures presented to them (i.e. they will be inclined to ignore new ideas). Another possible reason could be that complex software tools must undergo extensive commercialization before organizations will be willing to commit to using them. Not less important, introducing new technology into an organization must be accompanied by dedicated, internal or external, experts who can provide long-term, ongoing support in order to ensure that problems are resolved quickly and efficiently as they arise.

5.9 Bibliography

Alexandridis, A.N. (1986, February). Adaptable Software and Hardware: Problems and Solutions. *IEEE Computer* 19 (2): 29–39.

Ali, A.S., and Seifoddini, H. (2006). Simulation Intelligence and Modeling for Manufacturing Uncertainties, Proceedings of the 2006 Winter Simulation Conference, Monterey, California, December 3–6.

Altshuller, S.G. (1984). *Creativity as an Exact Science: The Theory of the Solution of Inventive Problems*. Boca Raton, FL: CRC Press.

Bejan, A., and Lorente, S. (2011). The constructal law origin of the logistics S curve. *Journal of Applied Physics*, 110, 024901.

Black, F., and Scholes, M. (1973). The pricing of options and corporate liabilities. *Journal of Political Economy* 81: 3.

Browning, T.R., and Honour, E.C. (2008). Measuring the life-cycle value of enduring systems. *Systems Engineering* 11 (3): 187–202.

Browning, T.R., and Honour, E.C. (2005). Measuring the Lifecycle Value of a System. INCOSE International Conference, Rochester, New York, July 10–14.

Christensen, C.M. (1992). Exploring the limits of the technology S-curve. *Production and Operations Management* 1 (4): 334–366.

CMU/SEI-2003-TN-017. (2003). Software Engineering Institute Carnegie Mellon University Pittsburgh, PA, 2003. http://www.sei.cmu.edu. Accessed: May 2017.

Coase, R. (1937). The nature of the firm. *Economica* 4 (16): 386–405.

Cooke, R.M. (1991). *Experts in Uncertainty: Opinion and Subjective Probability in Science*. Oxford: Oxford University Press.

Davey, B.A., and Priestley, H.A. (2002). *Introduction to Lattices and Order*. Cambridge University Press.

de Neufville, R., and Scholtes S. (2011). *Flexibility in Engineering Design*. Cambridge, MA: MIT Press.

de Neufville, R. et al. (2004). *Uncertainty Management for Engineering Systems Planning and Design*. MIT Engineering Systems Division.

de Weck, O., de Neufville, R., and Chaize, M. (2004). Staged deployment of communications satellite constellations in low earth orbit. *Journal of Aerospace Computing, Information, and Communication* 1 (4): 119–136.

Engel A., and Browning, T. (2006). Designing Systems for Adaptability by Means of Architecture Options, INCOSE-2006, the 16th International Symposium, Florida, USA, July 9–13.

Engel, A., and Browning, R.T. (2008). Designing systems for adaptability by means of architecture options. *Systems Engineering Journal* 11 (2): 125–146.

Engel, A., and Reich, Y. (2015). Advancing architecture options theory: Six industrial case studies, *Systems Engineering Journal* 18 (4): 396–414.

Engel, A., Browning R.T., and Reich Y. (2017). Designing products for adaptability: Insights from four industrial cases. *Decision Sciences Journal*, Online: September 28, 2016, Print: 48 (5): 875–917.

Engel, A., Reich, Y., Browning, T.R., and Schmidt D. (2012). Optimizing system architecture for Adaptability, International Design Conference – DESIGN-2012, Dubrovnik, Croatia, May 21–24.

Engel, A. (2010). *Verification, Validation and Testing of Engineered Systems* (Wiley Series in Systems Engineering and Management). Hoboken, NJ: John Wiley & Sons.

Eppinger, S.D., and Browning, T.R. (2012). *Design Structure Matrix Methods and Applications*. Cambridge, MA: The MIT Press.

Fortnow, L. (2013). *The Golden Ticket: P, NP, and the Search for the Impossible Hardcover*. Princeton, NJ: Princeton University Press.

Fricke, E., and Schulz, A.P. (2005). Design for Changeability (DfC): Principles to *Enable Changes in Systems Throughout Their Entire Lifecycle. Systems Engineering* 8 (4): 342–359.

Hanratty, M. (1999). Open systems and the systems engineering process. *Acquisition Review Quarterly*, Winter.

Hommes, Q.D.V.E., and Renzi, M.J. (2014). *Product Architecture Decision Under Lifecycle Uncertainty Consideration: A Case Study in Providing Real-time Support to Automotive Battery System Architecture Design, Technology and Manufacturing Process Selection* (pp. 1–20). London: Springer.

Hundal M., Ehrlenspiel K., Kiewert A., and Lindemann U. (2007). *Cost Efficient Design*. Berlin: Springer.

Karniel, A., and Reich, Y. (2009). From DSM-based planning to Design Process Simulation: A review of process-scheme logic verification issues. *IEEE Transactions on Engineering Management* 56 (4): 636–649.

Lin, J., de Weck, O., de Neufville, R., Robinson, D., and MacGowan D. (2009). Designing Capital-Intensive Systems with Architectural and Operational Flexibility Using a Screening Model, COMPLEX'2009, Shanghai, China, February 23–25.

Lindemann, U., Maurer, M., and Braun, T. (2008). *Structural Complexity Management: An Approach for the Field of Product Design*. Springer.

Loveridge, D. (2002). Experts and Foresight: Review and experience, Paper 02-09, PRES, The University of Manchester, UK. June.

Mann, D.L. (2003). Better technology forecasting using systematic innovation methods. *Technological Forecasting and Social Change*, 70 (8): 779–795.

Mathews, S., Datar, V., and Johnson, B. (2007). A practical method for valuing real options: The Boeing approach. *Journal of Applied Corporate Finance* 19 (2): 95–104.

Maurer, M., Deubzer, F., Kreimeyer, M., and Lindemann, U. (2005). MOFLEPS – Modeling Flexible Product Structure, The 7th International Dependency Structure Matrix (DSM) Conference, Seattle, Washington, USA, Oct 4–6.

Merton, R.C. (1973). Theory of rational option pricing. *Bell Journal of Economics and Management Science* 4 (1): 141–183.

Pimmler, U.T., and Eppinger, S.D. (1994). Integration analysis of product decompositions. Working Paper #3690-94-MS. MIT Sloan School of Management, Cambridge, MA.

Quan, N., and Kim, H.M. (2012). A Functionally Aware Product Schematic Clustering Algorithm, ASME, International Design Engineering Technical Conferences and Computers and Information in Engineering Conference (pp. 1011–1021). American Society of Mechanical Engineers.

Schilling, M.A. (2000). Toward a general modular systems theory and its application to interfirm product modularity *Academy of Management Review* 25 (2): 312–334.

Schrieverhoff, P. (2014). Valuation of Adaptability in System Architecture, Ph.D. dissertation, Technische Universität München.

Sered, Y., and Reich, Y. (2006). Standardization and modularization driven by minimizing overall process effort. *Computer-Aided Design* 38 (5): 405–416.

Shishko, R., Ebbeler, D.H., and Fox, G. (2004). NASA technology assessment using real options valuation. *Systems Engineering* 7 (1): 1–12.

Suh, E.S., Furst, M.R., Mihalyov, K.J., de Weck O.L. (2008). Technology infusion: An assessment framework and case study, DETC2008-49860, Proceedings of IDETC/CIE 2008, ASME 2008 International Design Engineering Technical Conferences, New York, New York, USA, August 3–6.

Wang, T., and de Neufville, R. (2005, June). Real options "in" projects. Real options conference. Paris, France.

Wang, T. (2005). Real Options "in" Projects and Systems Design Identification of Options and Solution for Path Dependency. PhD Dissertation, Massachusetts Institute of Technology, Cambridge, Massachusetts, USA.

Yassine, A.A. (2012). Parametric design adaptation for competitive products. *Journal of Intelligent Manufacturing* 23: 541–559.

Appendix A

Life Cycle Processes versus Recommended Creative Methods

Table A.1 depicts systems' life cycle processes and recommended creative methods.

Readers should note that most creative methods, described in Part III, could help systems engineers in performing the majority of these life cycle processes. Nevertheless, the book provides individual sets of recommended creative methods which are considered more appropriate for specific life cycle process.

Practical Creativity and Innovation in Systems Engineering, First Edition. Avner Engel.
© 2018 John Wiley & Sons, Inc. Published 2018 by John Wiley & Sons, Inc.

Table A.1 Systems life cycle processes versus recommended creative methods

(Continued)

Recommended creative methods	2.3.1.1 Acquisition process	2.3.1.2 Supply process	2.3.2.1 Life cycle model management process	2.3.2.2 Infrastructure management process	2.3.2.3 Portfolio management process	2.3.2.4 Human resource management process	2.3.2.5 Quality management process	2.3.2.6 Knowledge management process	2.3.3.1 Project planning process	2.3.3.2 Project assessment and control process	2.3.3.3 Decision management process
3.6.7 Conflict analysis					X	X				X	
3.6.6 Anticipatory failure								X			
3.6.5 FMEA analysis				X							
3.6.4 DSM analysis							X	X			
3.6.3 Technology forecasting			X		X	X					
3.6.2 Nine-screens							X				
3.6.1 Process map			X								X
3.5.6 Group decisions: Practice									X	X	
3.5.5 Group decisions: Theory									X	X	
3.5.4 Kano model analysis							X			X	
3.5.3 Cause-and-effect				X	X						
3.5.2 SAST analysis									X	X	
3.5.1 Delphi method			X		X	X					
3.4.5 Pareto analysis					X	X					
3.4.4 Value analysis / engineering			X				X	X			
3.4.3 Decision trees						X					
3.4.2 Morphological analysis									X		X
3.4.1 PMI analysis									X		
3.3.5 Focus groups	X	X		X			X				
3.3.4 SCAMPER analysis						X				X	
3.3.3 SWOT analysis							X			X	X
3.3.2 Six thinking hats			X	X	X						
3.3.1 Classical brainstorming	X	X		X		X					
3.2.4.3 Mind-mapping			X	X	X		X				
3.2.4.2 Concept fan	X	X									
3.2.4.1 Concept map	X	X			X						
3.2.3 Biomimicry innovation					X						
3.2.2 Resolving contradictions	X	X						X			
3.2.1 Lateral thinking					X		X				

Table A.1 (Continued)

(Continued)

Recommended creative methods \ Life cycle processes	2.3.3.4 Risk management process	2.3.3.5 Configuration management process	2.3.3.6 Information management process	2.3.3.7 Measurement process	2.3.3.8 Quality assurance process	2.3.4.1 Business or mission analysis process	2.3.4.2 Stakeholder needs and requirements definition process	2.3.4.3 System requirements definition process	2.3.4.4 Architecture definition process	2.3.4.5 Design definition process	2.3.4.6 System analysis process
3.2.1 Lateral thinking						X	X	X	X		
3.2.2 Resolving contradictions							X	X			
3.2.3 Biomimicry innovation	X		X			X	X		X		
3.2.4.1 Concept map								X		X	
3.2.4.2 Concept fan						X	X				X
3.2.4.3 Mind-mapping											
3.3.1 Classical brainstorming					X		X		X		
3.3.2 Six thinking hats				X	X		X				
3.3.3 SWOT analysis						X	X				
3.3.4 SCAMPER analysis						X	X				
3.3.5 Focus groups				X							
3.4.1 PMI analysis			X						X		X
3.4.2 Morphological analysis							X	X			
3.4.3 Decision trees	X	X					X		X		
3.4.4 Value analysis / engineering									X		
3.4.5 Pareto analysis						X	X				X
3.5.1 Delphi method			X						X		
3.5.2 SAST analysis				X							
3.5.3 Cause-and-effect	X								X	X	X
3.5.4 Kano model analysis	X	X			X	X	X				
3.5.5 Group decisions: Theory		X	X		X	X					
3.5.6 Group decisions: Practice		X			X						
3.6.1 Process map									X	X	
3.6.2 Nine-screens						X	X			X	X
3.6.3 Technology forecasting									X	X	X
3.6.4 DSM analysis		X		X							X
3.6.5 FMEA analysis	X			X	X						
3.6.6 Anticipatory failure	X			X					X		
3.6.7 Conflict analysis						X					

Recommended creative methods / Life cycle processes	2.3.4.7 Implementation process	2.3.4.8 Integration process	2.3.4.9 Verification process	2.3.4.10 Transition process	2.3.4.11 Validation process	2.3.4.12 Operation process	2.3.4.13 Maintenance process	2.3.4.14 Disposal process
3.6.7 Conflict analysis						X		
3.6.6 Anticipatory failure	X		X		X	X	X	
3.6.5 FMEA analysis			X		X		X	
3.6.4 DSM analysis		X						
3.6.3 Technology forecasting	X			X				X
3.6.2 Nine-screens								
3.6.1 Process map	X	X	X		X		X	
3.5.6 Group decisions: Practice			X		X			
3.5.5 Group decisions: Theory			X		X			
3.5.4 Kano model analysis			X		X			X
3.5.3 Cause-and-effect		X						
3.5.2 SAST analysis	X							
3.5.1 Delphi method				X			X	
3.4.5 Pareto analysis		X					X	
3.4.4 Value analysis / engineering		X	X	X	X	X		
3.4.3 Decision trees		X	X		X		X	X
3.4.2 Morphological analysis	X			X				
3.4.1 PMI analysis								
3.3.5 Focus groups						X		
3.3.4 SCAMPER analysis								
3.3.3 SWOT analysis								
3.3.2 Six thinking hats						X		X
3.3.1 Classical brainstorming	X					X		
3.2.4.3 Mind-mapping				X				
3.2.4.2 Concept fan	X							X
3.2.4.1 Concept map								
3.2.3 Biomimicry innovation								X
3.2.2 Resolving contradictions	X		X		X		X	X
3.2.1 Lateral thinking	X						X	X

Appendix B

Extended Laws of Technical Systems Evolution

An extended set of technical systems evolution laws[1] was used during the AMISA project. This extended set has also been embedded into the DFA-Tool. The appendix describes this set of 21 laws, in which each individual law defines unique mechanism along which systems increase their ideality (Figure B.1).

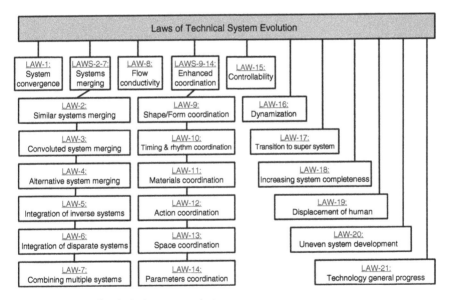

Figure B.1 Laws of technical system evolution

---

1 Adopted with modifications from: A. Lyubomirskiy and S.S. Litvin, Lows of technical systems evolution, GEN3 Partners, Feb. 2003 (А. Любомирский, С. Литвин, Законы развития технических систем, GEN3 Partners, Февраль 2003).

Practical Creativity and Innovation in Systems Engineering, First Edition. Avner Engel.

B.1 Law 1: System Convergence

The law of system convergence represents a pattern of systems evolution in which the number of elements from which the system is composed tends to decrease over time, without degradation in the performance of the system itself. This reduction in the number of components is often accompanied with system's cost reduction, both leading to improvement in the system's ideality.

The functional capacity of the system is often maintained by redistribution of the useful functions to the remaining elements of the system (i.e. this process is called system merging). In other cases, some inner system functionalities may disappear altogether without affecting the external capabilities of the system.

Example
A company manufacturing transmission systems with hydraulic torque converters was able to continually reduce the number of parts by 70% over a period of three decades (Figure B.2).

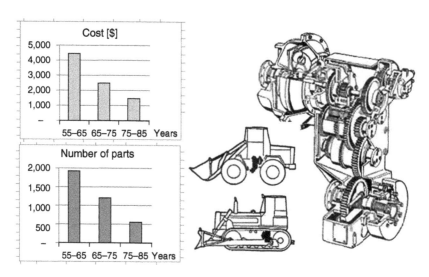

Figure B.2 Continual reduction in number of parts and costs

B.2 Laws 2 to 7: Systems Merging

The laws of systems merging represent a pattern of systems evolution in which several systems are merged into a single system that provides advantage over the original individual systems. These laws have several manifestations.

B.2.1 Law 2: Similar Systems Merging

The law of similar systems merging arises when two or more similar systems are merged together in the design of a new system.

Example
A catamaran combines two (or more) ship hulls to improve system stability and reduce overall system drag (Figure B.3).

Figure B.3 A catamaran

Photo: The Salem Ferry, Nathaniel Bowditch, approaches its dock in Salem, Massachusetts.

B.2.2 Law 3: Convoluted System Merging

The law of convoluted systems merging arises when two or more unrelated systems with shifted characteristics (i.e. features that are different from one another in one or more parameters) are merged together in the design of a new system.

Example
Two materials with different physical characteristics function together to produce a unique phenomenon. Here, a thermocouple composed of two dissimilar conductors produce electrical voltage when subjected to differing temperatures. Figure B.4 shows a thermocouple connection in a thermostat/gas valve.

Figure B.4 A thermocouple

B.2.3 Law 4: Alternative System Merging

The law of alternative systems merging arises when two or more systems having mutually opposite sets of strengths and weaknesses are merged together in the design of a new system.

Example

A common nail is easy to hammer, but it does not hold two wooden parts too well. A common screw holds two wooden parts much better, but it is relatively difficult to insert and tighten. The union of these two systems is the spiral nail, which has advantages of both the common nail and the screw (Figure B.5).

Common nail   Common screw   Spiral nail

Figure B.5 Nails, screws, and in-between

B.2.4 Law 5: Integration of Inverse Systems

The law of integration of inverse systems arises when two or more systems, having opposing characteristics, are combined in the creation of a new system. Such combined systems can increase the effectiveness, controllability, and operation range of the original individual systems.

Example

Combining a pencil and an eraser. Using the integrated system, one can draw or write on a paper and erase it as needed (Figure B.6).

Figure B.6 A pencil and an eraser

B.2.5 Law 6: Integration of Disparate Systems

The law of integration of disparate systems arises when two or more unrelated systems are combined in order to tap the resources of each other.

Example
A tower is designed to place communication antennas at a high elevation. In addition, the tower includes a restaurant located very high up. In this case, there is no functional connection whatsoever between the antenna tower and the restaurant, but the designers utilized the need to build a communication antenna at high altitude and combine it with customers' desire to see large portion of the area while eating.

Figure B.7 shows the Black Mountain Tower, located on Black Mountain in Canberra, Australian Capital Territory.

Figure B.7 The Black Mountain Tower

B.2.6 Law 7: Combining Multiple Systems

The law of combining multiple systems arises when several systems, each designed for a single application, are combined in order to create a system that has the combined capabilities of the original individual systems.

Example

Four related systems: Copier, Scanner, Printer and Fax are merged into a single system (Figure B.8).

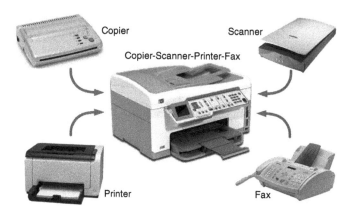

Copier   Scanner

Copier-Scanner-Printer-Fax

Printer   Fax

Figure B.8 Copier, scanner, printer, and fax – merged

B.3 Law 8: Flow Conductivity

The law of flow conductivity represents a pattern of systems evolution in which systems contain a flow or a stream of matter, energy or information, evolve such that it is implemented in a more effective and efficient way. There are two main streams of such systems evolution.

B.3.1 Law 8 A: Increasing the Positive Effect of Helpful Flows

Typically, each transformation of the flow (e.g. transfer of matter from one state to another, changing the types of energy, changing the ways of presenting information) is accompanied by some losses. Consequently, reducing the number of transformations often leads to greater system efficiency.

Example

In a diesel generator, the energy flows from chemical energy to mechanical energy and then to electricity. However, in a fuel cell the energy flows from chemical energy directly to electricity. Therefore, from an energy standpoint, a fuel cell is usually more efficient than a diesel generator (Figure B.9).

(a)   (b)

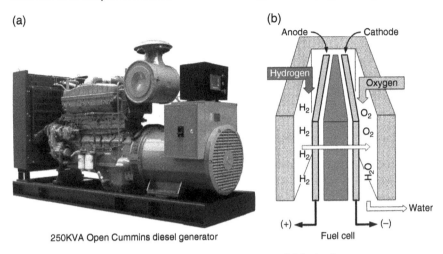

250KVA Open Cummins diesel generator   Fuel cell

Figure B.9 Electricity generation using (a) diesel generator (b) fuel cell

B.3.2 Law 8 B: Reducing the Negative Effect of Harmful Flows

Many systems flows are not desired or even harmful (e.g. waste and pollution generated by an automobile). According to the law of flow conductivity, such flows are reduced as systems evolve.

Example

A catalytic converter is a device used to convert toxic exhaust emissions from an internal combustion engine into nontoxic substances. Catalytic converters were installed in large numbers of gasoline-powered vehicles starting in 1975 (Figure B.10).

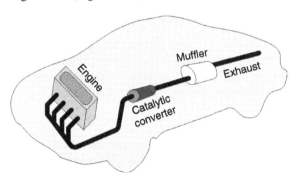

Figure B.10 Reducing automobile pollution

B.4   Laws 9 to 14: Enhanced Coordination

The laws of enhanced coordination represent a pattern of systems evolution in which, over time, certain characteristics of subsystems better match other subsystems or better match the super-system (i.e. the environment). The term *coordination* relates to the preference or the value of one parameter compared with the value of another. These laws have several manifestations, as described below.

B.4.1   Law 9: Shape and Form Coordination

The law of shape and form coordination arises when the shape or the form of parts of the system match better with other parts or with the super-system.

The shape of the system must be consistent with the form, features, and character of the internal system's elements in order to optimize the operations of the system.

Example
Standardization of bolt and nut thread (Figure B.11).

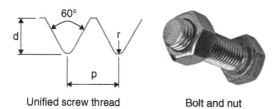

Unified screw thread   Bolt and nut

Figure B.11 Standardization

B.4.2   Law 10: Timing and Rhythm Coordination

The law of timing and rhythm coordination arises when the rhythm of different components becomes more synchronized with other components so the overall system begins to operate in step internally as well as with other systems. This can provide significant increases in efficiency and throughput.

Example
The Atacama Compact Array (ACA) produces multi-source information that must be perfectly synchronized in order to obtain a single image from the array. ACA, located on the ALMA high site in northern Chile, is composed of four 12-metre antennas and twelve 7-metre antennas (Figure B.12).

Figure B.12 European Southern Observatory

B.4.3   Law 11: Materials Coordination

The law of materials coordination arises when, over time, different materials of various components evolve to be more harmonized with one another.

The overall system tends to operate more harmoniously, which leads to increased efficiency and throughput.

Example
A crankshaft, located in internal combustion engines, is a subsystem that translates linear piston motion into reciprocal rotation movement. The crankshaft has a linear axis about which it rotates, typically with several bearings held inside the engine block. The crankshaft is made of steel and the bearings are made from bronze alloy. This combination of materials has evolved as the most optimal mechanical solution to withstand the forces applied and to minimize friction (Figure B.13).

Figure B.13 A crankshaft

B.4.4   Law 12: Action Coordination

The law of action coordination arises when the system evolves in such a way that several actions of the system are coordinated, producing more effective results and often utilizing less resources.

Example
Augmented reality (AR) is based on a live direct view of a physical, real-world environment on which appropriate computer-generated graphic elements are superimposed.

Figure B.14 is a LandForm video with map overlays showing landmarks and other indicators during a helicopter flight at Yuma Proving Ground.

Figure B.14 Augmented reality (AR)

B.4.5   Law 13: Space Coordination

The law of space coordination arises when a system evolves such that the space or locations of its parts are more suitable to fulfill its objectives in achieving more efficient processes.

Example
A production line is a set of sequential operations established in a factory where materials are put through a refining process to produce an end product that is suitable for onward consumption.

Figure B.15 shows a doughnut production line, Selfridges, Krispy Kreme Doughnut shop.

Figure B.15 Doughnut production line

B.4.6 Law 14: Parameters Coordination

The law of parameters coordination arises when systems evolve and progress into a more harmonized states. Such systems exhibit more harmonization among its components as well as between the system and its super-system.

Example
The physical parameters (height, width, weight, etc.) of a train must be harmonized with railways, railway electrical power distribution, bridges, railway platforms, etc.

Figure B.16 is an electrically hauled container-freight train on the West Coast Main Line near Nuneaton in Warwickshire, England.

Figure B.16 Container-freight train

B.5 Law 15: Controllability

The law of controllability arises when systems evolve and progress into a more controllable operation. Such evolving systems permit more and more external control, which modifies the system behavior to align it with either changing parameters of the system itself or the super-system or the environment.

Example
(1) The static (uncontrolled) Cleveland Bridge in Bath, England, built in 1830 (engraving and hand water-colored on print by FP Hay), and (2) the externally controlled, Tower Bridge, in London, England, built in 1892 (Figure B.17).

Cleveland Bridge, Bath, England

Tower Bridge, London

Figure B.17 Bridge design evolution

B.6   Law 16: Dynamization

The law of dynamization arises when systems evolve to encompass more and more behavioral states. Such evolving systems are able to exist in different modes of operations, adapted to different needs and environment conditions.

Typically, as systems evolve, they can transfer from one state to another, in order to take advantage of the characteristics exhibited by the system in each state.

Example
(1) An ancient Chinese parasol exhibiting a single, spread-out state; (2) A dual state collapsing umbrella exhibiting two states: spread out or collapsed; and (3) a compact mini umbrella, exhibiting three states: spread out, collapsed, or folded (Figure B.18).

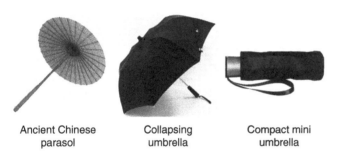

| Ancient Chinese parasol | Collapsing umbrella | Compact mini umbrella |

Figure B.18 Parasol & Umbrella evolution

B.7   Law 17: Transition to Super System

The law of transition to a super system arises when systems evolve and progress to the point where local resources are exhausted and the system integrates with other systems or with the super-system and continues to evolve in that environment.

Example
(1) A self-resourced, gyroscope-based, inertial navigation system (INS), (2) A LORAN (LOng RAnge Navigation) terrestrial radio navigation system using radio transmitters in multiple locations to determine the position and speed of aircrafts and ships, and (3) A satellite-based, global positioning system (GPS) (Figure B.19).

Gyroscope

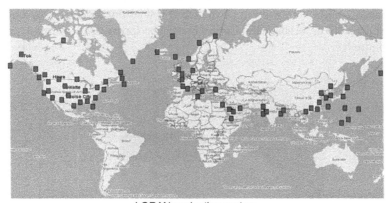

LORAN navigation system

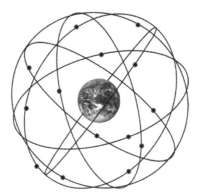

Global Positioning System

Figure B.19 System integrates with super-system

B.8   Law 18: Increasing System Completeness

The law of increasing system completeness arises when systems evolve such that they become less and less dependent on their environment.

Example

(1) In the early stages of aviation, aircraft flew at low altitude and fuel could easily combine with the oxygen in the air (Photo: De Havilland DH.90 at the 2008 "Flying Legends" air show in Duxford, UK.). Later, (2) when aircraft flew at higher altitude, early engines could not burn fuel in the thin atmosphere, so aircraft engines were provided with a compressor that compressed the air and provided sufficient concentration of oxidizers (Photo: Qantas Boeing 747-400ER). Later, (3) in order to fly beyond the atmosphere, specific engines and strategy evolved to abandon the use of air so the vehicle (rocket) itself would carry sufficient oxidizers on board. Thus, a system became independent of the environment for ongoing operations (Photo: The AS-203 rocket launched from NASA's Kennedy Space Center, 1966.) (Figure B.20).

De Havilland DH.90

Qantas Boeing 747-400ER

AS-203 NASA rocket

Figure B.20 Increased system completeness

B.9   Law 19: Displacement of Human

The law of displacement of human arises when systems evolve and progress to the point where they can govern and control themselves.

The system or super-system begins to make its own decisions and humans tend to either disappear completely from the system control cycle or assume general supervision positions.

Example
Driverless automobiles are prime example of a system displacing human at the control and working independently.

Volvo Car Group initiates world unique Swedish pilot project with self-driving cars on public roads (Figure B.21).

Figure B.21 Driverless automobile

B.10   Law 20: Uneven System Evolution

Ideally, all the components of a system should evolve at an equal pace. The law of uneven system evolution arises when different parts of a system evolve in an uneven pace. That is, some parts of the system evolve faster than other parts.

This may lead to contradictions between system components, which may reduce the efficiency of the system.

Example
Early automobile were produced in the early 1900. The evolution concentrated mostly on mechanical aspects of the automobile. Eventually, in the 1950s, electrical components were added to the system, and only in the late 1980s and onward were automotive electronic systems added to improve safety, comfort of driving, and reliability (Figure B.22).

Photo: Ford Model-T Pickup (1919).

Figure B.22 Old automobile

B.11 Law 21: Technology General Progress

Sometimes elements of a system improve along with other parts of the system.

Example
Within an electronic board, more unified resistors are produced, smaller electronic components are available in the market, and more reliable semiconductors are available. In such circumstances, improved systems can be manufactured and made available to users (Figure B.23).

Figure B.23 VLSI input-output chip for x86 computers

Appendix C

List of Acronyms

AADS	Analog Air Data System
AAV	Architecture adaptability value
ADC	Air data computer
ADT	Air data terminal
AFD	Anticipatory Failure Determination
AHRS	Attitude, Heading, Roll, System
AMISA	Architecting Manufacturing Industries and Systems for Adaptability
AO	Architecture option
API	Application program interface
APMI	Alternative Plus-Minus-Interesting
AR	Augmented reality
ASC	Average satisfaction coefficient
AUV	Autonomous underwater vehicle
AV	Air vehicle
AVB	Air vehicle bus
BD	BusinessDictionary
BDS	Biomimicry Design Spiral
BPS	Business process simulation
CAIB	Columbia Accident Investigation Board
CCB	Change Control Board
CEO	Chief executive officer
CIM	Cyclic Innovation Model
CM	Configuration management
CMM	Capability Maturity Model
CMMI	Capability Maturity Model Integration
COTS	Commercial off-the-shelf
CPU	Central processing unit
CSB	Control System Board

Practical Creativity and Innovation in Systems Engineering, First Edition. Avner Engel.
© 2018 John Wiley & Sons, Inc. Published 2018 by John Wiley & Sons, Inc.

DAESEO	Developing Adaptable Embedded Systems for Economic Opportunities
DCF	Discounted cash flow
DEC	Digital Equipment Corporation
DEC	Dissemination and exploitation committee
DEFS	Discrete Event Factory Simulation
DFA	Design for Adaptability
DI	Dissatisfaction index
DM	Decision maker
DoD	Department of Defense
DoE	Department of Energy
DPI	Dots per inch
DPT	Direct personalized treatment
DSM	Design Structure Matrix
EC	European Commission
EC	Engineering choice
ECU	Electronic control unit
EES	Economic, environmental, and social
EFR	External factors ratio
EGI	Embedded GPS-INS
EIRMA	European Industrial Research Management Association
EQ	Emotional quotient
ESA	European Space Agency
EW	Electronic warfare
FCS	Future combat systems
FDA	Food and Drug Administration
FDI	Foreign direct investment
FLIR	Forward-looking infrared
FM	Frequency modulated
FMEA	Failure mode effect analysis
FOT	Financial option theory
FP5	Fifth Framework Program
FTE	Full-time equivalent
GCS	Ground control system
GDM	Group decision-making
GDP	Gross domestic product
GDT	Ground data terminal
GERD	Gross expenditure on R&D
GII	Global Innovation Index
GMCR	Graph model for conflict resolution

GPS	Ground positioning system
GSM	Global System for Mobile Communications
HDD	Hard disk drive
HEV	High-energy visible
HGP	Human Genome Project
HUD	Head-up display
IAI	Israel Aerospace Industries
IAW	In accordance with
IBM	International Business Machine
IC	Integrated circuits
IC	Interface cost
ICD	Interface control document
ICMM	Innovation Capability Maturity Model
ICT	Information & communication technology
IEC	International Electrotechnical Commission
IEEE	Institute of Electrical and Electronics Engineers
IFF	Identification friend or foe
IFR	Ideal final result
IFR	Internal factors ratio
ILS	Integrated logistics support
IMS	Intelligent manufacturing system
INCOSE	International Council on Systems Engineering
INS	Inertial navigation system
IPR	Intellectual property right
ISO	International Organization for Standardization
IST	Information Society Technologies
JV	Joint venture
LASERS	Light Amplification by Stimulated Emission of Radiation
LORAN	Long-range navigation
LT	Lateral thinking
MCDM	Multicriteria decision-making
MDM	Multiple domain matrix
MIT	Massachusetts Institute of Technology
MMI	Man–machine interface
MOMP	Multiple objectives-multiple participants
MSI	Medium-scale integration
MTBF	Mean time between failures
MULE	Multifunctional utility/logistics and equipment

NASA	National Aeronautics and Space Administration
NCES	National Center for Education Statistics
NDI	Non developmental items
NPV	Net present value
NSF	National Science Foundation
NUF	Non useful functions
OECD	Organization for Economic Cooperation and Development
OV	Option value
PCM	Product configuration management
PCT	Patent Cooperation Treaty
PDP	Product development processes
PLPI	Product line practice initiative
PM	Person-months
PMI	Plus-Minus-Interesting
PPP	Purchasing power parity
PQA	Product quality assurance
QA	Quality assurance
QFD	Quality function deployment
QM	Quality management
R&D	Research and development
RF	Radio frequency
RFC	Requests for change
RFP	Request for proposal
RFV	Requests for variance
ROI	Return on investment
RPN	Risk priority number
RS	Rhizophora Stylosa
RTD	Research, technological development & demonstration
SAE	Society of Automotive Engineers
SAST	Strategic assumptions, surfacing and testing
SC	Steering committee
SCAMPER	Substitute, Combine, Adapt, Modify, Put, Eliminate, Reverse
SE	Systems engineering
SEI	Software Engineering Institute
SFIN	Skane Food Innovation Network
SI	Satisfaction index
SII	Standards Institution of Israel
SIPOC	Supplier, Input, Process, Output, Customer
SLS	Space launch system
SME	Small and medium enterprise

SMTF	Swedish Marine Technology Forum
SR	Short range
SSI	Small-scale integration
SSPA	Solid state power amplifier
STEM	Science, technology, engineering, and mathematics
STH	Six thinking hats
STS	Space Transportation System
SWOT	Strengths, weaknesses, opportunities, and threats
TAU	Tel-Aviv University
TCT	Transaction cost theory
TCU	Texas Christian University
TED	Technology entertainment and design
TRIZ	Theory of inventive problem solving
TRL	Technology readiness levels
TRW	Thompson Ramo Wooldridge
TUM	Technical University of Munich
UAV	Unmanned air vehicles
UF	Useful functions
UGV	Unmanned ground vehicle
UV	Ultraviolet
VA	Value analysis
VE	Value engineering
VE-QFD	Value engineering – Quality function deployment
VM	Virtual manufacturing
VR	Virtual reality
VVT	Verification, validation, and testing
WBS	Work breakdown structure
WP	Work package
WRT	With respect to
WSM	Weighted score matrix

Appendix D

Permissions to Use Third-Party Copyright Material

Book's front cover	Source: Image is part of a dry rock garden consisting of gravel and massive boulders placed by Hoichi Kurisu. Photo courtesy of Frederik Meijer Gardens & Sculpture Park, Grand Rapids, Michigan, USA.

D.1   Part I: Introduction

Figure 1.4 Moses, a creative and innovative giant	Source: Moses by Michelangelo Buonarroti, Tomb (1505–1545) for Julius II, San Pietro in Vincoli (Rome). Reproduced with permission of Wikimedia Commons, Author: Jorg Bittner Unna.

D.2   Part II: Systems Engineering

Figure 2.5 Concept testing at NASA's wind tunnel facility	Source: Media Invited to View Rocket Model Tests of NASA's Space Launch System, 1.5.2017. Image available in public domain (NASA).
Figure 2.12 The Creissels and Viaduct de Millau bridge	Adapted from: View of Creissels with the viaduct in the background. Reproduced with permission of Wikimedia Commons, author: Stefan Krause, Germany.
Figure 2.13 Aircraft system embedded in a national transport Supersystem	Adapted from: Model of an aircraft system as embedded in an air transport system, which is related to other transport systems. Reproduced with permission of Philosophy of Engineering, The Royal Academy of Engineering, London, UK, June, 2010.
Figure 2.14 The Millennium Dome London UK	Adapted from: The Millennium Dome, London, UK. Reproduced with permission of Wikimedia Commons, author: James Jin.

Practical Creativity and Innovation in Systems Engineering, First Edition. Avner Engel.
© 2018 John Wiley & Sons, Inc. Published 2018 by John Wiley & Sons, Inc.

D.3 Part III: Creative Methods

Figure 3.4 A rickety bridge example	Adapted from: Drahtsteg pedestrian hanging bridge, in the Zillertal Alps, Austria. Reproduced with permission of Wikimedia Commons, author: Bohringer Friedrich.
Figure 3.7 Water faucets	Adapted from: Tap mechanism. Reproduced with permission of Wikimedia Commons, author: Chabacano.
Figure 3.8 Safety matches "Weltholzer"	Source: Matchbox "Weltholzer", Reproduced with permission of Wikimedia Commons, author: Anagoria.
Figure 3.9 Motorcycle chain	Adapted from: Roller chain, used to transmit power in motorcycles and bicycles. Reproduced with permission of Wikimedia Commons, author: Ralf Roletschek (User:Marcela).
Figure 3.11 A rope	1) Adapted from: Hemp rope (Corderie Royale of Rochefort, Charente-Maritime). Reproduced with permission of Wikimedia Commons, author: Ji-Elle. 2) Adapted from: An unidentified competitor at the rope climbing competition at the shopping center, La Part Dieu, Lyon, 6.5.2006. Reproduced with permission of Wikimedia Commons, author: Stevage.
Figure 3.13 Phone for seniors	Source: Snapfon ezTWO3G Cell Phones for seniors. Reproduced with permission of Excellus Communications, LLC (Jim Tate, General Manager).
Figure 3.14 Example: UV-protected sunglasses	Adapted from: Ray-Ban Aviator sunglasses model RB3025 004/58. Reproduced with permission of Wikimedia Commons, author: Rich Niewiroski Jr.
Figure 3.17 Schematic of water filtration in mangrove roots	Source: Kim et al., 2016. Reproduced with permission of the author: Kim K.
Figure 3.33 Unmanned Ground Vehicle (UGV)	Adapted from: Future Combat Systems (FCS)-Multifunctional Utility/Logistics and Equipment (MULE) vehicle. Image available in public domain (US Army).
Figure 3.34 Autonomous Underwater Vehicle (AUV)	Adapted from: A AUV photo taken at the Student Autonomous Underwater Challenge Europe (SAUC-E) competition (2009). Reproduced with permission of Wikimedia Commons, author: Paul Esparon, Sunil Shah, Dr. Timothy Nickels and the 2008 CAUV team University of Cambridge Department of Engineering and Nokia.
Figure 3.36 The T-38 Talon aircraft	Adapted from: US Air force fact sheet, Article 104569/t-38-talon. Image available in public domain (U.S. Air Force, Photo/Steve White).

Figure 3.47 RSA Chemical reactor	Source: RSA Chemical reactor. Reproduced with permission of Wikimedia Commons, author: Echis.
Figure 3.51 Unmanned Air Vehicle MQ-5B Hunter	Adapted from: Inspecting an MQ-5B Hunter at Kandahar Airfield in Afghanistan. Image available in public domain (U.S. Navy. photo: Lt. Kristine Volk on Aug. 6, 2015).
Figure 3.52 Sinking of the *Titanic* (Engraving by Willy Stower)	Source: Engraving by Willy Stower (1912). Image available in public domain (Magazine Die Gartenlaube).
Figure 3.63 A rocket engine	Source: Liquid Hydrogen as a Propulsion Fuel, 1945–1959 Pratt & Whitney's RL-10 rocket engine. Image available in public domain (NASA).
Figure 3.64 A Jet engine	Source: Clay-cutaway-render-of-turbofan-jet-engine-isolated-on-white-background 687783895. Photo purchased from Shutterstock.
Figure 3.65 An interstellar ion propulsion engine	Adapted from: Bussard Interstellar Ramjet engine concept. Image available in public domain (NASA).

D.4   Part IV: Promoting Innovative Culture

Figure 4.6 Continual reduction in number of parts and costs	Adapted from: Ehrlenspiel et al. (2007). Image in public domain (AMISA).
Figure 4.7 1,024 bit core memory	Source: 1,024 bit core memory. Reproduced with permission of Wikimedia Commons, author: Konstantin Lanzet
Figure 4.8 IBM 737, Magnetic core storage	Source: IBM 737 Magnetic Core Storage Unit. Reproduced with permission of the International Business Machines Corporation ©.
Figure 4.11 System integrates with super-system	1) Adapted from: Litton LN3-2A Inertial Navigation Platform of the F-104G Super Starfighter. Reproduced with permission of Wikimedia Commons, author: HvO / Klu andre at en.wikipedia. 2) Adapted from: Loran stations. Reproduced with permission of Wikimedia Commons, author: Zeitan. 3) Adapted from: GPS constellation visible from Golden CO. Reproduced with permission of Wikimedia Commons, author: Paulsava.

Figure 4.16 Older cellphone versus iPhone	1) Source: Nokia 8210 in light cover. Reproduced with permission of Wikimedia Commons, author: krystof.k (Twitter) & nmuseum.
Figure 4.18 Milling machining	Adapted from: Milling machine (Vertical, Manual). Reproduced with permission of Wikimedia Commons, author: Tosaka.
Figure 4.24 Electricity generation using (a) Diesel generator (b) Fuel cell	1) Source: The electric generating unit, with the fuel tank, the mounting base and the radiator. Reproduced with permission of Wikimedia Commons, author: Abdul Aziz Abdo.
Figure 4.27 IBM 350 Disk Storage Unit	Source: IBM 350 Disk Storage Unit. Reproduced with permission of the International Business Machines Corporation ©.
Figure 4.28 Container-freight train	Source: An electrically hauled container freight train. On the West Coast Main Line near Nuneaton in Warwickshire, England. Reproduced with permission of Wikimedia Commons, author: G-Man.
Figure 4.29 Metal cutting laser head	Source: Laserhead of an AMADA FO-4020NT industrial laser installed at Metaveld BV. Reproduced with permission of Wikimedia Commons, author: Metaveld BV.
Figure 4.39 Common obstacles to innovation	Source for background image: William Blake, John Bunyan, Christian Reading in His Book (Courtesy: Frick Collection, New York). Image available in public domain.
Figure 4.45 Evaluating production line concept using virtual reality tool	Source: Designing manufacturing floor plan in 3D, Lake Engineering and Fabrication. Reproduced with permission of Brian D. Lake (Lake Engineering and Fabrication).
Figure 4.48 IBM flag products (1965 and 2012)	1) Source: Bundesarchiv, B 145 Bild-F038812-0014 / Schaack, Lothar / CC-BY-SA 3.0. Reproduced with permission of Wikimedia Commons, German Federal Archive (Photo: Schaack, Lothar). 2) Source: The IBM Blue Gene/Q installed at Argonne National Laboratory, near Chicago, Illinois. Reproduced with permission of Wikimedia Commons, Argonne National Laboratory (Flickr) / U.S. Department of Energy.
Figure 4.49 DEC flag products (Mid 1970s)	1) Adapted from: PDP11/40 exhibited in Vienna Technical Museum. Reproduced with permission of Wikimedia Commons, author: Stefan Kogl. 2) Adapted from: Digital Equipment Corporation (DEC) VAX 780 computer, Living Computer Museum, Seattle, Washington, USA. Reproduced with permission of Wikimedia Commons, author: Joe Mabel.
Figure 4.50 Quickly rejecting new ideas	Adapted from: Business-cartoon-about-job-search-the-job-candidate-has-some-requirements-for-his-potential, 362027465. Drawing purchased from Shutterstock.

Figure 4.51 Human diversity - Asiatiska folk	Source: Nordisk familjebok (1904), vol.2, Asiatiska folk. Image available in public domain (Nordisk Familjebok has credited the image to Bibliographisches Institut, Leipzig, G. Mutzel).
Figure 4.56 Typical small leisure boat	Source: 80 foot motor yacht Alchemist photo D Ramey Logan. Reproduced with permission of Wikimedia Commons, author: Don Ramey Logan (1961–).
Figure 4.57 Ovadia Harari and the Lavi fighter aircraft	Source: Ovadia Harari and the Lavi fighter aircraft. Reproduced with permission of the Israel Society of Aeronautics & Astronautics.
Figure 4.58 (a) Space shuttle Columbia and (b) Flight STS-107 crewmembers (left to right): Brown, Husband, Clark, Chawla, Anderson, McCool and Ramon (NASA pictures)	1) Source: A close-up camera view shows Space Shuttle Columbia as it lifts off from Launch Pad 39A on mission STS-107. Image available in public domain (NASA). 2) Source: The crew of the final ill-fated flight of the Space Shuttle Columbia, mission STS-107, in October 2001. Image available in public domain (NASA).
Figure 4.59 Lincoln's US Cabinet (1864)	Adapted from: The first reading of the emancipation proclamation before the cabinet. Engraving by Alexander Hay Ritchie, painting by Francis Bicknell Carpenter, 1864. Image available in public domain (United State Senate).

D.5 Part V: Creative and Innovative Case Study

Figure 5.2 View from mount Floyen, Bergen, Norway	Source: Floybanen – Funicular. Reproduced with permission of Wikimedia Commons, author: Svein-Magne Tunli.
Figure 5.5 Cyclical architecture improvement	Adapted from: Hellenize et al. (2007). Image in public domain (AMISA).
Figure 5.16 Cap Applicator Machine (CAM)	Source: Cap Applicator Machine (CAM). Image in public domain (AMISA).
Figure 5.17 Fiber Placement System (FPS)	Source: Fiber Placement System (FPS). Image in public domain (AMISA).
Figure 5.18 Power Train (PT) of a standard MAN TGM truck	Source: Power Train (PT). Image in public domain (AMISA).
Figure 5.19 Vehicle Localization System (VLS) for an Unmanned Ground Vehicle	Source: Vehicle Localization System (VLS). Image in public domain (AMISA).

Figure 5.20 Solid State Power Amplifier (SSPA)	Source: Solid State Power Amplifier (SSPA). Image in public domain (AMISA).
Figure 5.21 Hologram Production Line (HPL)	Source: Hologram Production Line (HPL). Image in public domain (AMISA).
Figure 5.31 Two types of architectures: (a) Traditional; (b) Adaptable1.	Adapted from: Browning and Honour, 2005, 2008. Reproduced with permission of the authors.
Figure 5.46 Merging the Driver and the HPA in the SSPA amplifier stage	Source: Merging the Driver and the HPA in the SSPA amplifier stage. Image in public domain (AMISA).

D.6 Appendices

Figure B.2 Continual reduction in number of parts and costs	Adapted from: Ehrlenspiel et al. (2007). Image in public domain (AMISA).
Figure B.3 A catamaran	Adapted from: The Salem Ferry, Nathaniel Bowditch, approaches its dock in Salem, Massachusetts. Reproduced with permission of Wikimedia Commons, author: Fletcher6.
Figure B.4 A thermocouple	Source: Thermocouple connection. Reproduced with permission of Wikimedia Commons, author: Z22.
Figure B.7 The Black Mountain Tower	Source: Black Mountain Tower located on Black Mountain in Canberra, Australian Capital Territory. Reproduced with permission of Wikimedia Commons, author: Bidgee.
Figure B.9 Electricity generation using (a) Diesel generator (b) Fuel cell	1) Source: The electric generating unit, with the fuel tank, the mounting base and the radiator, Reproduced with permission of Wikimedia Commons, author: Abdul Aziz Abdo.
Figure B.12 European Southern Observatory	Source: Atacama Compact Array (ACA) on the ALMA high site at an altitude of 5000 meters in northern Chile. Reproduced with permission of Wikimedia Commons, author: ESO.
Figure B.13 A crankshaft	Derived from: Solid model assembly created in en:NX (Unigraphics). Reproduced with permission of Wikimedia Commons, author: Freeformer.
Figure B.14 Augmented reality (AR)	Source: LandForm video with map overlays showing landmarks and other indicators during helicopter flight at Yuma Proving Ground. Reproduced with permission of Wikimedia Commons, author: Winged1der.

Figure B.15 Doughnut production line	Source: Selfridges has a Krispy Kreme Doughnut shop which has its own doughnut production line. Reproduced with permission of Wikimedia Commons, author: Neil T.
Figure B.16 Container-freight train	Source: An electrically hauled container-freight train. On the West Coast Main Line near Nuneaton in Warwickshire, England. Image in the public domain (G-Man).
Figure B.17 Bridge design evolution	1) Source: The New Bridge at Bathwick, Bath, England. Reproduced with permission of Wikimedia Commons, author: Unbekannt. 2) Source: Tower Bridge, London getting opened. Reproduced with permission of Wikimedia Commons, author: Mvkulkarni23.
Figure B.19 System integrates with super-system	1) Adapted from: Litton LN3-2A Inertial Navigation Platform of the F-104G Super Starfighter. Reproduced with permission of Wikimedia Commons, author: HvO / Klu andre 2) Adapted from: Loran stations. Reproduced with permission of Wikimedia Commons, author: Zeitan. 3) Adapted from: GPS constellation visible from Golden CO. Reproduced with permission of Wikimedia Commons, author: Paulsava.
Figure B.20 Increased system completeness	1) Adapted from: De Havilland DH.90 at the 2008 'Flying Legends' air show in Duxford, UK. Reproduced with permission of Wikimedia Commons, author: Rror. 2) Source: Qantas Boeing 747-400ER; VH-OEI@ LAX;18.04.2007/463fi. Reproduced with permission of Wikimedia Commons, author: Aero Icarus from Zurich, Switzerland. 3) Source: The Apollo second stage AS-203 rocket launched from NASA's Kennedy Space Center (1966). Image in public domain (NASA).
Figure B.21 Driverless automobile	Source: Volvo Car Group initiates world unique Swedish pilot project with self-driving cars on public roads, Dec. 02, 2013, ID: 136186. Reproduced with permission of the Volvo Media Relations.
Figure B.22 Old automobile	Source: 1919 Ford Model T pickup. Reproduced with permission of Wikimedia Commons, author: Writegeist.
Figure B.23 VLSI Input-Output Chip	Source: The VLSI VL82C106 is a I/O chip for x86 computers. Reproduced with permission of Wikimedia Commons, author: Appaloosa.

Index

Practical Creativity and Innovation in Systems Engineering, First Edition. Avner Engel.
© 2018 John Wiley & Sons, Inc. Published 2018 by John Wiley & Sons, Inc.

Wiley Series in Systems Engineering and Management

William Rouse, Series Editor
Andrew P. Sage, Founding Editor

ANDREW P. SAGE and JAMES D. PALMER
Software Systems Engineering

WILLIAM B. ROUSE
Design for Success: A Human-Centered Approach to Designing Successful Products and Systems

LEONARD ADELMAN
Evaluating Decision Support and Expert System Technology

ANDREW P. SAGE
Decision Support Systems Engineering

YEFIM FASSER and DONALD BRETINER
Process Improvement in the Electronics Industry, Second Edition

WILLIAM B. ROUSE
Strategies for Innovation

ANDREW P. SAGE
Systems Engineering

HORST TEMPELMEIER and HEINRICH KUHN
Flexible Manufacturing Systems: Decision Support for Design and Operation

WILLIAM B. ROUSE
Catalysts for Change: Concepts and Principles for Enabling Innovation

UPING FANG, KEITH W. HIPEL, and D. MARC KILGOUR
Interactive Decision Making: The Graph Model for Conflict Resolution

Practical Creativity and Innovation in Systems Engineering, First Edition. Avner Engel.
© 2018 John Wiley & Sons, Inc. Published 2018 by John Wiley & Sons, Inc.

DAVID A. SCHUM
Evidential Foundations of Probabilistic Reasoning

JENS RASMUSSEN, ANNELISE MARK PEJTERSEN, and LEONARD P. GOODSTEIN
Cognitive Systems Engineering

ANDREW P. SAGE
Systems Management for Information Technology and Software Engineering

ALPHONSE CHAPANIS
Human Factors in Systems Engineering

YACOV Y. HAIMES
Risk Modeling, Assessment, and Management, Third Edition

DENNIS M. SUEDE
The Engineering Design of Systems: Models and Methods, Second Edition

ANDREW P. SAGE and JAMES E. ARMSTRONG, Jr.
Introduction to Systems Engineering

WILLIAM B. ROUSE
Essential Challenges of Strategic Management

YEFIM FASSER and DONALD BRETTNER
Management for Quality in High-Technology Enterprises

THOMAS B. SHERIDAN
Humans and Automation: System Design and Research Issues

ALEXANDER KOSSIAKOFF and WILLIAM N. SWEET
Systems Engineering Principles and Practice

HAROLD R. BOOHER
Handbook of Human Systems Integration

JEFFREY T. POLLOCK and RALPH HODGSON
Adaptive Information: Improving Business Through Semantic Interoperability, Grid Computing, and Enterprise Integration

ALAN L. PORTER and SCOTT W. CUNNINGHAM
Tech Mining: Exploiting New Technologies for Competitive Advantage

REX BROWN
Rational Choice and Judgment: Decision Analysis for the Decider

WILLIAM B. ROUSE and KENNETH R. BOFF (editors)
Organizational Simulation

HOWARD EISNER
Managing Complex Systems: Thinking Outside the Box

STEVE BELL
Lean Enterprise Systems: Using IT for Continuous Improvement

J. JERRY KAUFMAN and ROY WOODHEAD
Stimulating Innovation in Products and Services: With Function Analysis and Mapping

WILLIAM B. ROUSE
Enterprise Transformation: Understanding and Enabling Fundamental Change

JOHN E. GIBSON, WILLIAM T. SCHERER, and WILLAM F. GIBSON
How to Do Systems Analysis

WILLIAM F. CHRISTOPHER
Holistic Management: Managing What Matters for Company Success

WILLIAM B. ROUSE
People and Organizations: Explorations of Human-Centered Design

MO JAMSHIDI
System of Systems Engineering: Innovations for the Twenty-First Century

ANDREW P. SAGE and WILLIAM B. ROUSE
Handbook of Systems Engineering and Management, Second Edition

JOHN R. CLYMER
Simulation-Based Engineering of Complex Systems, Second Edition

KRAG BROTBY
Information Security Governance: A Practical Development and Implementation Approach

JULIAN TALBOT and MILES JAKEMAN
Security Risk Management Body of Knowledge

SCOTT JACKSON
Architecting Resilient Systems: Accident Avoidance and Survival and Recovery from Disruptions

JAMES A. GEORGE and JAMES A. RODGER
Smart Data: Enterprise Performance Optimization Strategy

YORAM KOREN
The Global Manufacturing Revolution: Product-Process-Business Integration and Reconfigurable Systems

AVNER ENGEL
Verification, Validation, and Testing of Engineered Systems

WILLIAM B. ROUSE (editor)
The Economics of Human Systems Integration: Valuation of Investments in People's Training and Education, Safety and Health, and Work Productivity

ALEXANDER KOSSIAKOFF, WILLIAM N. SWEET, SAM SEYMOUR, and STEVEN M. BIEMER
Systems Engineering Principles and Practice, Second Edition

GREGORY S. PARNELL, PATRICK J. DRISCOLL, and DALE L. HENDERSON (editors)
Decision Making in Systems Engineering and Management, Second Edition

ANDREW P. SAGE and WILLIAM B. ROUSE
Economic Systems Analysis and Assessment: Intensive Systems, Organizations, and Enterprises

BOHDAN W. OPPENHEIM
Lean for Systems Engineering with Lean Enablers for Systems Engineering

LEV M. KLYATIS
Accelerated Reliability and Durability Testing Technology

BJOERN BARTELS, ULRICH ERMEL, MICHAEL PECHT, and PETER SANDBORN
Strategies to the Prediction, Mitigation, and Management of Product Obsolescence

LEVANT YILMAS and TUNCER OREN
Agent-Directed Simulation and Systems Engineering

ELSAYED A. ELSAYED
Reliability Engineering, Second Edition

BEHNAM MALAKOOTI
Operations and Production Systems with Multiple Objectives

MENG-LI SHIU, JUI-CHIN JIANG, and MAO-HSIUNG TU
Quality Strategy for Systems Engineering and Management

ANDREAS OPELT, BORIS GLOGER, WOLFGANG PFARL, and RALF MITTERMAYR
Agile Contracts: Creating and Managing Successful Projects with Scrum

KINJI MORI
Concept-Oriented Research and Development in Information Technology

KAILASH C. KAPUR and MICHAEL PECHT
Reliability Engineering

MICHAEL TORTORELLA
Reliability, Maintainability, and Supportability: Best Practices for Systems Engineers

DENNIS M. BUEDE and WILLIAM D. MILLER
The Engineering Design of Systems: Models and Methods, Third Edition

JOHN E. GIBSON, WILLIAM T. SCHERER, WILLIAM F. GIBSON, and MICHAEL C. SMITH
How to Do Systems Analysis: Primer and Casebook

GREGORY S. PARNELL
Trade-off Analytics: Creating and Exploring the System Tradespace

CHARLES S. WASSON
Systems Engineering Analysis, Design and Development

AVNER ENGEL
Practical Creativity and Innovation in Systems Engineering